AF545128

APPLICATIONS OF HOLOGRAPHY

Edited by

Euval S. Barrekette

Manager, Electro-Optical Technologies
IBM Research Division
Thomas J. Watson Research Center
Yorktown Heights, New York

Winston E. Kock

Vice President and Chief Scientist
Bendix Corporation
Southfield, Michigan

Teruji Ose

Institute of Industrial Science
University of Tokyo
Tokyo, Japan

Jumpei Tsujiuchi

Imaging Science and Engineering Laboratory
Tokyo Institute of Technology
Tokyo, Japan

George W. Stroke

Professor of Electrical Sciences
and Medical Biophysics
and
Head, Electro-Optical Sciences Center
State University of New York at Stony Brook
Stony Brook, New York

APPLICATIONS OF HOLOGRAPHY

Proceedings of the United States–Japan Seminar on Information Processing by Holography, held in Washington, D. C., October 13-18, 1969

PLENUM PRESS • NEW YORK–LONDON • 1971

Library of Congress Catalog Card Number 76-148415

SBN 306-30526-7

A Division of Plenum Publishing Corporation
227 West 17th Street, New York, N. Y. 10011

United Kingdom edition published by Plenum Press, London
A Division of Plenum Publishing Company, Ltd.
Davis House (4th Floor), 8 Scrubs Lane, Harlesden, NW10 6SE, England

Printed in the United States of America

PREFACE

Holography has matured over the years since it was first invented by Dennis Gabor and further developed by many workers during the last decade. With some applications a commercial reality and many more continually appearing, it is now possible to compile a volume devoted to the applications of holography.

This book consists primarily of papers presented at the second U.S. Japan Seminar on holography, held in Washington, D.C., from October 20-27, 1969, under the auspices of the National Academy of Sciences, National Science Foundation, and the Japanese Society for the Promotion of Science. In addition to most of the papers presented at the seminar, several others are included because they cover important applications which are appropriate to this volume.

Attendance at the seminar was limited to ten delegates from each country; these delegates being chosen so as to permit the seminar to include as wide a range of subjects as possible, with each delegate being recognized as an authority in his particular field of holography. In addition, Dr. Gilbert Devey (Program Director for Engineering Systems for the National Science Foundation, and the one most responsible for making the seminar possible), Professor Wolf F. Berg, Director of the Photographic Institute, ETH, Zurich, Switzerland (who presented a tutorial paper on the subject: "The Photographic Material as a Three-dimensional Recording Medium" on one of the free evenings), two interpreters, and an occasional invited visitor participated in the seminar. Since the seminar extended over a five-day period, only four papers were presented each day, with the rather extensive remaining time being given to lengthy discussions. The suggestions given to the speakers by other members of the delegations have been, in most cases, incorporated into the papers so that many concepts arising out of the discussions are now contained in this book and published for the first time.

For historical purposes it is to be noted that the first U.S.-Japan seminar on holography took place in Tokyo and Kyoto, Japan, on October 2-10, 1967. In that seminar only six delegates from each country were involved in presentations, with the arrangements being similar to those described above for the second seminar. A report on the 1967 seminar was published in the journal Applied Optics, vol. 7, no. 4, April, 1968, p. 22.

The editors, Dr. E. S. Barrekette of the IBM Corporation, and one of the U.S. delegates, and Dr. Winston E. Kock of the Bendix Corporation, co-chairman of the U.S. delegation wish to express their appreciation for the help rendered by the assistant editors Prof. Teruji Osė of the Institute of Industrial Science, University of Tokyo, who was chairman of the Japanese delegation, Professor Jumpei Tsujiuchi of the Applied Optics Department of Tokyo Institute of Technology, who was assistant chairman of the Japanese delegation, and Professor George Stroke of the State University of New York at Stony Brook, co-chairman of the U.S. delegation.

Thanks are also due to Robert Ubell of Plenum Press for his saintly patience and to Toni Gallelli who nearly lost her sanity typing and correcting many of the manuscripts herein.

Finally we wish to thank Applied Optics, the IBM Journal of Research and Development, the Journal of the Optical Society of America and Optica Acta as well as Cambridge University Press for permission to reprint, as part of some of the papers in this volume, material previously included in the following: R. V. Pole et al, App. Opt. 6(1967)1571-5, H. Wieder et al, App. Opt. 6(1967)1761-5, B. R. Brown et al, IBM J. Res. Dev. 13(1969)160-168, H. Wieder et al, IBM J. Res. Dev. 13(1969)169-171, T. Tsuruta, JOSA 60(1970)44-48, G. Stroke, Optica Acta 16(1969)401-422, T. Tsuruta et al, Optica Acta 16(1969)723-733, E. R. Robertson et al, Engineering Uses of Holography, Cambridge U Press (1970).

Euval S. Barrekette
Winston E. Kock

October 1970

CONTENTS

TWO CHALLENGES TO HOLOGRAPHY

1. HOLOGRAPHIC ELECTRON MICROSCOPY 2. SPECKLE-FREE ILLUMINATION

Dennis Gabor, F.R.S.
Staff Scientist, CBS Laboratories, Stamford, Conn.

I want to challenge experimental physicists to tackle two important and difficult problems. In both cases, a solution is theoretically possible, but the experimental realization will require skills of a very high order.

HOLOGRAPHIC ELECTRON MICROSCOPY

Holography was started in 1948 when I proposed to overcome the limitations of electron microscopy by taking first a confused,"coded" picture with coherent electrons, and decoding it by light. In the last six years, holography had a spectacular revival, and has scored many notable successes; but no significant progress appears to have been made beyond what M. E. Haine, J. Dyson, and I have achieved in a premature attack, 1950-1953.[1 2 3]

All aberrations of electron objectives can be corrected or made negligibly small except the spherical aberration. In modern electron microscopes, this limits the resolution to 3.6 Angstroms, though 2.8 Å have also been claimed. Reducing this to 2 Å would be sufficient for the resolution of almost all atomic lattices, and phase contrast would make it just possible to see single atoms of carbon. This was, and remains a great challenge. One could never go much further, because at about 1.5 Angstroms, de Broglie's "destruction limit" presents an insuperable obstacle. If one wants to see material structures in too fine detail, one destroys them.[4]

Haine, Dyson and I started much too early. At that time, the resolution limit stood at about 10 Angstroms. This limit was set by "trivial" errors; by the astigmatism (imperfect roundness) of the electron objectives, by the vibrations of the column, by stray magnetic fields, by the creep of the stage, and by "contamination;" the phenomenon that objects grow under the electron beam, by the decomposition of residual vapours.

In the meantime, every one of these disturbances has been eliminated by the patient work of electron microscopists, and the resolution limit has been reduced to that set by the spherical aberration. Now is the time to review the situation. Can we reduce the resolution limit to 2 Å by halving the spherical error at twice the aperture?

In 1953, the analysis of M. E. Haine had shown that the main obstacle was the long exposure required in a coherent electron beam, at least of minutes. At that time, the creep of the stage was of the order of an Angstrom per second, and contamination thickened the object by at least five Angstroms in the course of an exposure, thus completely frustrating the higher resolution. The long exposure time had another fatal effect. At the high resolution, the focal depth becomes very small, and the current in the magnetic objective, as well as the voltage, had to be kept constant within $\pm 10^{-6}$ ($\pm$ 0.1 volt in 10^5!) if the interference fringes in the hologram were not to be wiped out. M. E. Haine solved this problem by an ingenious regulator, but by that time the project had to be abandoned.

In modern electron microscopes, the creep of the stage is of the order of 0.1 Å or even less, and contamination has been as good as completely eliminated by cooling of the object chamber with liquid nitrogen; hence, we could afford exposures of 10 seconds. Can we obtain holograms in such times?

I have shown in 1951, that for every type of electron emitter there exists a certain coherent current; a limit which can be reached but not bettered by any careful design of the electron gun.[2] This limit, I_{coh} is

$$I_{coh} = (\pi h^2/32mkT)i_{em}$$

where h, m, k are the fundamental constants, T is the absolute temperature of the emitter, and i_{em} its emission current at T. For tungsten cathodes, I_{coh} is $4.3.10^{-14}$ amp. at T = 2800°K, $8.55.10^{-14}$ at T = 2900°K. These are about the highest safe operating temperatures, and the currents are much too small.

This was the situation in 1951. I believe that it has now appreciably improved by two developments. The first is the development of certain composite cathodes. Michael Albert, Mansur Atta and I have discovered in 1966, the curious fact that the emissivity of rare earth metals in a tungsten matrix can be improved by at least an order of magnitude by adding to them small quantities of a "mobilizer," which has the curious effect of enhancing the flow of the rare earth metals to the surface.[5] The best mobilizers were titanium, zirconium, and hafnium. In a tungsten-thorium-zirconium cathode, the emission density was 50 amperes/cm^2 at T = 1900°K, more than 100 amperes/cm^2 at T = 2000°K, and could be maintained for hours in a good vacuum. The last data correspond to $I_{coh} = 1.6.10^{-12}$ amperes, about twenty times more than obtainable with tungsten; hence, an exposure of five minutes could be reduced to 15 seconds, and would give about 1.5×10^8 electrons.

This is of the order of 100 bombarding electrons per resolution element in a 1000 x 1000 image, which can be considered as sufficient to overcome the shot noise.

Some patient work may be necessary to adapt our composite emitters to electron guns, but it is well possible that small emitter spots could be found which give densities even higher than what we have found in the average over large cathodes of more than one cm^2 area.

The second important development which has taken place in the meantime, and which may ease the problem of the coherent current, concerns field emitters. After decades of trying, several authors have been able to report steady currents of the order of 10^{-9} amps from point-emitters, so small that the emission density can be estimated to at least several thousand amperes/cm^2. These field-emitters require extremely good vacua because they are easily ruined by ion-bombardment, and I am not sure whether such vacua can be maintained in an electron microscope, even with the modern means of vacuum technique, such as titanium ion pumps near the electron gun; but if this could be done, there would be no difficulty with the coherent current, and holograms could be taken in a flash.

Another advantage which the electron-microscopist of 1970 would have over his colleagues of almost 20 years ago is the improvement in photographic material. The Agfa-Gevaert Scientia emulsions contain about 80% of silver bromide, in very fine grains of nearly equal size. If used in layers of two micron thickness, the resolution could be about 500 lines/mm, and a magnification of about 20,000 would be sufficient instead of 100,000 or more, which we had to use in 1950-1953.

In the case of electrons, in-line holography is very much simpler than holography with skew reference beams. In principle, the second method has now also become available, since G. Moellenstedt invented his "biprism," but the experimental complications are considerable, and the coherence requirements are prohibitive. Nor is it necessary, because I have shown[2] that in the reconstruction with the "in-line" method, the "conjugate image" is almost completely wiped out. In this reconstruction process, the investigator of 1970 has again a great advantage: When I made my suggestions, I thought only of using aspherical optical elements. But E. N. Leith and his collaborators, in particular, A. Vander Lugt, have already shown in 1964 that this can be achieved also with holographic filters, and more recently, G. W. Stroke and his collaborators have brought the correction of lens errors by photographically produced filters to a fine art.

Summing up, I believe that the way to a resolution limit of two Angstroms is now open for experimenters with great skill and patience.

SPECKLE-FREE ILLUMINATION

My second challenge has much in common with the first one, because it requires patient and skillful work. The great enemy of holography is laser speckle. Diffuse illumination of the object was a great progress in holography, but it brought us face to face with this phenomenon. If one illuminates an object with a single coherent wave (plane or spherical), there will be no speckle. But take two beams, and the object will appear hatched with interference fringes. Take many beams, for instance, issuing from a frosted glass, and it will appear badly speckled, owing to the random intereferences of the many waves whose amplitude sometimes add, sometimes subtract. If the hologram is sufficiently large, it will only faithfully reproduce the original speckle. But if the hologram is small, the effect will be enhanced, ultimately to the point where there is no likeness whatever between the reconstruction and the original.

On the other hand, diffuse illumination has great advantages. It was this that in the hands of E. N. Leith and J. Upatnieks first allowed the production of three dimensional reconstructions which could be viewed with two eyes. In a diffuse hologram, the information on any small part of the object is distributed almost evenly over the whole plate. Consequently, such holograms are very insensitive to scratches and dirt; even small parts of them can reproduce the whole object, though with a greatly increased "noise;" that is to say, speckle.

The challenge that I want to throw out is this: I can see a way theoretically for producing speckle-free reconstructions from a speckle-free illuminated plane object, from holograms which retain to a great extent, the advantages of diffuse holograms, because the information is widely (if not quite evenly) distributed over the plate. As in the case of the first challenge, the realization of the theoretical idea will require much skillful and patient work.

It is not possible to illuminate a plane evenly with more than one beam. There will always be interferences. But I will show that it is possible to illuminate a plane with many plane beams, at equal angular spacing from one another, so that all interferences vanish, with the exception of those of the two outermost beams; and these fringes are harmless if they are finer than the finest detail which we want to resolve. The effect will then be merely as if we had placed a grid over the plane object, with a spacing too fine to be seen.

Consider first the one-dimensional case. We illuminate the plane z = 0 with N plane waves, with the resulting amplitude

$$A(X) = \sum_n a_n e^{inKx} \qquad (1)$$

The resulting intensity is

$$I(x) = A.A^* = (\Sigma a_n e^{inKx})(\Sigma a_m^* e^{-imKx}) \qquad (2)$$

This sum contains N terms of the form $a_n a_n^*$ which represent <u>uniform</u> illumination, two terms,

$$a_1 a_N^* e^{-(N-1)Kx} + a_1^* a_N e^{(N-1)Kx}$$

which represent the interference fringes of the two extreme waves, and for the rest, fringes with wave numbers K, 2K...(N-2)K, which can all be suppressed if the following equations are satisfied:

$$
\begin{aligned}
a_1 a_2^* + a_2 a_3^* + a_3 a_4^* + \dots\dots\dots\dots + a_{N-1} a_N^* + \text{conj.} &= 0 \\
a_1 a_3^* + a_2 a_4^* + \dots\dots + a_{N-2} a_N^* + \text{conj.} &= 0 \\
\dots\dots\dots\dots\dots\dots\dots\dots\dots\dots & \\
a_1 a_{N-1}^* + a_2 a_N^* + \text{conj.} &= 0
\end{aligned}
\qquad (3)
$$

These are N-2 equations for the N-1 essential amplitudes, a_n. From a practical point of view, the <u>symmetrical</u> solutions are of interest, with N odd, so that <u>n</u> runs from $-\frac{1}{2}(N-1)$ to $+\frac{1}{2}(N-1)$. For N = 3 and N = 5, the solutions are easily found. The amplitudes are

N = 3		1	ik	1	
N = 5	1	ik	$\frac{1}{2}k^2$	ik	1

This shows a simple regularity. The coefficients are alternately real and purely imaginary, and in progressing from N to N + 2, only one new coefficient has to be calculated. The rule for calculating this from the preceding can be seen as follows: In passing from the solution for N to N + 2, only the equations for the first and second differences change. The others are solved by shifting the previous amplitudes to the right and left. In the equation for the first differences, three new terms arise and one drops out, while the rest is already zero. Denoting the new amplitudes (for N + 2) by underlining, this gives an equation

$$a_o a_o^* - a_1 a_1^* + \underline{a}_o a_1^* + \underline{a}_o a_1 = 0$$

In the equation for the second differences, there are two new terms:

$$\underline{a}_o a_o^* + \underline{a}_o a_o = 0$$

i. e., the new central amplitude is orthogonal to the previous one.

Solving these two equations (both real, so there is no need to add the conjugate), we obtain for the new central amplitude

$$\underline{a}_o = - a_o \frac{a_o a_o^* - a_1 a_1}{a_o a_1^* - a_o^* a_1} \tag{4}$$

With this rule, we now obtain the family of symmetrical solutions

N = 3				1	ik	1			
N = 5			1	ik	$-\frac{1}{2}k^2$	ik	1		
N = 7		1	ik	$-\frac{1}{2}k^2$	$-\frac{1}{8}ik(k^2-4)$	$-\frac{1}{2}k^2$	ik	1	
N = 9	1	ik	$-\frac{1}{2}k^2$	$-\frac{1}{8}ik(k^2-4)$	$-\frac{1}{64}(k^4-24k^2+16)$	$-\frac{1}{8}ik(k^2-4)$	$-\frac{1}{2}k^2$	ik	1

and so on.

The result is so far, that <u>if</u> we could illuminate a plane with such a group of plane waves, it would appear evenly illuminated, apart from the fine marginal fringes of the extreme waves. If we now take a hologram of it, it will be a multiple hologram (N overlapping holograms) with the advantage of diffuse holograms; <u>distributed information</u>. But evidently, this is not yet very practical advice, because producing these waves with the accurately prescribed amplitudes and phases is not an easy matter.

There is, however, a practical way of realizing these waves; let us make a <u>filter which has an amplitude transmission function</u>

$$t(x) = A(x) = \sum_n a_n e^{inKx}$$

<u>and illuminate it with a single, collimated beam, at right angles to it.</u> The amplitude and phase distribution behind this filter will be just $A(x)$. We have then only to put this filter in contact with the plane to be imaged. It will be speckle-free, and its image, or holographic reconstruction, will also be speckle-free, provided of course, that we do not cut out too many of the waves at some point of the optical tract.

In order to realize such a filter, write $t(x) = A(x)$ in the form $R.e^{i\phi}$. We can then realize this filter by putting together a photographic plate with an amplitude transmission proportional to R, and a pure phase filter with phase shift $\phi(x)$.

Using our previous equations, we obtain R in the form

$$R = (A.A^*)^{\frac{1}{2}} = \{\sum a_n a_n^* + 2a_1 a_N^* \cos (N-1)Kx\}^{\frac{1}{2}} = t_1(x)$$

because all other terms have dropped out.

We have only a uniform ("d.c.") term and the interference fringes of the two extreme waves. Only these will be visible in the plane z = 0 or its image, because $t_2(x)$ is a pure phase filter, invisible in its own plane. For this second filter, we obtain

$$e^{i\phi} = t_2(x) = \frac{\Sigma a_n e^{inKx}}{\{\Sigma a_n a_n^* + 2a_1 a_N^* \cos(N-1)Kx\}^{\frac{1}{2}}}$$

This is a rather complicated expression which can hardly be realized photographically. I propose to realize it photomechanically. The profile of t_2, which is proportional to $\phi(x)$ has to be calculated, drawn out on a larger scale with sufficient accuracy, and reduced to produce a shaping tool which can produce the phase filter in a ruling engine. The width (or period) of this shaping tool is a multiple of the resolution limit. For instance, in the case, N = 9, it is 8 times the period of the transparency t_1, and on this scale, the required accuracy need not be prohibitive.

We can now go a step further. We can cross at right angles, two such filters, and we again obtain one with the required properties, but one which now spreads out the hologram in two dimensions instead of one. The transmission of such a crossed filter is

$$t(x,y) = \sum_{n_x} \sum_{n_y} a_{n_x} a_{n_y} e^{i(n_x x + n_y y)K}$$

Working out the transmitted intensity, proportional to $t.t^*$, one finds that this will again consist of d.c. terms, plus the two marginal fringe systems, at right angles, plus two new systems proportional to

$$\cos (N-1)K(x+y) \text{ and } \cos (N-1)K(x-y),$$

at 45°, which are also harmless, as they are beyond the resolution.

I am throwing out the challenge of realizing this filter, which, I am convinced, would represent an important step in the art of holographic information storage in small sizes.

REFERENCES

1. D. GABOR, Proc. Roy. Soc. A. 197, 454, 1949

2. D. GABOR, Proc. Phys. Soc. B. 64, 244, 1951

3. M. E. HAINE AND J. DYSON, Nature, 166, 315, August, 1960 (Basic Idea)

M. E. HAINE AND MULVEY, Optical Soc. of America, 42, No. 10, PP 763-77 October 1952 (Analysis and Application)

M. E. HAINE AND JERVIS, Proceedings of IEE, 102, Part b, No. 3, May, 1955 (H.V. Stabilizer)

4. D. GABOR, Louis de Broglie et les Limites du Monde Visible, in Louis de Broglie, Physicien et Penseur, Paris, 1953

5. M. J. ALBERT, M. A. ATTA AND D. GABOR, A New Type of Composite All-metal Electron Emitter for Valves and Electron-optical Devices, Brit. J. Appl. Phys. 18, 627, 1967

M. J. ALBERT AND M. A. ATTA, Thermionic Electron Emitters Based on the Alloys of Rare Earth Metals, Metals and Materials, February 1967, P. 43

THE PHOTOGRAPHIC EMULSION LAYER AS A THREE-DIMENSIONAL RECORDING MEDIUM

W.F. Berg

Photographic Department, Swiss Federal Institute of Technology, Zürich

INTRODUCTION

If we want to understand how the basic characteristics of a photographic picture come about, we have to take the depth of the image-recording and -bearing layer into account. The characteristics of interest are the density (D), the granularity (G) and the modulation transfer function MTF. Here MTF is concerned with the distribution of the exposing light in the emulsion layer, whereas D and G describe the nature of the resulting developed granular silver image, as a rule referring to uniformly exposed areas.

The MTF is a complete description of the distribution of the exposing light in three dimensions arising from sinusoidal distributions of light of various spatial frequencies incident on the surface of the emulsion layer; processing then results in a three-dimensional distribution of density, from which, if the density-exposure relationship is known, the MTF can be worked out. This is one of the major difficulties in determining MTF: the light-sensitive material itself has to be used to record the distribution of light, and the density-exposure relationship ought to be the "micro-characteristic curve" applying to uniformly exposed areas of size close to the inverse of the spatial frequency for any one point of the MTF. Ideally, these micro-characteristic curves should be determined by varying the exposures to the sinusoidal test object in a known manner by an amount corresponding to their modulation[1]; in practice it is common to utilize the "macro-characteristic curve" (spatial frequency zero) in the hope that it will not differ materially from the micro-curve, which will be true if development is full, thus avoiding neighbourhood effects.

The density distribution in depth is a special case of the application of the concept of MTF to uniform exposures of large areas, i.e. spatial frequency zero, and is singled out for its importance in tone reproduction. For full development, again from the density distribution can be worked out the exposure distribution; it is necessary, however, to prove in each case that development was even in depth.

It can be shown that D, G, and MTF are additive if emulsion layers are placed on top of one another, and if exposure and read-out are in essentially parallel light. If therefore, one wishes to investigate the characteristics of a given emulsion layer, one has to subdivide it into a number of elementary layers (ELs). Additivity means that if we want to learn the characteristics of E_i, all we have to do is to wear down the emulsion layer to E_i, measure the property in question, remove E_i, measure again and take the difference.

Ideally, the ELs should be infinitely thin; in practice, since they should also be independant of one another, and since the grains are of finite size, they will have to be fairly thick. Useful data are obtained by a compromise, subdividing a layer into rather fewer than 10 ELs.

In many instances, exposure and/or read-out are not in parallel light. If the depth-of-field of the optical system is smaller than the depth of the layer in which the image is recorded, the optical system will have different transfer functions according to the focusing position. Fig.1 shows the MTF curves of a diffraction-limited lens for different focusing positions as worked out by Wollenmann[2] from formulae by Hopkins[3]. The way in which images are recorded and read-out under these conditions is poorly understood.

EXPERIMENTAL REALIZATION OF ELEMENTARY LAYERS

Two experimental methods have been established for realizing ELs[4,5]. In one, which is specially suited for uniformly exposed and developed emulsion layers, an oxidizing agent is produced electrolytically; this reacts layerwise and quantitatively with the silver[6], and hence not only the density is reduced, but the amount of silver associated with it is determined. We can thus find the photometric constants of the individual ELs.

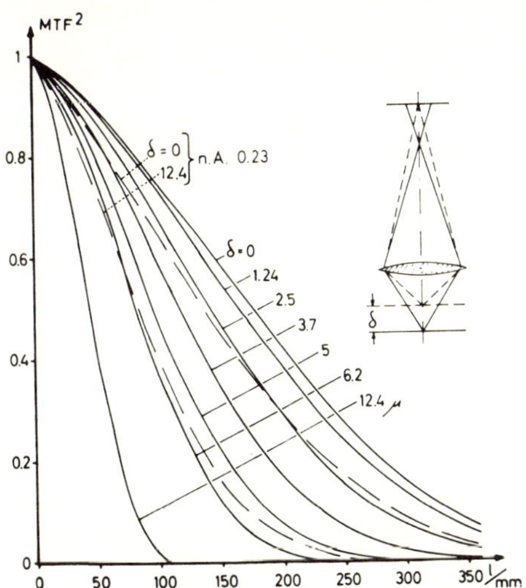

Fig.(1) MTF curves for off-focus positions for a diffraction-limited lens of numerical aperture 0.46, in conjunction with a 2-μm wide slit. Dotted curves for n.-a. 0,23.

The other method is an enzyme attack on the gelatine, which seems to be fairly independant of the presence or absence of silver in the layer. This method is specially useful for structured images.

DISTRIBUTION OF EXPOSURE IN DEPTH

The original assumption was that the distribution of exposure in depth was exponential. Theoretical work by Metz[7] has shown that this is but approximately correct and only for the case of strongly absorbed light. For less strongly absorbed and more strongly scattered light, the distribution falls off essentially linearly in depth, and for both conditions there is an initial rise.

The relationship of outer exposure (in mcs) and the inner exposure (total light flux through a volume element of an emulsion layer) is remarkable; for a film for pictorial work inner exposure is approximately 8 times the outer[8,9,10]. This is due to scatter of light within the emulsion layer; the effective sensitivity of a film is achieved at the expense of structural image quality.

Recent experimental work by Strübin[4] has confirmed Metz' data. The density distribution for different exposures enabled the characteristic curves for the various ELs to be obtained. From their position on the log exposure axis the exposure distribution in depth

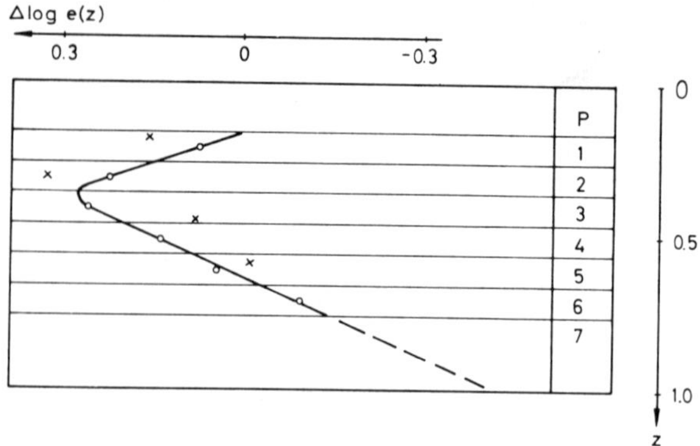

Fig.(2) Exposure distribution in depth with a Process Film (Typon FRP), after Strübin and Franchetti[4,5]. The circles relate to enzyme attack, the crosses to electrolytic oxidation.

was derived, as shown in fig.2. The initial rise in exposure was greater than that found by Metz but exposure conditions were very different.

DISTRIBUTION OF GRANULARITY IN DEPTH

Here the most significant information is in the form of the power spectrum[11], here written down as one-dimensional

$$N(\nu) \cong \left| \int \Delta T(x) e^{-\iota 2\pi\nu x} dx \right|^2$$

in effect a Fourier analysis of the transparency deviations $\Delta T(x)$ from their mean. Riva[12] determined this from a microdensitometric scan, recording directly on magnetic tape and carrying out the analysis on a digital computer.

Fig.3 shows noise power curves for a film carrying a silver image for a uniformly exposed and fully developed process film, after removal of various fractions by electrolytic oxidation. The shape of the curves for the lower layers differs materially from that for the bulk layer; the lower layers make a relatively large contribution to the lower spatial frequencies. The interpretation is not clear. The effect may be due to an artefact: finer grains may be oxidized preferentially; this is not too likely in view of the filamentary nature of developed silver. A more fundamental reason might be that exposure, being attenuated in depth, will

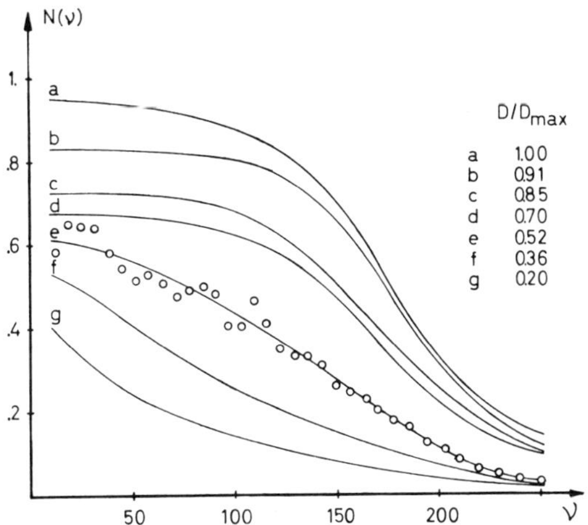

Fig.(3) Power spectra of the developed silver image in an emulsion layer (Typon FRN), from which various fractions have been removed as indicated. The original density was 1.18. For the lower layers, the noise power is relatively high for the lower spatial frequencies.

pick out the most sensitive i.e. the largest grains in the lower layers.

Another manifestation of the effect of depth on granularity arises when one attempts to obtain a measure of granularity through the Fraunhofer spectrum[13]. The granular deposit is simply set up as the diffracting element in a spectrograph, a scanning slit in the Fourier plane acting as a band-pass filter. The resulting spectrum is that of fluctuation of amplitude $\Delta a(x)$ - as against transparency = amplitude^2 in the power spectrum

$$W(\nu) \sim \left|\iint \Delta a(x)\, e^{-i2\pi\nu x} dx\right|^2$$

Although physically different from the power spectrum, this amplitude spectrum would be an attractive measure of granularity and very simple to determine. However, the method does not work, as shown by Hautot et al[14] and more recently by Riva[12]. Fig.4 shows measurements indicating that all one sees is evidence for scatter of light, the blue light being deviated more strongly than

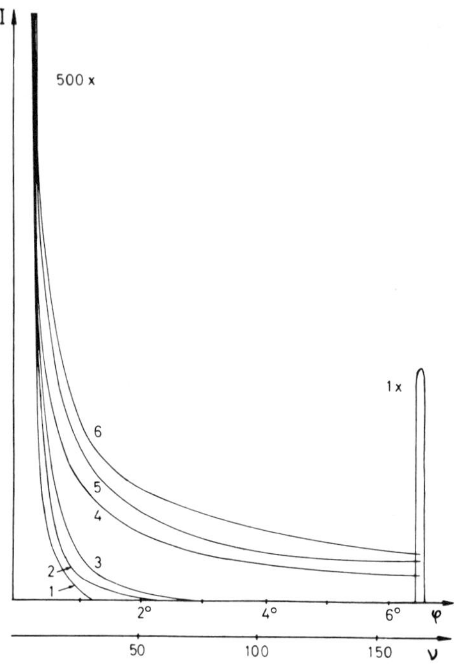

Fig.(4) Amplitude "spectra" from a processed process film Typon FRN, D = 0.4. ν represents a frequency scale for λ = 546 nm (1) was measured without film, (2) with film base, both for the blue and the green Hg-line; (3) for the whole Hg spectrum. (4),(5),(6) are measurements with the green line, the whole spectrum and the blue line respectively.

the green - just the opposite of what would occur with diffraction.

One concludes that there must occur not only amplitude changes, but also shifts in phase, due to the arrangement of the grains in depth and to local tanning effects on the gelatin by the development of the grains.

DISTRIBUTION OF MTF IN DEPTH

The Fraunhofer diffraction method provides a simple scheme for determining MTF of emulsion layers[15]. A coarse (2 or 5 lines/mm) square-line grating is printed on the material under test; the resulting grating is set up as the diffracting element in a spectrograph. If light and dark lines are equal in width, the even orders in the spectrum will be missing, provided the recording process is linear: transparency is a linear function of exposure. This was fulfilled up to densities of the order of 0.8.

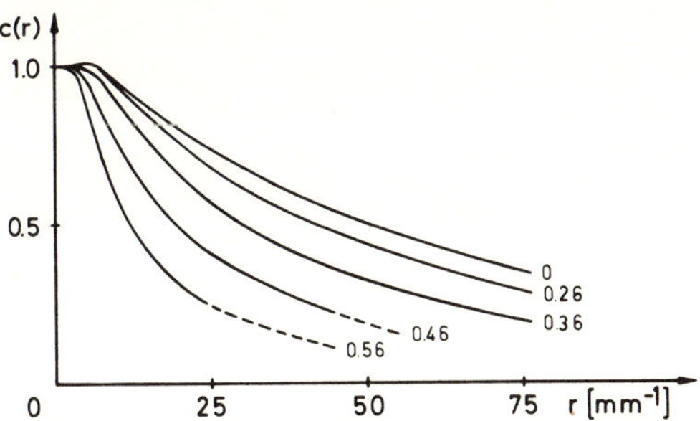

Fig.(5) MTF curves of a process film, (Typon FRP) after the fractions indicated have been removed. Normalization: MTF = 1 for $\nu = 0$.

Fig.5 shows MTF curves for emulsion layers from which the fractions indicated had been removed by enzyme action[4]. The normalization adopted was to give the MTF the value of 1.0 for $\nu = 0$. The MTF for the lower layers is found to be strongly degraded.

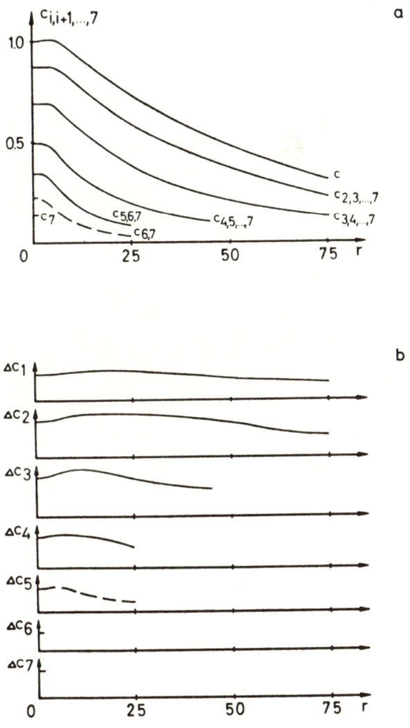

Fig.(6) Top: The same data as fig.(5), but normalization taken from fig.(2). Bottom: Difference between any two adjacent curves above give MTF curves for the individual ELs.

Fig.6, top, shows the same data replotted, using the light distribution in depth for $\nu = 0$ (fig.2) as basis for normalization. The difference between any two adjacent curves represents the MTF for one of the individual ELs. The MTF for the top layer is flat up to high spatial frequencies; that for the lower layers is progressively degraded.

SUMMARY AND CONCLUSIONS

Methods are described for investigating the distribution of the exposing light and that of the resulting density, as well as that of granularity in the depth of an emulsion layer. Experimental methods for studying the elementary layers of an emulsion coating have been established. The MTF for the lower emulsion layers was found to be degraded for high spatial frequencies; at zero frequency exposure first rises below the surface, to fall off exponentially further down into the coating. The power spectrum of the lower layers makes a relatively large contribution at low spatial frequencies.

These preliminary results were obtained on process films. No results are available yet on emulsion layers of the type used in holography.

LITERATURE REFERENCES

(1) De Belder, M., Jespers, J., Verbrugghe, R., Phot.Sci.Engng.9, 314 (1965).

(2) Wollenmann, H.P., (ETH, Zürich) unpublished calculations.

(3) Hopkins, H.H., Proc.Roy.Soc. (London) 231, 91 (1955).

(4) Strübin, H., Phot.Korr. 104, 5, 26, 53 (1968).

(5) Franchetti, F. (ETH, Zürich), unpublished work in part fulfilment of the requirements for a Ph.D. degree of the University of Padova, Italy.

(6) Causer, R.L., Smith, C.H. "Photographic Science", ed. J. Pouradier, p.495 (Focal Press, London 1967).

(7) Metz, H.-J., "Die Tiefenabhängigkeit der Belichtung im Inneren einer photographischen Emulsion", Diploma Thesis, T.H. München (1964).

(8) Frieser, H., Klein, E., Mitt.Forschungslaboratorien Agfa, Leverkusen-München, 2, 1948 (1958).

(9) Farnell, G.C., J.phot.Sci.7, 83 (1959).

(10) Spühler,A., Trautweiler, F., J.phot.Sci.12, 57 (1964).

(11) Jones, R.C., J.Opt.Soc.Amer. 45, 799 (1955).

(12) Riva, R., Phot.Korr.105, 111, 128, 143, 159 (1969).

(13) Thiry, H., Phot.Sci.Engng.4, 19 (1969); J.phot.Sci.11, 121 (1963).

(14) Hautot, A., Ooue, S., "Photographic Science", ed. G.Semerano, U.Mazzucato, p.148 (Focal Press, London 1965).

(15) Thiry, H., Hautot, A., Berwart, L., Phot.Korr. 104, 64 (1968).

3-D CONSTRUCTION OF IMAGINARY OBJECTS BY THE METHOD OF HOLOGRAPHIC STEREOGRAM

Tadashi Kasahara, Yoshiaki Kimura and Masanori Kawai

Research & Development Laboratory

Konishiroku Photo Ind. Co., Ltd. Tokyo Japan

1. INTRODUCTION

The methods of reconstruction of three-dimensional objects may be divided into the following three groups. The first includes the ordinary holographic reconstruction[1)2)] of objects which are illuminated with coherently diffused light. The second utilizes the so-called integral photography[3)4)] which was studied initially during the first decade of this century. Recently, it has been developed by the introduction of holographic techniques, based on studies carried out by Pole[5)], Collier[6)] and his co-workers, and many others. The third utilizes "Stereogram".[7)] Similar but imperfect reconstruction device for the third method is called commercially: "Lenticular Stereogram".

The reconstruction of three-dimensional objects which utilize "Stereogram" has the following characteristics:

(1) It makes use of rather small information on images, compared with the hologram or the integral photography.

(2) Reconstruction of the object spacing may be not always approximated to the original object spacing. Furthermore, there may be no three-dimensional information along vertical direction especially.

(3) In spite of the imperfect spatial reconstruction as mentioned above, it results in sufficient three-dimensional sensations. Sometimes, it makes it possible to get numerical data concerning the spatial distribution of reconstructed objects.

When characteristics of the stereogram mentioned above is combined with holographic techniques, there seem to be several methods of three-dimensional reconstruction, which cannot be obtained by the ordinary hologram or the integral photography.

Some of these methods were discussed by McCrickerd[8] about a year ago. Applications of such methods to radiographic diagnosis were reported by Redman[9] at about the same time and one of us[10] after that. These may be considered as one of the modified holocoder holograms originated by Pole,[5] from which various developments have been carried out by him.

Originally, three-dimensional reconstructions are so inevitably connected with video communication facilities that many authors have talked about the bandwidth compression of 3-D images and the holography when they are transmitted by communication lines. Leith and Upatnieks reported[11] the necessary bandwidth to transmit the holographic information and Burckhardt calculated[12] information contents of the integral photography. Of course, results of these calculation require a quite broad bandwidth and some suggestions on the bandwidth compression have been proposed by DeBitetto[13] and other authors.[14]

The holographic stereogram has special capability of constructing the three-dimensional image of an imaginary object, although it has been developed on the same line as the holocoder hologram and the integral photography. In case of the coherent illumination, a perfect three-dimensional reconstruction of the object can be achieved holographically, and in case of the incoherent illumination, the method of the integral photography might be used as a substitute. If the reconstruction of objects is desired at various magnifications, methods mentioned above are affected by intermediate optical systems and suffered from some sort of spatial distortion.

In other cases, it is often desired to reconstruct the 3-D object which cannot be seen by naked eyes. For example, it is desirable to look into the human body internally by X-ray radiation, to magnify a micro structure of substance by electron radiation, or to clarify a purely imaginary subject by the aid of displaying devices of an electronic computer.

The method of holographic stereograms would be applied advantageously to solve these problems instead of using the ordinary hologram or the integral photography.

2. DISPLAYING THROUGH A WINDOW

An object is reconstructed through the plane of the holographic stereogram. This situation is treated as the reconstruction through a window, which is similar to the integral photography or the ordinary hologram. Instead of reconstructing a wavefront by a hologram,

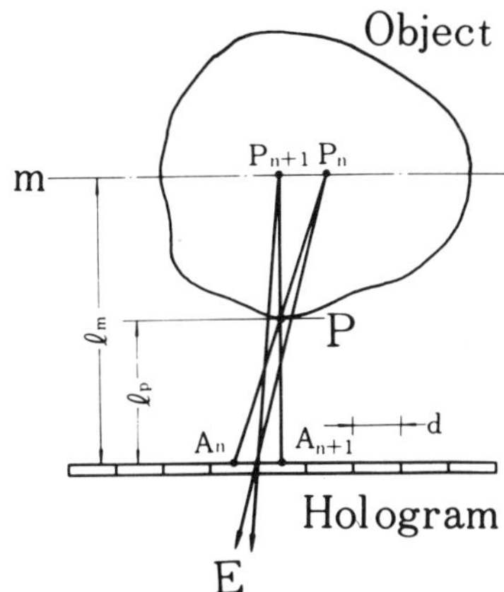

Fig. 1. 3-D Reconstruction of the Object by the Method of Holographic Stereogram.

a geometrical ray distribution at the plane of the window may be constructed approximately by a plate of the holographic stereogram.

The holographic plate includes small elementary holograms which usually have a shape of vertically elongated rectangular strips and are arranged contiguously. Each elementary hologram contains the image information characterized by the perspective, and its projection center is located at the midpoint of the elementary hologram. Boundaries of the elementary holograms act as an aperture for imaging. This means that the reconstruction property of the holographic stereogram may be treated in the same way as the integral photography except for the information storing mechanism for sensitized materials and an aberration of imaging systems. Assuming that a photographic emulsion of the holographic plate has enough quality, each elementary hologram works as the combination of a lens system with the emulsion.

In Fig. 1, each elementary hologram reconstructs a 2-D image of the object observed from the mid point of the elementary hologram onto about the middle plane of the object, which is marked as m. The eye of an observer, positioned at E, will be able to see the fragment of the image through each elementary hologram. As a result, synthesized sensations will be given as an image of the whole object. When the observer moves his eye horizontally, he will be able to find different aspects of the object and will have a full three-dimensional sensation when he observes binocularly. Since the width of the elementary hologram is sampled by finite interval d, the object point P located at some distance from m will be projected as different points P_n and P_{n+1} on the plane m by the contiguous holograms A_n and A_{n+1}. This might cause some discontinuity or flickering when projected images are quite sharp and the observer distinguishes P_n from P_{n+1} as a separate point.

In order to reduce laborious work for synthesizing each elementary hologram, it is desirable to minimize the number of sampling points. Sometimes one might be obliged to reduce the dose of X-ray radiation for safety of patients so that a critical sampling interval should be required. On the other hand, quite a narrow sampling interval for high fidelity reconstruction of the object can be impractical because of diffraction of each edge of the elementary hologram. Consequently, the sampling interval of the width of elementary holograms should be chosen so as to secure image sharpness and at the same time to avoid any sensation of discontinuity. According to various construction procedures of the holographic stereogram, the quality of the projected image will be suffered from some sort of degradation. In order to calculate the optimum sampling interval, the point spread function of the holographically reconstructed image has to be examined.

Various types of the point spread function can be considered. For instance, Airy disk is important in the case of an ideal construction procedure. The circular disk type distribution function is necessary for treating out-of-focus images which have been recorded by an ordinary photographic objective. In some complicated construction procedure, the Gaussian distribution function plays an important role. Here, a diffraction limited image is treated below as an example.

Assuming that no serious degradation affects a projected image quality of the point P and that the pupil diameter of the eye of the observer is rather small, reconstructed images P_n and P_{n+1} cannot be distinguished from each other, when

$$\overline{P_n P_{n+1}} \leqq \frac{1.22\lambda}{\alpha / \ell_m} \quad , \tag{1}$$

where α is the diameter of the pupil of the observer's eye and λ is the wavelength of the reconstructing radiation. While, the distance $\overline{P_n P_{n+1}}$ is given as

$$\overline{P_n P_{n+1}} = d \left| \left(\frac{\ell_m}{\ell_p} - 1\right) \right| \quad , \tag{2}$$

and consequently, we obtain

$$d \leqq 1.22 \frac{\lambda}{\alpha} \left| \left(\frac{1}{\ell_m} - \frac{1}{\ell_p}\right)^{-1} \right| . \tag{3}$$

In this expression, $1.22\frac{\lambda}{\alpha}$ corresponds to the angular resolution of the eye and is estimated to be about 1/3000 radian in ordinary condition, and (3) is simplified to

$$d \leqq \frac{1}{3000} \left| \left(\frac{1}{\ell_m} - \frac{1}{\ell_p}\right)^{-1} \right| \quad , \tag{4}$$

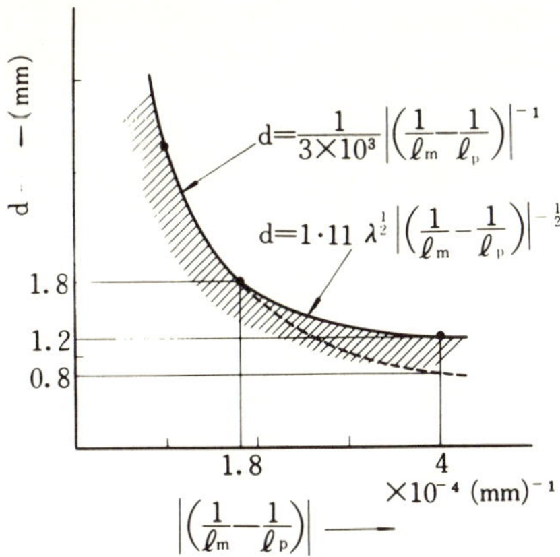

Fig. 2. Permitted Width of the Elementary Hologram for Avoiding Discontinuous Sensation.

where absolute value is adopted as the point where P is located before or behind the plane m.

When the sampling interval is reduced below certain value, e.g. about 1.8mm, the width of each elementary hologram acts as a finite aperture and α in the expression (3) has to be substituted by d, we obtain

$$d < 1.11\lambda^{\frac{1}{2}}\left|\frac{1}{\ell_m}-\frac{1}{\ell_p}\right|^{-\frac{1}{2}} . \tag{5}$$

This means that, when the width of the elementary hologram becomes narrower than the pupil diameter of the eye, the observed image quality is blurred by the diffraction of the edge of each hologram and, correspondingly, sampling intervals larger than (4) are permitted. (Fig. 2.)

3. RECONSTRUCTION OF ORDINARY SCENES

The procedure for producing the holographic stereogram of an ordinary scene is illustrated in Fig. 3. 2-D images of the object are recorded on the photographic film by the ordinary camera which can be moved in succession by interval d. After processing as a positive, each image on the photographic film is illuminated by coherent light and recorded as the elementary hologram having the width d'. If the object should be reconstructed at infinity or around a plane far apart, a projecting lens system must be placed between the photographic film and the holographic plate, and the aperture of the

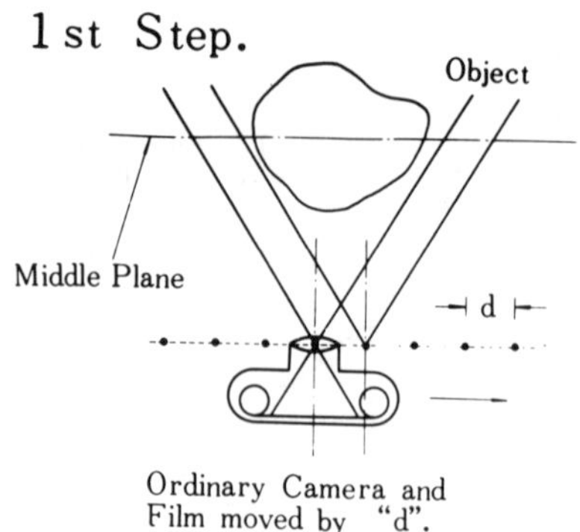

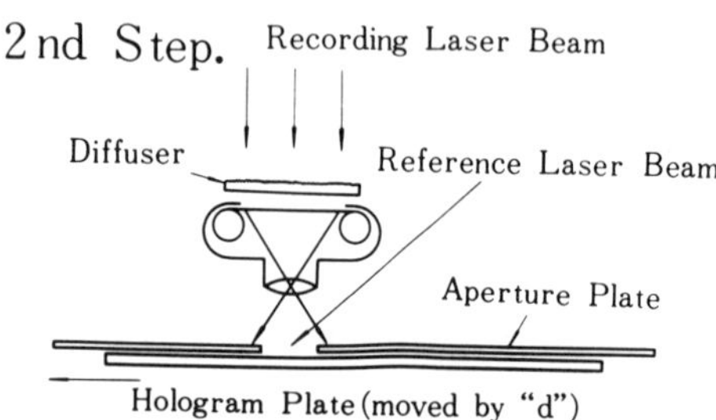

Fig. 3. Typical Procedure for Fabricating Holographic Stereogram

lens should have enough diameter to cover the area of the elementary hologram. Many variations of the constructing procedure can be considered. In other words, the designing of each step is affected by the following factors.

(1) Magnification of the reconstructed object space.
(2) Distance of the reconstructed object from the holographic stereogram.
(3) Allowable distortion of the reconstructed image space.
(4) Presence or absence of the 3-D information along with the vertical direction.
(5) Required image sharpness of the most important object point.
(6) Depth of the reconstructed object space.
(7) Allowable image blur associated with the depth of focus of the photographic objective at the marginal plane of the object space.
(8) Final configuration of the visual display system.

In these factors, the image sharpness, the depth of the reconstructed space, and the magnification relate one another. Therefore, the systematic discussion is necessary. However, discussion here is limited to the life size reconstruction of the ordinary scene.

In the first step, the photographic objective is focused onto the plane containing the most important subject. This plane has to be spatially reconstructed precisely at the same position by each elementary hologram. That is the reason why this plane is designated as the middle plane. Assuming that the resolving power of the photographic film used in the first step is large enough and the aberration of the photographic objective can be neglected, the image quality recorded on the photographic film depends only on the aperture of the photographic objective. As is well known, resolving distance obtained in the object space is given, by

$$\delta = \frac{1.22}{D/\ell_m} \quad , \tag{6}$$

where D is the diameter of the photographic objective and λ_o is the wavelength of the radiation used at the first step. For the object point which is positioned at distance ℓ from the photographic objective, the image quality is not seriously degraded when the wave aberration associated with the defocusing does not exceed certain value, $\lambda_o/4$. This condition can be formulated as the expression

$$\frac{\lambda_o}{4} > \frac{D^2}{8} \left| \left(\frac{1}{\ell_m} - \frac{1}{\ell}\right) \right| \quad . \tag{7}$$

If it is necessary to reconstruct object points which are located at distances between ℓ_1 and ℓ_2 in the equal sharpness, equations

$$\ell_m = \frac{2\ell_1 + \ell_2}{\ell_1 + \ell_2} \quad , \tag{8}$$

$$D = 2\sqrt{\frac{\ell_1 \ell_2 \lambda_o}{\ell_1 - \ell_2}} \, , \tag{9}$$

should be satisfied. Equations (8) and (9) can be obtained by substituting ℓ_1 and ℓ_2 with (7). The necessary width of the elementary hologram can be calculated by (7) and (3), in which ℓ is substituted for ℓ_p, that is

$$d < 0.61\frac{D^2}{\alpha} \quad . \tag{10}$$

When the quite sharp reconstruction of the whole subject is desired, the expression (10) gives an important relation among the diameter of the photographic objective, the width of the elementary hologram, and the pupil of the observer.

If the object is large in depth, and the equally sharp reconstruction of the whole object is required, the value D given by (9) is reduced below α. In this case, the expression (5) must be used to calculate the necessary width of the elementary hologram and

$$d < 0.785D \tag{11}$$

is obtained. The sharpness of the reconstructed object is inferior to the original scene observed by a naked eye.

Some special displaying techniques can be considered. When the depth of the object is too large, the out-of-focus image blur could be permitted at marginal parts of the object. As is well known, the point spread function of the out-of-focus image has the form of a circular disk and an easy calculation gives the width of the elementary hologram as follows,

$$d < \frac{\sqrt{3}}{2} D \quad = 0.87D \quad . \tag{12}$$

When the photographic objective has a rectangular aperture of horizontal width D_R, the similar result is obtained as (12),

$$d < D_R \quad . \tag{13}$$

Based on expressions (10), (11), (12), and (13), it is found that when the diameter of the photographic objective has nearly the same value as the width of the elementary hologram, the observer feels neither discontinuity nor flickering at any condition. This is the case of a 3-D reconstruction employing the integral photography. In other words, the integral photography can be treated as the special case of the holographic stereogram.

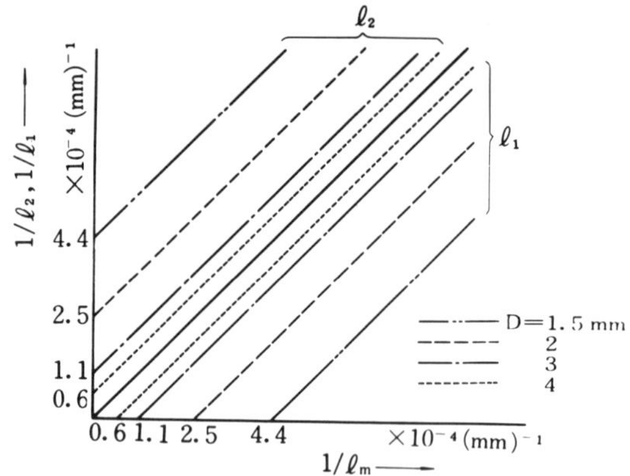

Fig. 4. Range of Object Space Where Enough Sharpness is Secured.

Fig. 4 shows the range of the object space which is secured by an ordinary naked eye to obtain the sufficient sharpness of the reconstruction.

4. MAGNIFIED AND DEMAGNIFIED OBJECT SPACE

In order to reconstruct the magnified or demagnified object space, an assumption is made that the angular equality of the space should be maintained, or some distortion of the space is introduced and no faithful reconstruction could be expected. This means that the perspective of the original object at the principal point of the photographic objective has to be definitely kept at the middle point of each elementary hologram. When the middle plane which contains the most interesting point of the object is magnified photographically by M, the distance from the reconstructed middle plane to the hologram should be selected to be $M\times\ell_m$, where ℓ_m is the distance of the main object point measured from the principal point of the objective.

Resolvable distance of the objective at the first step is given as (6) and this can just be resolved by the eye of the observer at the final step, when the condition of (14) is satisfied:

$$M\delta = \frac{1.22\lambda}{\alpha/M\times\ell_m} \quad . \tag{14}$$

This results in

$$D = \frac{\lambda_o}{\lambda}\alpha \quad . \tag{15}$$

The depth of focus given by the $\lambda/4$ criterion is already shown in (7). Considering in the same way as applied for (3), the allowable width of the elementary hologram is given as

$$d < 1.22\frac{\lambda}{\alpha}\left|\left(\frac{1}{M\cdot\ell_m}-\frac{1}{M\cdot\ell}\right)\right|$$

$$= 1.22\frac{\lambda}{\lambda_o}\frac{D^2}{2}M = 0.61\frac{\lambda_o}{\lambda}M \quad . \tag{16}$$

The implication of (15) and (16) will be examined as follows, applying to the case of 3-D reconstruction of the microscopic object.

Assuming that the object is reconstructed at the distance of 250mm behind the holographic stereogram, that is, $M\times\ell_m=250$, and $\lambda=\lambda_o=0.5$micron, the range of the object space where sufficient sharpness is secured by (7) is plotted in Fig. 5. In this case, $\alpha=D=1.5$ mm. According to this diagram, the range of focus at M=10 is about 0.28mm and is only 0.028mm at M=100. In the case of M=10, the magnified reconstructed object will be sharply displayed within 2.8mm but only 0.28mm for M=100. The corresponding width of the elementary hologram is 9.2mm for M=10, and 92mm for M=100. The latter case is meaningless as the value is far beyond the length of human binocular base.

It may be concluded that the holographic stereogram has little practical significance, when the magnification is larger than 10. For this magnification one would be able to observe only the depth of 2.8mm sharply and this value diminishes in inverse proportion to the magnification. This situation can be improved, however, when the radiation having different wavelength is employed to record the original object. For example, an electron microscope might be used to reconstruct a 3-D micro structure with enough depth.

In order to reconstruct the demagnified object space, some care should be taken to avoid the discontinuity or the flickering, because in ordinary case, the depth of focus of the photographic objective is so large that the blur of the out-of-focus image could hardly be expected. For such case, the width of the elementary hologram is calculated by (3) or (5). If sufficient sharpness of the reconstruction is necessary, the extent of the object has to be confined as follows,

$$\alpha^2 < 1.22M\lambda \left| \frac{1}{\ell_m} - \frac{1}{\ell} \right|^{-1} \tag{17}$$

where α is the diameter of the pupil of the observer's eye.

Fig. 5 shows the effective range of the magnified or demagnified object space which can be reconstructed by the holographic

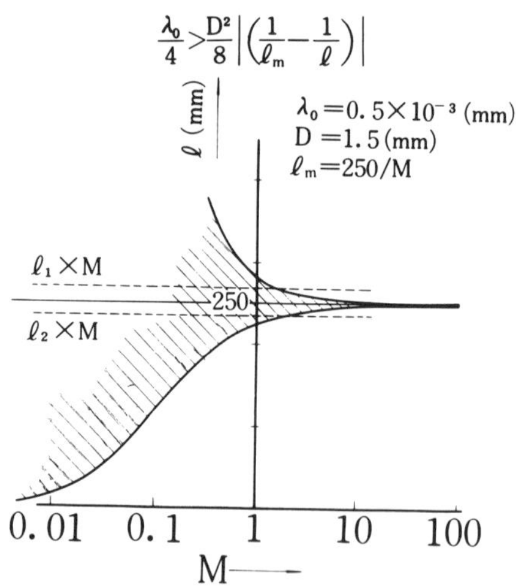

Fig. 5. Range of Sharp Focus When the Object is Magnified "M" Times at a Distance of 250mm.

stereogram. When the whole object point is contained in the shaded area of this diagram, the sharpness of the reconstruction is satisfactory, while the points included within the two broken lines will give no sensation of the discontinuity. These conditions seem to be rather severe and in practice the slightly larger depth of the object space may be permitted. In this diagram, the distance of the reconstructed middle plane from the holographic stereogram is given as 25cm. If the different values were selected as that distance, the effective range of the demagnified object space changes, but the range of magnified object space will keep constant.

5. RECONSTRUCTION OF A RADIOGRAPHIC IMAGE OF THE OPAQUE OBJECT

In the case mentioned above, the characteristics of the holographic stereogram have been affected remarkably by the nature of the optical system employed at the first step. When the holographic stereogram is applied to radiography, the difference of the first picture taking process should be examined.

A typical example of the constructing procedure is shown in Figs. 6 and 7. These figures imply that, when the life size and undistorted reconstruction is required, the plane of the stereogram has to coincide with the plane of the source of the radiation. The middle plane which is mentioned in the previous section may not

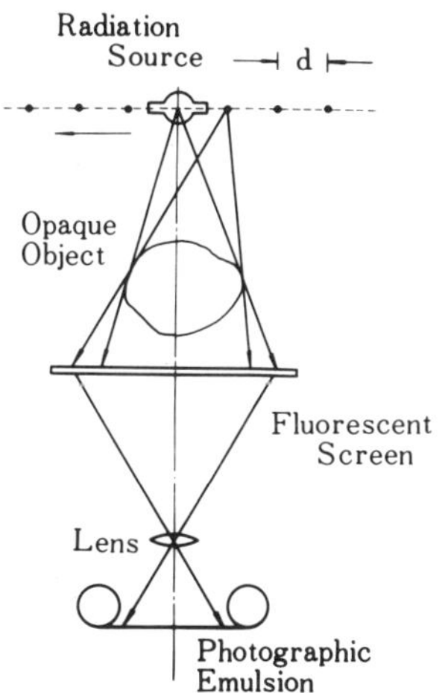

Fig.6. Taking of a Sequence of Ordinary Radiographs.

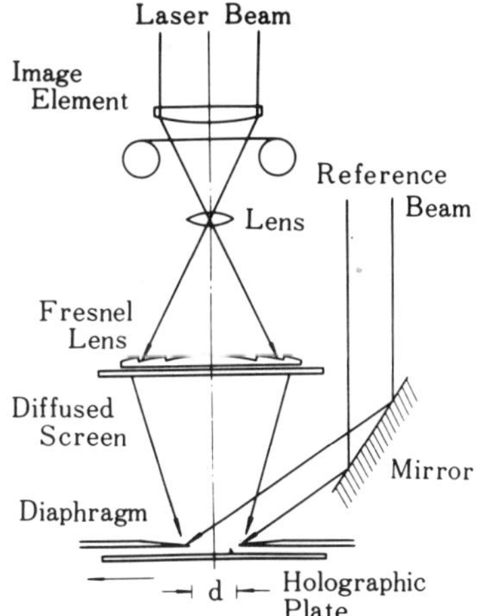

Fig.7. Synthesis of Holographic Stereogram.

be selected distinctly in the first radiographic process, because this imaging system has no focusing action. This plane must be selected so as to coincide with the most interesting part of the object and usually is located at about the center of whole object points. The discontinuity and the flickering of the reconstructed object could be minimized only when the object point contained in this plane is projected on the same position.

The image quality projected on this plane is determined by the geometrical dimension of the radiation source a, the distance between the radiation source and the object point ℓ, a selected distance between source and the middle plane ℓ_m, and the distance between the source and the fluorescent screen ℓ_s. These are shown in Fig. 8. The resulted image blur γ of the point P projected on the middle plane is given as

$$\gamma = a\ell_m \left(\frac{1}{\ell} - \frac{1}{\ell_s}\right) \quad . \tag{18}$$

When the value γ does not exceed the resolving limit of the observer's eye positioned behind the radiation source, the displayed image could be sharp enough and the expression (4) might be used to avoid the flickering. If the values of γ given by the expression exceed the resolving limit of the eye, a slightly different rule could be applied to select the position of the middle plane and to calculate the width of the elementary hologram.

Fig. 9 shows another setup for producing the 3-D radiographic display. In this instance, the radiant point source of the refer-

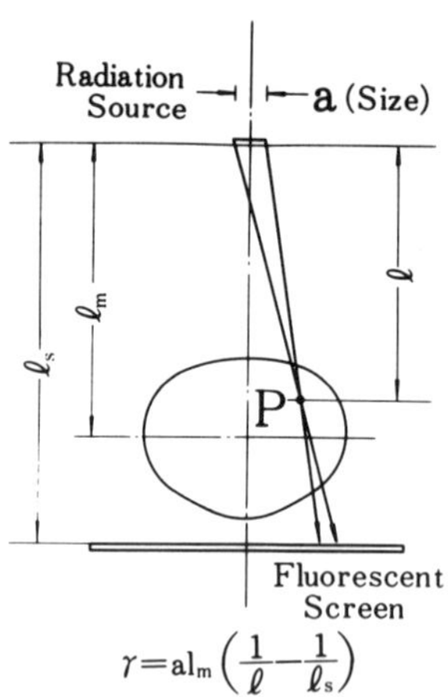

Fig. 8. Typical Radiographic Imaging System

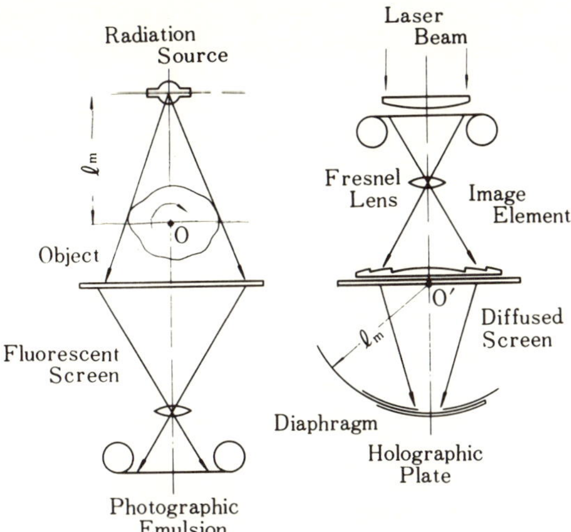

Fig. 9. Holographic Stereogram System Which Permits 360° View of the Opaque Object.

ence coherent beam is located on the vertical axis around which the object is rotated and thus the reconstruction of 360° view of the object will be given. Radiographic images are projected on the plane which includes the vertical axis and this can be treated as the middle plane.

6. RECONSTRUCTION OF A PURELY IMAGINARY OBJECT

The three-dimensional reconstruction of the purely imaginary object can easily be achieved using a cathode ray tube computer display device. When each coordinate of the imaginary object point is given to the computer, the appearance observed from many different viewpoints can be displayed according to the law of the perspective projection, and then images on the display device are photographed and synthesized to the holographic stereogram. In this case, the position of the middle plane should also be selected so as to minimize the flickering.

The sharpness of the reconstructed image is wholly affected by the point spread function obtained by the cathode ray tube, and this can be considered as the Gaussian distribution function. Assuming that, in the reconstructed object space, the diameter of the distribution function is much larger than the resolving distance of the observer's eye, for example 0.5mm, and that the distance between the reconstructed middle plane and the observer's eye is 500mm, the imaginary object in depth of 10cm can be well reconstructed. In this example, the optimum width of the elementary hologram can be easily calculated as 5mm.

7. EXPERIMENT

As already mentioned, many variations of the experimental setup can be considered. The first trial was to reconstruct rather small objects illminated by incoherent light and located at about 70cm from the ordinary 35mm Camera. After each exposure the camera was moved by 4mm, and 20 pictures which were seen from the different view points were recorded in succession on an ordinary 35mm reversal color film. The photographic objective of 58mm focal length and 1.4 maximum aperture was used. However, the aperture of F 11 was used, because this value corresponded to about 5mm of the aperture diameter. This objective was used again in the step of exposing the holographic plate. The reason why the objective with such small F number was used will be explained later. Fig. 3 shows diagramatically the constructing process of the holographic stereogram, in which the camera body and the objectives are the same as those used in the first step and focused at the distance of 70cm. The processed color film is loaded again in this camera and is illuminated from the backside by coherent diffused light. A holographic plate is located in front of the objective, and after recording each image holographically, the camera and the plate is relatively moved by 4mm at each time. The thin metal plate which has a 4x40mm rectangular aperture to eliminate the overlapping of the elementary hologram is placed almost touching the sensitized material. The parallel reference laser beam which makes 60° with the optical axis is used at each exposure. To reconstruct thus-produced holographic stereograms, the anti-parallel reconstructing laser beam is used. The virtual image will then be observed. In this stereogram, the vertical movement of the camera has been neglected at all, and yet it results in a satisfactory 3-D sensation.

In order to reconstruct the object at a short distance, the direct holographic recording of the projected original image is recommended as shown in Fig. 7 because in this process the lens system can be excluded, and vertically quite-long elementary holograms can be produced. In this case, however, some care should be taken to reduce the vibration of the projection screen and the air turbulance derived from the rather long optical path.

Other kinds of objects, such as landscapes, indoor scenes, radiographic images, and phantoms of crystal structure drawn by hand have been tried and the satisfactory 3-D reconstruction has been obtained.

Fig. 10 shows a typical holographic stereogram and reconstructed

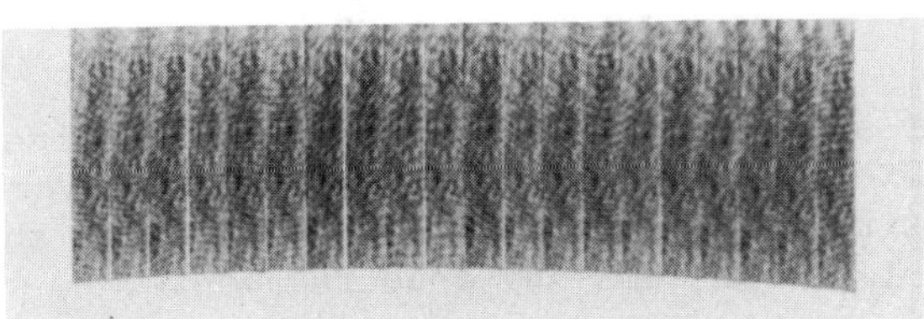

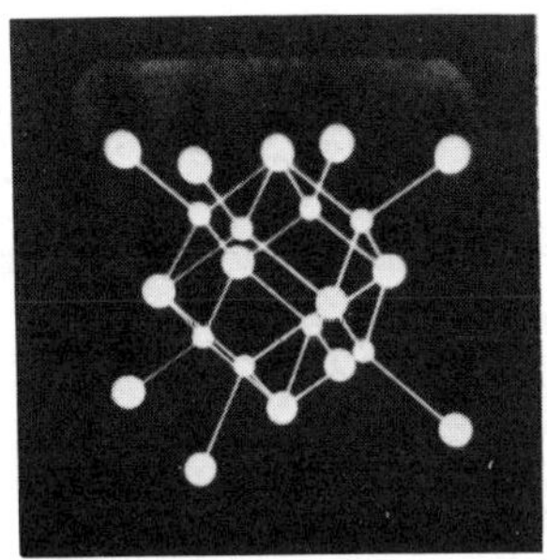

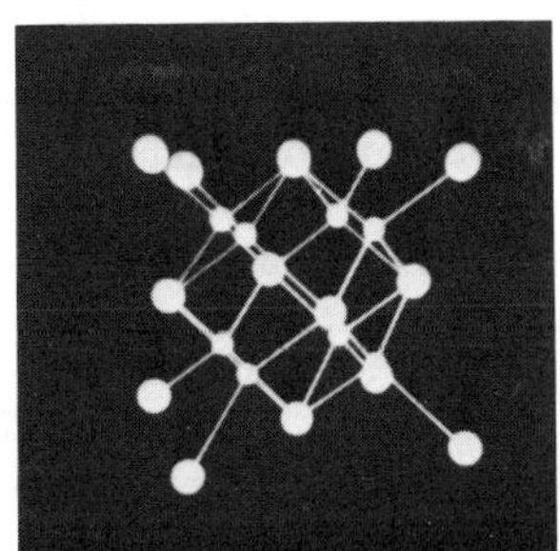

Fig. 10. Typical Holographic Stereogram and Reconstructed Virtual Images.

virtual images of a purely imaginary object viewed from two horizontally separated positions.

This method reconstructs the image just as if it were reconstructed by the conventional hologram.

The most serious problem for practical application of the holographic stereogram is the speckle noise, especially when the contrast of the object is rather low. As is well known, the speckle pattern is not localized in space and renders the granular sensation of rather high contrast and fine structure to the eye. When the observer gazes at the object having low contrast, the accommodation of the eye is varied from time to time owing to the confusing action of speckle noise, and he will not be able to have the acute visual sensation at all. Many suggestions[15)16)] have been proposed to eliminate the speckle noise for practical purposes. This obstacle will remain as a future problem, however.

ACKNOWLEDGEMENT

The earlier stage of this work was conducted under the leadership of Dr. R. Hioki, who was formerly at Tokyo University and now at Chiba University. The authors wish to thank him and Dr. S. Tanaka for their helpful discussions and encouragements.

REFERENCES

1. E. N. Leith and J. Upatnieks, J. Opt. Soc. Am. 54, 1295(1964).
2. K. S. Pennington and L. H. Lin, Appl. Phys. Letters 7, 56 (1965).
3. G. Lippmann, J. Phys. théorique et appliquée 7, 821(1908).
4. H. E. Ives, J. Opt. Soc. Am. 21, 171(1931).
5. R. V. Pole, Appl. Phys. Letters 10, 20(1967).
6. R. J. Collier, Physics Today 21, 54(1968).
7. L. P. Dudley, Stereoptics (Macdonald, London, 1951) P. 96.
8. J. T. McCrickerd and N. George, Appl. Phys. Letters 12, 10 (1968).
9. J. D. Redman, W. P. Wolton and E. Schuttleworth, Nature 220, 58(1968).
10. T. Kasahara, Y. Kimura, R. Hioki and S. Tanaka, Japan. J. Appl. Phys. 8, 124(1969).
11. E. N. Leith, J. Upatnieks, B. P. Hilderbrand and K. Haines, J. Soc. Motion Picture Televis. Engrs 74, 893(1965).
12. C. B. Burckhardt, J. Opt. Soc. Am. 58, 71(1968).
13. D. J. De Bitetto, Appl. Phys. Letters 12, 176(1968).
14. C. B. Burckhardt, J. Opt. Soc. Am. 58, 241(1968).
15. E. N. Leith and J. Upatnieks, Appl. Opt. 7, 2085(1968).
16. H. J. Gerritsen, W. J. Hannan and E. G. Ramberg, Appl. Opt. 7, 2301(1968).

OPTICAL APERTURE SYNTHESIS PRODUCING HIGH-RESOLUTION PHOTOGRAPHS BY INCOHERENT SUPERPOSITION OF LOW-RESOLUTION PARTIAL-FREQUENCY RANGE COMPONENT PHOTOGRAPHS

G. W. Stroke

State University of New York

Stony Brook, New York 11790

ABSTRACT

The long-sought extension to the optical domain of the Ryle (1957) computer-synthesis 'synthetic-aperture radio telescope' principle has been obtained by simple photographic synthesis of the 'high-resolution' image in a single photograph, exposed successively through sets of small apertures, placed successively to generate the spatial frequency components of the desired large aperture.

The high-resolution "full spatial-frequency range" optical image may be synthesized by superposing in a single plate a suitable set of low-resolution partial frequency-range photographs, obtained separately or simultaneously, with "low-frequency redundancy" supressed by spatial filtering.

Very considerable attention has been recently devoted to possibilities of transposing into the optical domain the remarkable 'aperture synthesis' principle first described in radio astronomy by Sir Martin Ryle and his colleagues [1-4] (for a general background see e.g. ref. [5]). One aim in optics is to truly synthesize a posteriori the equivalent of the image which would be obtained by a large-aperture telescope, in cases when only small-aperture optics may be usable. This may be achieved, for instance, by using a pair of successively spaced small-aperture telescopes and by synthesizing the desired image from the set of successively recorded small-aperture images. In effect, as far as the final image is concerned, one aims to 'simulate' a large telescope with the aid of the pair of successively spaced small telescopes, notably in cases when sufficient light is available. Our method consists simply in superposing the several small-aperture "partial frequency range" component images directly in a single photographic plate. The result is an image equivalent to that which would be obtained by the large-aperture "full frequency range" optical system.

For the purpose of clarity, it may be desirable to precede the presentation of our new method with three preliminary remarks. We may first briefly recall the radio-astronomical form of the 'aperture synthesis' principle, which has become well established as a result of the work of Ryle and his colleagues [1-5]. In a simple analysis, the principle consist of using a pair of small antennas in such a way that one of the antennas is moved successively to sample the currents (induced by a radio source) which would have existed in each of the elements of a hypothetical array of large physical extent. The required synthesis is then achieved with the aid of electronic digital computers. The method thus replaces the problem of antenna construction with one of computation. Next we may mention another famous radio-astronomical image-synthesis method, now well-known as the 'Culgoora annular-aperture radioheliograph', in which J. Paul Wild has pioneered by showing how an 'on-line' electronic analogue computation method may be used to synthesize an image (of the sun) which would have been obtained by a fully-filled aperture, in the case when the images (a three thousand point array, imaged almost simultaneously) are actually obtained through only the annular aperture [6]. Even though not directly related to the Ryle aperture synthesis method, the Wild radio astronomical method may be considered either as an 'aperture synthesis' or as an 'image deconvolution' method. Finally, we have ourselves similarly introduced several image 'deconvolution' and 'aperture synthesis' methods for use in the optical and X-ray imaging domains (see e.g. refs. [7-10]). However, in their initial forms, our optical image deblurring and aperture synthesis methods were not primarily aimed at the a posteriori purely photographic 'aperture synthesis' resolution-increasing principles, which we introduce here. Rather they used holographic analogue computing methods [11] for two purposes. On one hand they aimed at attaining the maximum 'diffraction-limited' resolution available within the aperture used, starting from a 'blurred' (e.g. out-of focus, moved, etc.) photograph obtained with that aperture. On the other hand, they aimed a synthesizing into a single image (without increase of resolution) the multiplicity of images obtained simultaneously with a number of small apertures [9], notably in view of achieving the significant signal-to-noise ratio gain in X-ray astronomy predicted by Dicke [12]. With the preceding remarks, we may now introduce the new result of our work.

In order to extend to the optical domain the remarkable advantages of the Ryle radio-stronomical aperture-synthesis principle, as we do here, it had heretofore appeared necessary to use either extensive and very lengthy electronic digital Fourier-transform computation methods [13], or , at best, as first proposed by J. Wilczynski [14], it appeared necessary to synthesize the final image with the aid of the holographic intensity-superposition method which we ourselves first described with D. Gabor in view of such 'image synthesis' applications in 1965[15,11]. The use of holographic aperture synthesis, according to Wilczynski, held out the promise of being the most likely one to succeed in practice, notably because of its ability to synthesize simultaneously the very considerable number of 'resolved' image points which characterize optical images. But no experimental Ryle-type 'aperture synthesis' results are known in optics, even though we had ourselves previously described several applications of our method for other purposes [11, 15, 16]. However, in further thinking about applying the holographic image-synthesis method [11, 15] in view of the 'aperture synthesis' problem which we consider here, it suddenly became clear to us that the best result of the holographic image synthesis, under the most favorable 'linearity' conditions, would in fact, be identical to that which we could achieve simply by successively superposing the component small-aperture pair images directly in a single photographic plate. This will be especially true, in general, with the use of 'extended dynamic range film' now under development [10], or with equivalent photo-transistor arrays, among others.

Moreover, it also became clear that an experimental verification in the optical domain could be very readily carried out simply with the aid of a set of suitable 'masking' apertures, used successively in front of a single 'large-aperture' "full frequency range" photographic lens, when exposing the single photographic plate, successively to the set of 'low-aperture' "partial frequency range" component images. In this way, we aimed to convincingly demonstrate in a model the important new optical principles involved, without the need of resorting to the rigorously accurate (wavelength precision) positioning of the movable small aperture, relative to the fixed aperture, which may be required (and achieved see e.g. ref. [17]) in the most exacting practical applications. Such experimental proof has appeared necessary at this stage because of some apparent general difficulty of accepting the possibilities and implications of using 'aperture synthesis' principles in optics, for the purpose of increasing a posteriori the resolution obtainable through each of the small-aperture telescopes (objectives) when used alone.

Because of the perhaps at first surprising conclusion which we presented in simple physical terms above, it may be desirable to precede the presentation of our experimental results by their theoretical basis. We shall limit ourselves here to the briefest possible description, compatible with clarity. More extensive analysis including that of signal-to-noise considerations will be presented in a future publication, together with the results of our further experiments now under way. Let f(x,y) be the intensity distribution in the 'diffraction-limited' image which would be obtained with the 'large' aperture. Let h(x,y) be the corresponding spread function (i.e. image of a point). The intensity distribution in the image obtained with the large aperture is given by the convolution integral

$$I(x',y') = \iint f(x,y)h(x'-x,y'-y)dxdy \tag{1}$$

under the usual assumptions. We may write eq. (1) in the symbolic form

$$I(x,y) = f(x,y) \otimes h(x,y) \tag{1'}$$

Eqs. (1), respectively (1') may be written in the spatial frequency domain in the form

$$T[I] = T[f]T[h], \tag{2}$$

where T[...] indicates a spatial Fourier transformation

$$T[...] = \iint [...]\exp[2\pi i(ux+vy)]dxdy \tag{3}$$

and u and v are coordinates in the aperture (spatial frequency) domain. It is well known (see e.g. refs. [11, 17, 18]) that T[h] is nothing but the spatial frequency transfer function (i.e. the MTF function, for short). It is also known that T[h] = MTF is equal to the spatial auto-correlation function of the field produced by a plane wave (more specifically by point-source illumination) in the aperture (exit pupil) of the optical image-forming system. In the case of real positive (partially null) aperture fields, the auto-correlation function, and therefore the MTF - T[h] functions are real and positive (partially null). This will be the case, for instance for well corrected systems, as commonly used in optics. Indeed, a field E(u,v) in the aperture produces in the image plane a field

$$E(x,y) = \iint E(u,v)\exp[2\pi i(ux+vy)]dudv \tag{4}$$

such that the spread function is

$$h(x,y) = E(x,y)E^*(x,y) \tag{5}$$

with the exponential time factor exp(iwt) and time averaging omitted as usual. It immediately follows from eq. (5)(see e.g. ref. [11], page 174) and from the definition of the MTF function that

$$\text{MTF} = T[h] = \iint E(u,v)E^*(u+u',\ v+v')\,du\,dv \tag{6}$$

For the complete large aperture, to be synthesized, we thus have

$$I_{OC} = f \otimes h_{OC} \tag{7}$$

and

$$T[I_{OC}] = T[f]T[h_{OC}]. \tag{8}$$

Next, we assume that a photograph is exposed successively through only suitably placed small apertures, placed within the 'large' aperture to be synthesized, say the apertures [1] and [n] (as shown in the figures illustrating our experimental results). Let T_{On} be the corresponding MTF function, h_{On} the corresponding spread function and I_{On} the corresponding 'component' image. We have for instance

$$I_{On} = f \otimes h_{On} \tag{9}$$

and

$$T_{On} = T[f]T[h_{On}]. \tag{10}$$

Let us further assume that we may "synthesize" the MTF h_{OC} of the complete aperture in the form of a sum of the MTF functions $T[h_{On}]$ that we may write

$$T[h_{synth}] = \sum_n w_n T[h_{On}] \tag{11}$$

where w_n are suitable "weighting" factors. Clearly [1-5, 11, 13, 14, 17, 18] such a synthesis may be readily achieved to a very good approximation in the case of the real positive (partially null) aperture functions which we consider here. (The case of complex MTF function synthesis is mentioned below.) Accordingly we shall have the equation

$$T[I_{synth}] = T[f] \sum_n w_n T[h_{On}) \tag{12}$$

i.e.

$$T[I_{synth}] = T[f]\ \{w_1 T[h_{01}] + w_2 T[h_{02}] + \ldots\} \tag{13}$$

which immediately gives

$$T[I_{synth}] = \sum_n w_n \, T[f]T[h_{On}] \tag{14}$$

that is

$$I_{synth} = \sum_n w_n [f \otimes h_{on}] \tag{15}$$

Eq. (15) may also be written in the form

$$I_{synth} = \sum_n I_{On} \tag{16}$$

It is eqs. (15) and (16 which permit us to conclude simply that the desired synthesized "full-frequency range" image I_{synth} (equivalent to the image I_{Oc} obtainable with a large aperture) may be obtained by a simple summation $\Sigma_n I_{On}$ of the component images I_{On}, each recorded successively with the suitable "partial frequency range" small apertures. Moreover, it is immediately clear also that the required summation, according to eq. (16), may be carried out directly, by successive exposure of a single photographic plate to the component images I_{On}, of course under suitable photographic linearity and weighting conditions. Clearly also the direct photographic superposition, obtained in coherent light, avoids the considerable "laser speckle" and other difficulties, such as spurious interference patterns which tend to plague and to limit resolution and image quality in coherent optical processing systems, including holography.

Another remarkable conclusion, which directly follows from the preceding equations, is that the "synthesized" high-resolution diffraction pattern h_{SYNTH} is indeed obtained by incoherent superposition of the component partial-resolution diffraction patterns h_{On} according to the equation

$$h_{SYNTH} = \sum_n w_n h_{On} \tag{17}$$

where h_{On} are the spread functions of the component images. The "weighting" factors w_n may be readily realized, in part by exposure time duration and suitable photographic processing, and, in part, by suitable spatial filtering of the component photographs, before (or during) superposition (for instance by partially attenuating their "low-frequency" region in the spatial Fourier-transform domain, e.g. as in Fig. 10 of our paper on "Optical Image Deblurring", this volume).

We have carried out a series of exhaustive experiments which have fully borne out our theoretical predictions. In our original experiments [*], carried out as a model, we verified the new optical image-synthesis principle (also known as "optical aperture synthesis") with the aid of "one-dimensional" varying frequency

* G.W. Stroke, Optics Comm. Vol. 1, No. 6 (January 1970) pp. 283-90.

test chart. We have since successfully verified: 1) that two-dimensional image synthesis is readily achievable with available photographic methods notably with "half-tone" scenes, 2) that the "low-frequency" "background" (redundancy) associated with each of the component photographs can be readily removed by spatial filtering, when necessary.[**]

It is essential to note (as we have **) that the desired synthesis may be achieved from a set of component photographs obtained with a set of pre-set partial-resolution cameras operating simultaneously, side by side. This arrangement may present very desirable grain-averaging advantages, of well-known interest in astronomy.

In conclusion, we may further note that our method has in effect permitted us to realize in incoherent light an aperture synthesis of a type which had recently been considered as readily realizable in optics only in coherent light (and then only for small-scale laboratory-size objects, for instance in a scaled-down version of the holographic method of image synthesis used in synthetic-aperture coherent 'side-looking' radar, refs. [19-21]). Applications of our method clearly exist throughout the electromagnetic wave domain, and indeed in ultrasonic imaging, among others. We shall report the results of our further experimental work, now under way, in a future publication, together with the details of further analysis of signal-to-noise theory, necessary in cases when a great number of images are being superposed [13,14]. The use of well-known spatial-filtering, scanning and subtracting schemes, photographic as well as photoelectric and holographic, for the purpose of reducing 'background', when required, forms a direct part of our method (see e.g. ref. [22]). However, it may again be in order to stress the important fact that the simple photographic superposition of the component "partial frequency range" images in a single photographic plate, in order to form a "full frequency range" image, does in practice work to a remarkable degree, even <u>without</u> "low frequency redundancy as we describe.

Our experimental results (several of which are given for the first time here) are shown in Fig. 1 to Fig. 6.

One additional remark may be in order. The experimental arrangement illustrated in Fig. 1 may also be particularly useful for the purpose of rapidly assessing, in the laboratory, the quality of images which may be expected from such "optical aperture synthesis" schemes and arrangements which appear to be particularly desirable, without the need of other "scaled-down" experimentation or computation.

**G.W.Stroke, Physics Lett. Vol. 38A, No.9(23 Dec. 1969) pp.485-6.

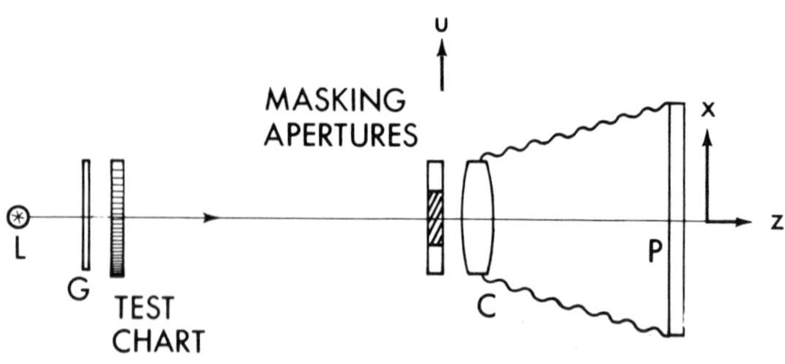

Fig. 1 -- Schematic diagram of the experimental arrangement used to verify the principle of "optical aperture synthesis" by superposition of component 'small-aperture' images in a single photographic plate. L: white-light source, C: photographic camera, using a f=240mm Schneider Kreuznach Symmar lens. The "test chart" consists both of a television test chart, including half-tone scenes, and also of half-tone scenes (see figures 4, 5 and 6). Dimensions of "test chart": 70x90 mm, reporduced in entirety in the experimental results in Fig. 4 and Fig. 6. We also show an enlarged central 23x23mm section in the experimental result in Fig. 5. Dimensions of image of 70x90 mm area on plate: 10x13 mm. Details of the masking apertures used are given in Figures 2 and 3. The apertures are used successively in front of the lens to record the images according to the details given in Figure 2 and in Fig. 3. Naturally, the arrangement represents a model for the actual "optical aperture synthesis" arrangement. In an actual arrangement, several schemes are possible, for instance the following two schemes. In both schemes component systems used to superpose the several 'low-resolution' (partial spatial-frequency resolution) photographs in the same plate may consist of pairs of lens (or mirror) systems virtually 'cut out' of the part of the large (full spatial-frequency resolution) lens (or mirror) system, for instance, according to Fig. 3. In scheme A the various pre-set component systems are successively brought in front of the same plate, as in the model shown here. In scheme B the various pre-set component systems operate simultaneously side by side, exposing separate component plates which are subsequently superposed in the same plate. Spatial frequency filtering (see Fig. 3) for 'low-frequency redundancy'("background") suppression may be used either during superposition of the component photographs, as in scheme B, or to operate on the synthesized photograph, both in scheme A and in scheme B. In any event, it should be clear that optical aperture synthesis may be of interest primarily when a large-diameter system may not be physically or economically realizable, It should however also be clear that the model arrangement, using the masking of a conveniently small-diameter lens, as here, is of very considerable practical interest during the design stages of an actual system, since it permits one to assess by actual photograph experiment, under ideal condiations, the quality of images to be obtained by optical aperture synthesis under the best possible conditions.

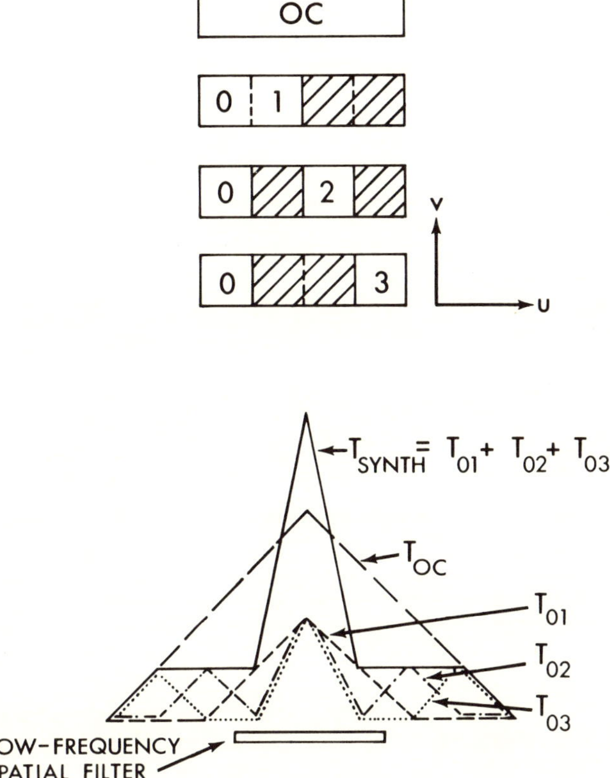

Fig. 2 -- Masking apertures and corresponding MTF (spatial frequency transfer functions) for the one-dimensional experiments.

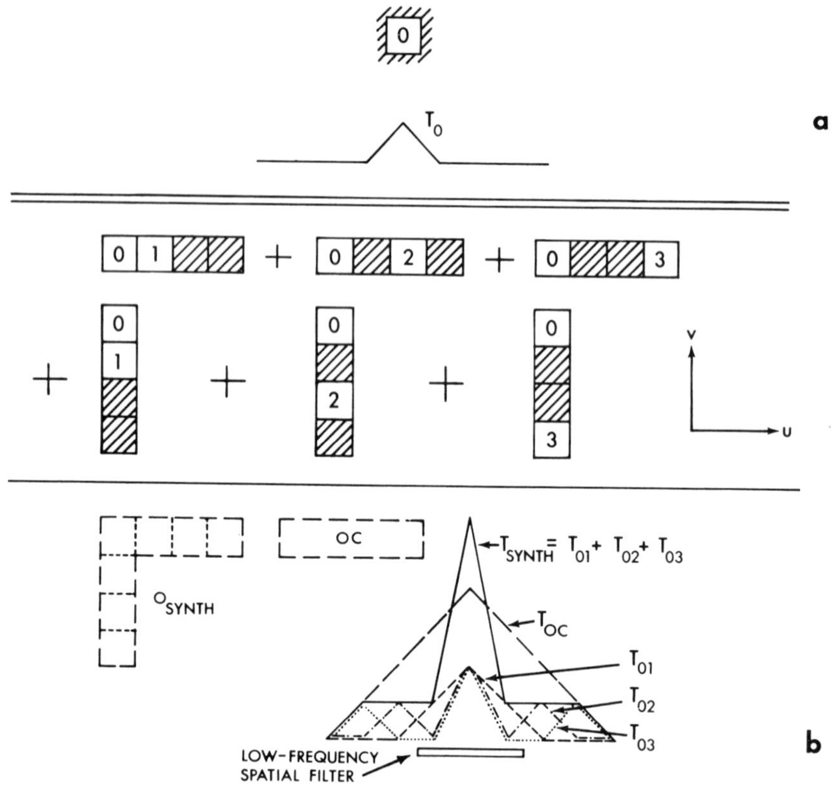

Fig. 3.

a. 1 mm aperture (and corresponding MTF function) used with f=240mm Schneider Kreuznach Symmar lens at f/240 to photograph 'low-resolution' images of Figures 4.a., 5.a., and 6.a.

b. Aperture-pair sets [0+1, 0+2, 0+3] horizontal+ [0+1, 0+2, 0+3] vertical used in synthesizing the 'high-resolution' photographs of Figures 4.b., 5.b., and 6.b. The component apertures in each pair were such that not one of the small apertures was larger than that used to obtain the 'low-resolution' photographs in Figures 4.a., 5.a., and 6.a. The "low-frequency" filter (shown schematically in position in the spatial-frequency domain) was actually not needed in the synthesis shown. In practice, the component photographs may be taken simultaneously by separate suitably pre-set component sub-systems (of which the component spacings and orientations may be maintained by suitable interferometric servo-control [according to the interferometric control principles described e.g. in ref. 17]. The synthesized aperture 0_{SYNTH} (sufficient in this case for illustration of the new principle with the horizontal and vertical bars in the test chart and of the corresponding frequencies in the continuous tone photo) and the aperture OC are shown for comparison, and the MTF functions are drawn slightly separated for clarity.

Fig. 4.

a. Component low-resolution photograph of 70x90mm test chart recorded through component f/240 aperture (Fig. 3.a.). [Best possible photo].

b. Synthesized 'high resolution' photograph using aperture synthesis scheme according to Fig. 3.b.

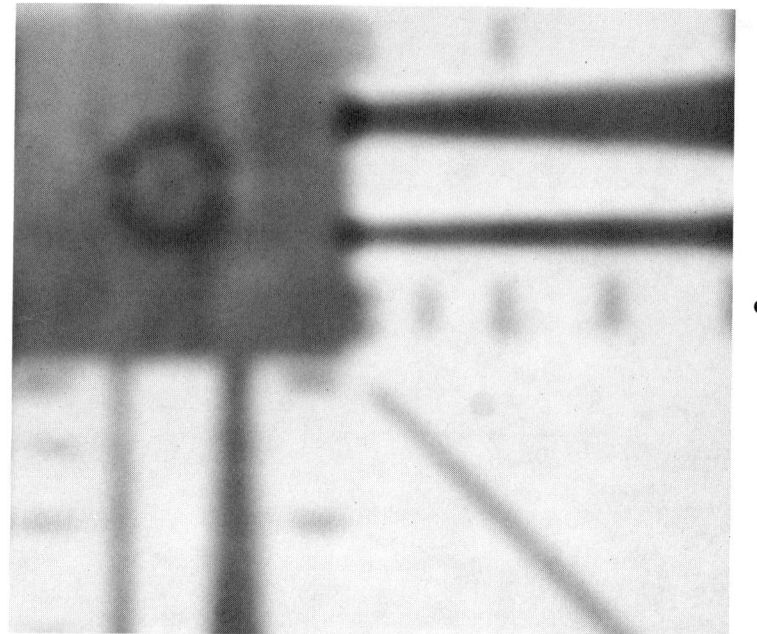

a

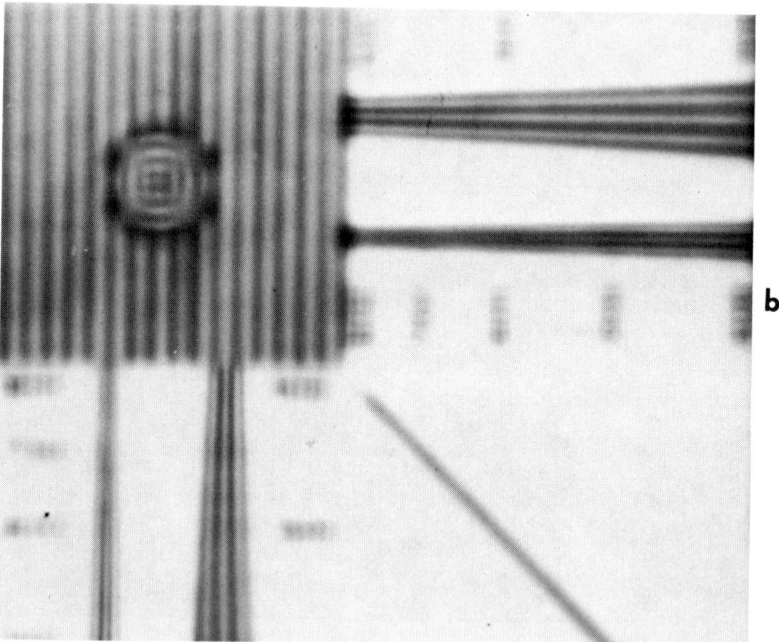

b

Fig. 5. Enlarged portions of central 23x23mm of photos shown in Fig. 4.

a. Component low-resolution photograph recorded through component f/240 aperture (according to Fig. 1. and 3.a.). [Best possible photo]

b. Synthesized 'high-resolution' photograph using aperture synthesis scheme according to Fig. 3.b.

a

b

Fig. 6.

a. Component low-resolution photograph recorded through component f/240 aperture (according to Fig. 1 and Fig. 3.a.) [Best possible photo].

b. Synthesized 'high-resolution' photograph using aperture synthesis scheme according to Fig. 3.b.

REFERENCES

[1] J. H. Blythe, Monthly Notices Roy. Astron. Soc. 117(1957) 644.
[2] M. Ryle, A. Hewish and J. R. Shakeshaft, IRE Trans. Antennas Propagation, Spec. Suppl. 7(1959)120.
[3] M. Ryle and A. Hewish, Monthly Notices Roy. Astron. Soc. 120(1960)220.
[4] A. Hewish, Proc. IRE(Australia) 24(1963)225.
[5] R. C. Hansen, ed. Microwave scanning antennas, Vol. 1 (Academic Press, New York, 1964).
[6] J. P. Wild, Proc. Roy. Soc. (London) A286(1965)499.
[7] G. W. Stroke and R. G. Zech, Phys. Letters 25A(1967)89.
[8] G. W. Stroke, Opt. Acta 16(1969)401.
[9] G. W. Stroke, G. S. Hayat, R. B. Hoover and J. H. Underwood, Opt. Commun. 1(1969)138.
[10] G. W. Stroke, F. Fürrer and D. Lamberty, Opt. Commun. 1(1969)141.
[11] G. W. Stroke, An introduction to coherent optics andnd holography, 2nd ed. (Academic Press, New York, 1969).
[12] R. H. Dicke, Astrophys. J. 153(1968)L101.
[13] Synthetic aperture optics, Proceedings of August 1967 Woods Hole Summer Study, Vol. I and Vol. II (U. S. National Academy of Sciences, Washington D.C. 1968).
[14] J. W. Wilczynski, IBM Res. Repts. RC-1988 and RC-2034 (1968) unpublished; private communication to the author from E. S. Barrekette, IBM 10 Sept. 1969.
[15] D. Gabor, G. W. Stroke, R. Restrick, A. Funkhouser and D. Brumm, Phys. Letters 18(1965)116.
[16] G. W. Stroke, F. H. Westervelt and R. G. Zech, Proc. IEEE 55(1967)109.
[17] G. W. Stroke, in: Handbuch der Physik, Vol. 29, ed. S. Flügge (Springer Verlag, Berlin, 1967) p. 426.
[18] A. Maréchal and M. Françon, Diffraction structure des images (Revue d'Optique, Paris, 1960).
[19] L. J. Cutrona, E. N. Leith, L. J. Porcello and W. E. Vivian, Proc. IEEE 54(1966)1026.
[20] W. E. Kock, Proc. IEEE 56(1968)238.
[21] E. N. Leith and A. L. Ingalls, Appl. Opt. 7 (1968)539.
[22] G. Nomarsky and Y. Levy, paper presented at I.O.C. Symp. on the applications of coherent light, Florence (1968)

IMAGING WITH LOW-REDUNDANCY ARRAYS*

J. W. Goodman

Department of Electrical Engineering

Stanford University, Stanford, California 94305

INTRODUCTION

A conventional imaging system, consisting of a collection of lenses or mirrors with clear apertures, provides non-uniform coverage of the spatial spectrum of an incoherent object.[1,2] Stated another way, a certain spatial-frequency redundancy is inherent to such a system, with low spatial frequencies being weighted more heavily than high spatial frequencies. In this paper we propose that it may sometimes be desirable to destroy this redundancy by inserting a complicated mask in the exit pupil of the system, thereby allowing the images to be formed with a multi-element interferometer. With a proper choice of mask geometry, the spatial frequency redundancy can be minimized or eliminated. The destruction of redundancy will be shown to be advantageous if the optical elements are aberrated and if post-detection image processing may be applied to improve image quality. We emphasize at the start that all arguments to follow apply only for spatially *incoherent* objects.

THE TWO-ELEMENT INTERFEROMETER

The simplest type of interferometric imaging system is the two-element interferometer (sometimes called the Fizeau stellar interferometer) shown in Fig. 1. A mask is inserted in the optical system such that light from two small regions of the primary collector is caused to interfere in the focal plane of the instrument.

*Work supported by the Office of Naval Research

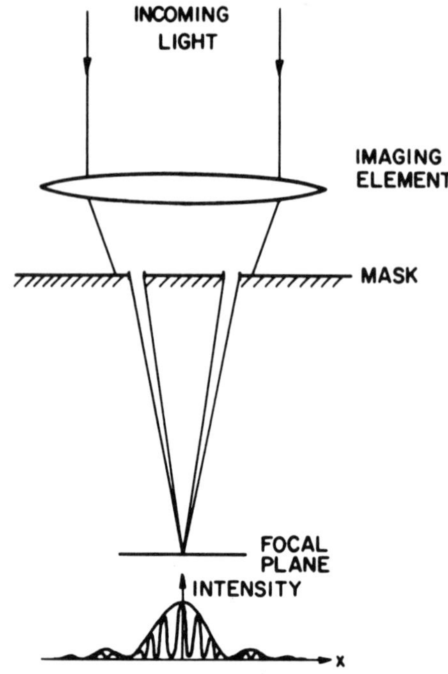

Figure 1. Two-element interferometer.

If the (identical) openings in the mask are too small to resolve the angular dimension of the object individually, then the pattern of intensity observed in the focal plane of the instrument is a Young's fringe pattern lying under a broad spatial envelope. In the absence of aberrations, the intensity distribution near the center of the envelope may be written

$$I(\vec{r}) = 2I_o\left\{1 + \mu(\vec{\nu}_{12}) \cos\left[2\pi\vec{\nu}_{12} \cdot \vec{r} + \psi(\vec{\nu}_{12})\right]\right\} \qquad (1)$$

Here I_o represents the intensity produced by each of the small apertures individually near the center of their diffraction patterns; $\vec{\nu}_{12}$ is a vector spatial frequency given by $\vec{s}_{12}/\lambda z_i$, where $\vec{s}_{12}$ is the vector separation of the centers of the two small openings on the exit pupil, λ is the mean wavelength of the light, and z_i is the image distance; $\vec{r}$ is a position vector in the image plane; and μ and ψ represent the contrast and spatial phase of the sinusoidal intensity distribution. The sinusoidal distribution so-recorded is in fact simply one Fourier component of the object brightness distribution; thus μ and ψ are determined by the object itself.

If the openings of the interferometer are so large as to resolve the object of interest, then the fringe pattern loses its crisp sinusoidal character. In this case the interferometer collects more than one spatial frequency component of the object. The optical transfer function (OTF) of the interferometer, as shown in Fig. 2, consists of a low-frequency passband and two high-frequency passbands, centered at frequencies $\pm\vec{\nu}_{12}$. In the material

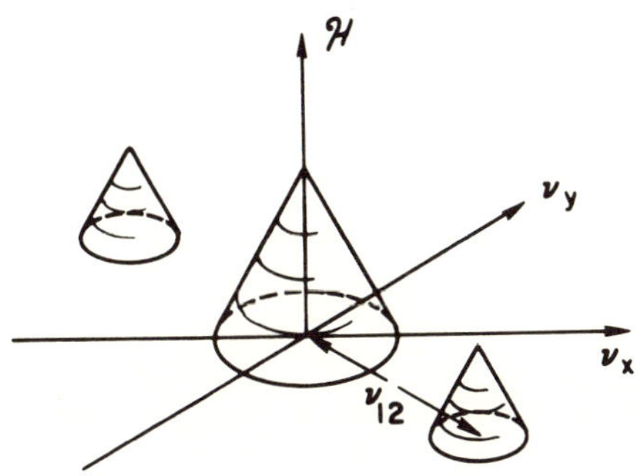

Figure 2. OTF of the two-element interferometer.

to follow we shall assume that the individual interferometer elements (i.e., the small apertures) are too small to resolve the object; however, this assumption is not always necessary.

We can now understand the behavior of the full-aperture system in terms of the two-element interferometer. Any two elementary areas separated by vector spacing $\vec{s}$ on the face of the exit pupil of the system contribute to the image an elementary Young's fringe pattern of vector spatial frequency $\vec{s}/\lambda z_i$. Since small spacings are embraced by the full (unobstructed) aperture more redundantly than are the large spacings, we find the usual drop of the OTF at high spatial frequencies. For a circular, aberration-free system of diameter d, the redundancies present are such as to produce the familiar OTF[3]

$$\mathcal{H}(\rho) = \begin{cases} \frac{2}{\pi}\left[\cos^{-1}\left(\frac{\rho}{\rho_o}\right) - \frac{\rho}{\rho_o}\sqrt{1-\left(\frac{\rho}{\rho_o}\right)^2}\,\right] & \rho \leq \rho_o \\ 0 & \text{otherwise} \end{cases} \tag{2}$$

where $\rho = |\vec{s}|/\lambda z_i$ and $\rho_o = d/\lambda z_i$.

EFFECTS OF ABERRATIONS

When the optical elements of an imaging system are imperfect, the wavefronts produced suffer from aberrations. The aberrations in turn affect the OTF of the system. It can be shown[4] that aberrations never increase the modulus of the OTF; generally they severely suppress the system response at high spatial frequencies.

The physical reason for a loss of high-frequency response can be understood by first examining the performance of the two-element interferometer in the presence of aberrations. Let the object be a point source and suppose that the aberrations are described by an "optical-path-difference function" $W(x,y)$, which represents the difference between the optical path length traveled by the ray leaving the exit pupil at (x,y) and the path length appropriate for a perfect spherical wave converging towards the gaussian image point. We suppose that the openings of the interferometer mask are sufficiently small so that W has only one constant value in each opening, and we represent these two values by W_1 and W_2. With this assumption the Young's fringe pattern produced by the interferometer becomes

$$I(\vec{r}) = 2I_o \left\{ 1 + \mu(\vec{\nu}_{12}) \cos\left[2\pi\vec{\nu}_{12}\cdot\vec{r} + \psi(\vec{\nu}_{12}) + k(W_1 - W_2)\right]\right\} \tag{4}$$

where $k = 2\pi/\lambda$ and all other symbols have been defined previously. Comparison of Eqs. (1) and (4) shows that the aberrations of the system have introduced a <u>phase error</u> in the OTF of the interferometer but have not changed the amplitude of the frequency response.

Why is it, then, that when the full aperture of the system is used for image formation the high-frequency response is severely reduced? The answer becomes clear when we recall that, for the full-aperture system, each spatial frequency is embraced redundantly by the aperture. Thus for each different pair of elementary areas separated by a given spacing $\vec{s}$ the values of W are in general different, and the various elementary Young's fringe patterns at a given frequency $\vec{\nu}$ are added with different spatial phases. In the presence of aberrations, the phasor contributions to that frequency component add in such a way that they partially cancel each other, resulting in a suppression of that spatial frequency component in the image. Since the optical-path-difference function $W(x,y)$ generally changes most rapidly near the edge of the aperture, and since the high spatial frequency components in the image are produced by elementary areas near the edge of the aperture (i.e. for large separations $\vec{s}$), the high-frequency response is most seriously affected.

We have postulated at the start that image processing may be used to restore the frequency response of the system a posteriori if necessary. Why, then, should a loss of high-frequency response be of concern when we can, after the fact, restore the suppressed frequency components to their proper amplitudes? Unfortunately the most common image detector, photographic film, has a limited dynamic range, and we find that for severe aberrations the high-frequency components are so strongly suppressed as to be irretrievably lost in the grain noise. Thus we require some method for eliminating the strong suppression of high-frequency components if successful restoration is to be achieved in the presence of severe aberrations. One such method is offered by the use of array technique, as we now explore in some detail.

LOW-REDUNDANCY ARRAYS

Let the exit pupil of the system be obstructed by a mask containing K separate small openings. The geometrical configuration of the mask is dictated by the following considerations. If the maximum angular dimension of the desired image field is Ω radians, then the sampling theorem implies that sampling of the Fourier components need only be performed on a rectangular lattice in the frequency domain, with the lattice constant no greater than $\Delta\nu = (z\Omega)^{-1}$. Equivalently, the vector separation of the centers of the openings of the mask in the exit pupil should cover a rectangular lattice with lattice constant $\Delta s = \lambda/\Omega$.

When the centers of the openings in the mask are arranged such that a single vector separation is embraced only once by the array, we refer to this structure as a non-redundant array. We shall assume further that all vector separations (on the required lattice) out to some maximum separation $|\vec{s}_{max}|$ are contained. If we arbitrarily number the openings in the mask 1 through K, the first opening generates K-1 distinct spacings with its neighbors, the second generates K-2 new spacings, etc. The total number of different spacings generated by the array is thus

$$(K-1) + (K-2) + \dots + 1 = \frac{K(K-1)}{2} \quad . \tag{5}$$

Therefore a non-redundant array containing K elements will pass K(K-1)/2 Fourier components of the object.

The zero-frequency component is, of course, collected by each of the elements. Hence if we normalize the OTF to unity at zero frequency, the value of the modulus of the OTF at all other frequencies on the lattice covered is K^{-1}. By destroying the redundancy of the optical system we have reduced the low-frequency response, but when aberrations are present the high-frequency response will be better than that obtained by the full aperture. This improvement comes about by virtue of the fact that only one Young's fringe pattern component contributes to the image at each spatial frequency, and therefore the possibility for destructive interference of several components has been eliminated.

We note in addition that some redundancy can be allowed in the array at small separations without destroying our argument, for the high spatial frequencies are generally the ones most severely affected by aberrations. Hence the array need not be non-redundant, but instead can be a low-redundancy array with the redundancies confined to the low spatial frequencies.

The problem of finding low-redundnacy one-dimensional arrays has been studied by radio astronomers,[5] but to date there has been relatively little attention devoted to the two-dimensional case. Examination shows that there is more flexibility in the two-dimensional case than in the one-dimensional case; the problem is also simplified when some redundancy at low frequencies is allowed. There is little doubt that appropriate arrays exist, but finding the best one remains a difficult problem. Some progress in this area will be reported in the future, but for the present purposes we simply assume that a suitable structure can be found.

IMAGE PROCESSING

The reader may well wonder what relevance holography has to the preceding discussion, but it is in the image processing step that holography plays a crucial role. The image recorded through the array in the presence of aberrations suffers several defects which must be corrected. For the moment we suppose that the array is non-redundant, returning later to the effects of redundancy.

For a non-redundant array of K elements, the zero-frequency component is passed with K times the weighting of other components. Thus it is necessary to attenuate the zero-frequency component, although this must be done with some moderation to avoid introducing "negative intensities". The best solution is probably to reduce the zero-frequency component by a factor K^{-1}, and then to apply a further frequency-dependent attenuation of the general form described by Eq. (2).

When a non-redundant array is used, the sole effect of aberrations is to change the phase of the image Fourier components. The required phase corrections can be accomplished by making a Fourier-transform hologram of the point-spread function of the aberrated array. The hologram is then placed in the frequency plane of a coherent optical processing system. The conjugate transmittance term of the hologram will correct the spectrum of any general image recorded through the array and placed at the input plane of the processor.

When some redundancy exists in the array, aberrations can introduce both amplitude and phase errors at those frequencies collected redundantly. In this case simple phase-correction filtering is generally not adequate, although it may suffice if the redundancy is very low. More generally the spectrum can be corrected by "inverse filtering" in a coherent optical processor, restoring both the amplitude and phase distributions to their proper form. Methods for performing inverse filtering are discussed in the literature[6,7] and will not be treated in detail here. It suffices to say that two separate frequency-plane filters are sandwiched, one a conjugate filter to correct the phase, and the second an attenuating filter to correct the amplitude.

CONCLUSION

We have proposed above that, in the presence of strong aberrations, an image formed through a low-redundancy array is easier to correct by image processing than is an image formed through a full aperture. When a proper array is used, aberrations are exhibited solely as errors of the phase of various image Fourier components, whereas when the full aperture is used the high-frequency components are often suppressed irretrievably.

The price paid by the use of a low-redundancy array is a loss of sensitivity. The size of the individual array elements must be limited to assure that the optical path length through the imaging system is constant over any single element. Since only a fraction of the total aperture collects energy, the sensitivity of the instrument is reduced accordingly. In some applications, however, this may be a tolerable price to pay.

REFERENCES

1. E. L. O'Neill, Introduction To Statistical Optics, Addison-Wesley Publishing Compnay, Inc., Reading, Mass. (1963)
2. J. W. Goodman, Introduction to Fourier Optics, McGraw-Hill Book Co., New York, N.Y. (1968), Chapter 6.
3. Ibid, p. 120
4. Ibid, p. 122
5. A. T. Moffet, IEEE Trans. Ant. and Prop., AP-16, 172 (1968)
6. G. W. Stroke and R. G. Zech, Phys. Lett. 25A, 89 (1967)
7. G. W. Stroke, Phys. Lett. 28A, 252 (1968)

CORRECTION OF LENS ABERRATION BY HOLOGRAPHY

Teruji Ose, Masaru Noguchi and Toshihiro Kubota
Institute of Industrial Science, University of Tokyo
Roppongi, Minato-ku, Tokyo, Japan

1. Introduction

There have been several studies on the application of holography to improve the quality of an image which is formed by an optical system. Some of them have been done to improve a lens image correcting the lens aberration with a hologram. These studies are classified into two groups according to the use of holograms. The first group is the method using a conjugate wavefront reconstructed from a hologram in which the wavefront at the pupil of a lens is recorded. Upatnieks[1] and Tsuruta[2] have studied this method. A hologram is used in this case as an optical spatial filter; therefore this method is referred to hereafter as a holographic filtering method for the correction of lens aberrations. The second group is the method carried out by the authors[3]. In this case both reference and signal wavefronts are affected by the same amount of phasic deformation caused by lens aberrations, and with these reference and signal waves a hologram is photographed. The interference pattern in this hologram suffers no serious effects from the lens aberrations. As shown later, the properties of a hologram are well-employed in this method. Therefore, this method is referred to hereafter as a holographic correction method of lens aberrations.

Both methods mentioned above are available for a correction of turbulence effect on a wavefront of light when passing through a turbulent medium as well as for correction of lens aberration. Kogelnik[4] and others[5-6] have used a holographic filtering method and Goodman[7] and Gaskill[8] have used holographic correction method.

At first the principle of a holographic filtering method will be described and next the principle and the results of experiments of holographic correction method for lens aberrations will be shown.

2. Principle of Holographic Filtering Method

The pupil function f of a lens is regarded as the product of the complex amplitude of aberration free wavefront f_L and that of aberrated wavefront f_A which is caused by the lens aberration. If the spatial filter whose complex transmittance is f_A^*, that is, the complex conjugate of amplitude f_A, is placed in front of the lens, then the resultant pupil function f_R of the lens is given by

$$f_R = f_A^* f = |f_A|^2 f_L. \quad (1)$$

Since the image of the lens is obtained using the Fourier transform of f_R, the phase of complex amplitude f_A does not seriously affect the image. If the complex amplitude f_A is recorded in a hologram, the amplitude f_A^* is obtained at the reconstruction with this hologram. Thus, the lens aberration is corrected with such a hologram as a filter. This method of correction has been carried out by Upatnieks et al[1].

Let the complex amplitude of the wavefront diffracted from an object be f_B at the entrance pupil of the lens, then the amplitude of the wavefront at the exit pupil is given by

$$f_E = f_B f', \quad (2)$$

where f' is the pupil function of the lens when the object is a point source.

When f_E is recorded in a hologram and the hologram is illuminated with f', then f_B and its conjugate amplitude f_B^* are reconstructed from the hologram. This method was carried out by Tsuruta et al[2]. The aberrated lens itself is used as a filter in this case.

In this holographic filtering method, the relative position of the hologram should be strictly fixed with respect to the lens during both procedures of production and reconstruction of the hologram.

3. Principle of Holographic Correction Method

The hologram, which is made on the condition that an object and a reference light source are located in a plane as is shown in Fig.1(a), is called a lensless Fourier transform hologram, and its theory of image formation has already been derived by Stroke[9].

Let t_s and t_R be the Fourier spectra of the amplitude of object T_S and the reference light source T_R respectively. According to his theory, the amplitude transmittance of this hologram is given by

$$|t_S|^2 + |t_R|^2 + t_S^* t_R + t_S t_R^* \qquad (3)$$

When the hologram is illuminated with a plane wavefront as is shown in Fig.1(b), the reconstructed wavefronts are $t_S t_R^*$ and $t_S^* t_R$. These are the spectra of the direct and conjugate images

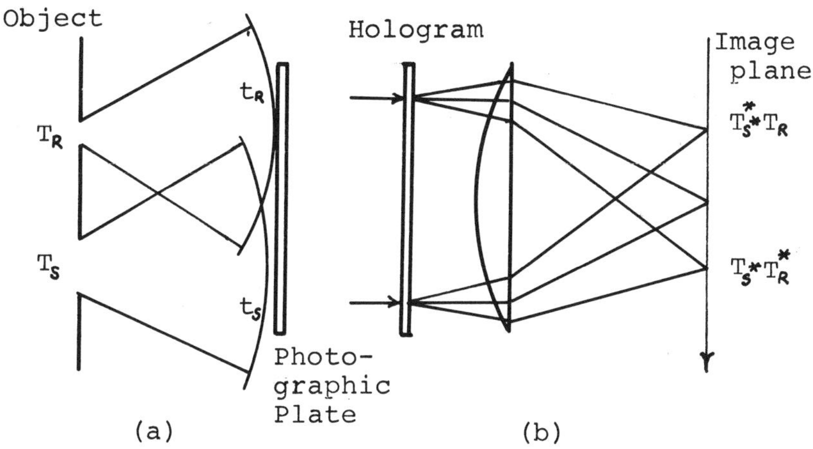

Fig.1

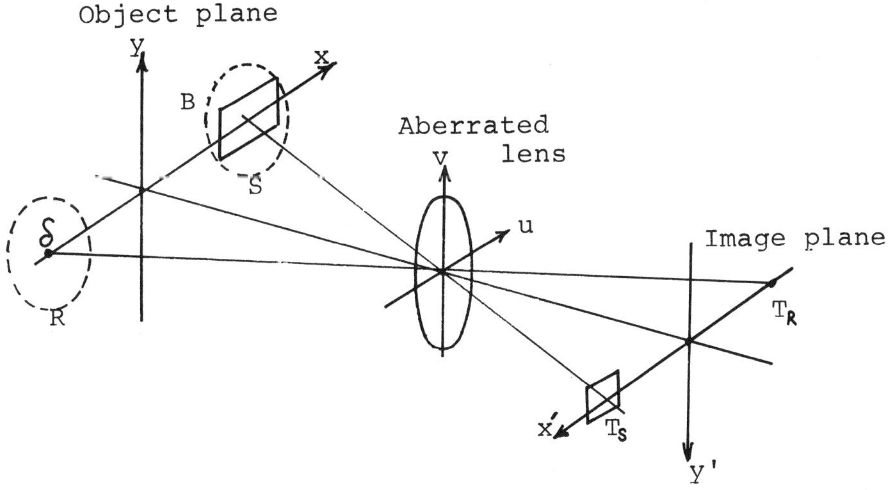

Fig.2

which are reconstructed from the hologram.

Now consider the lens images of an object $B(x,y)$ and a reference point light source $\delta(x-a,y-b)$ located in the x-y plane as is shown in Fig.2, where $\delta(x,y)$ denotes Dirac's delta function. We assume that the object and the point source are within different isoplanatic regions S and R of an image-forming lens, respectively. An image $T(x',y')$ of the object and an image $T(x,y)$ of the point source are given by

$$\begin{aligned} T_S(x',y') &= \iint F_S(x'-x,y'-y)\ B(x,y)dxdy \\ T_R(x',y') &= \iint F_R(x'-x,y'-y)\ \delta(x-a,y-b)\ dxdy \\ &= F_R(x'-a,y'-b) \end{aligned} \tag{4}$$

where $F_S(x,y)$ and $F_R(x,y)$ denote the point spread functions within corresponding isoplanatic regions S and R of the lens.

A photographic plate is placed just behind the lens, but away from its image plane, and a hologram is made. Since the amplitude transmittance of this hologram is proportional to

$$|t_S|^2 + |t_R|^2 + t_S t_R^* + t_S^* t_R \tag{5}$$

as is mentioned above, where t_S and t_R denote the spectrum of T_S and T_R which are given by equation(4).

Let the spectrum of $B(x,y)$, $F_S(x,y)$ and $F_R(x,y)$ be $b(u,v)$, $f_S(u,v)$ and $f_R(u,v)$, where u and v denote spatial frequencies along the direction of coordinates x and y. By the Fourier transform of the equation(4), t_S and t_R are given as follows.

$$\begin{aligned} t_S &= f_S b \\ t_R &= f_S \exp\left[-2\pi i(au+bv)\right] . \end{aligned} \tag{6}$$

Since this optical system is illuminated with a coherent light, f_S and f_R, which are defined as the spectra of point spread functions F_S and F_R at the beginning of this section, are the pupil functions of this lens.

Let the wavefront aberrations corresponding to isoplanatic regions S and R be W_S and W_R, the pupil functions f_S and f_R are written as

$$\begin{aligned} f_S &= \exp\left[ikw_S\right] \\ f &= \exp\left[ikw_R\right] \end{aligned} \tag{7}$$

where $k=2\pi/\lambda$, λ denotes the wavelength of light. Using equations(6) and(7), the amplitudes of reconstructed wavefronts with the hologram are given from equation(5) as follows.

$$t_S t_R^* = b \exp\left[ik(W_S - W_R)\right] \exp\left[2\pi i(au+bv)\right]$$
$$t_S^* t_R = b^* \exp\left[-ik(W_S - W_R)\right] \exp\left[-2\pi i(au+bv)\right]. \quad (8)$$

Since these are the spectra of reconstructed images, if they could become just the same as the spectra of object b and its complex conjugate b*, the hologram produces the reconstructed image, which is the same as the original object. From this consideration, the following condition is derived in order to correct the lens aberrations.

$$W_S - W_R = 0 . \quad (9)$$

This condition means that both wavefronts from object and point source should be deformed by the same amount of lens aberration.

If the object and the reference light source could be located in the same isoplanatic region of the lens, this condition would be perfectly satisfactory. However, an isoplanatic region is usually very small; then the size of object and also a space between the object and the reference light source should be as small as possible.

It is necessary to devise an optical system which satisfies the condition (9) against a large size of object and to make the carrier frequency of hologram high.

4. Correction of Spherical Aberration and Astigmatism

In the case when the object and the point reference light source are located with symmetry about the optical axis of a lens as is shown in Fig.3, the wavefront aberrations W_S and W_R

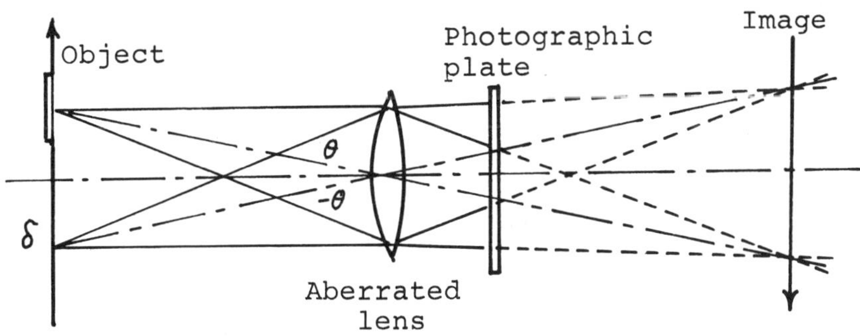

Fig.3

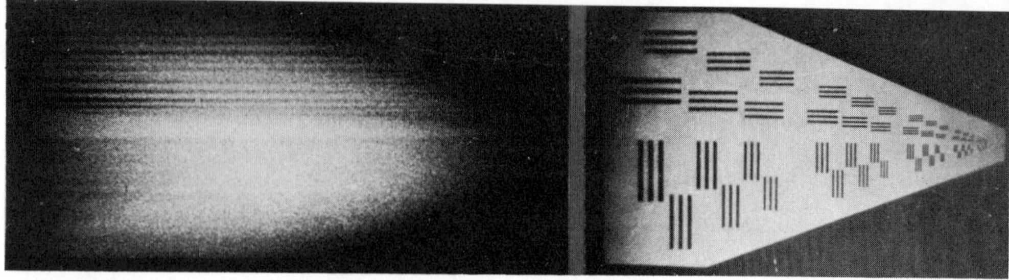

Fig.4 Correction of astigmatism.

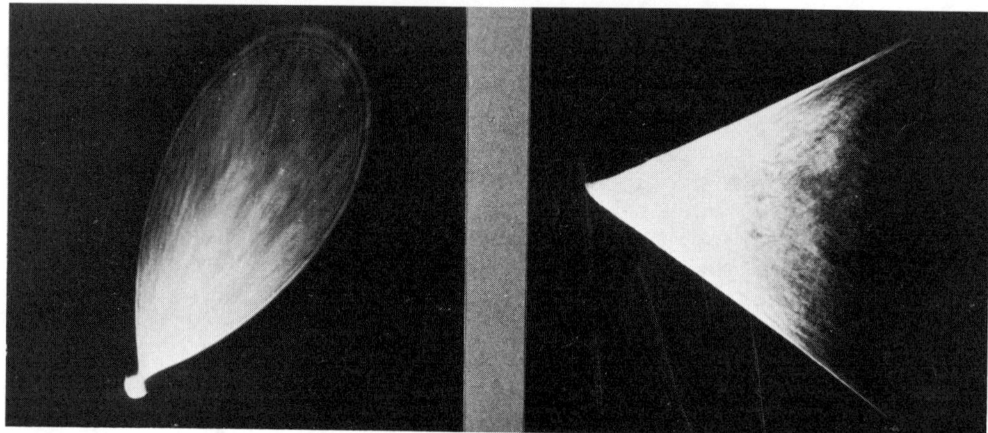

Fig.5 Correction of coma and astigmatism.

become $W(\theta)$ and $W(-\theta)$, where θ denotes the field angle of the lens. Then the condition (9) is written as follows in this case,

$$W(\theta) - W(-\theta) = 0. \tag{10}$$

The wavefront aberrations, which are proportional to the even powers of such spherical aberrations and astigmatism, can satisfy the above condition.

As an example, the result of an experiment on the correction of astigmatism is shown in Fig.4. The lefthand photograph shows the blurred image formed by the lens with astigmatism, that is, a cylindrical lens. The righthand photograph shows the corrected image using this method. Fig.5 shows an example of the correction of the aberration mixed with coma and astigmatism. The lefthand photograph shows the blurred point image made by the lens with coma and astigmatism. The righthand photograph shows the corrected image using this method. In this case the pure coma becomes dominant since the astigmatism can be eliminated.

5. Correction of Coma

The wavefront aberrations which are proportional to the odd powers of field angle, such as coma and distortion, can not only satisfy the condition (10) but also become twice as much as initial values. Such aberrations can be corrected by 180° rotation of the reference point image using a rotator prism which is inserted into the optical path of the reference beam between the lens and hologram as is shown in Fig.6. However in this case, the aberrations which are proportional to the even powers of field angle cannot be corrected.

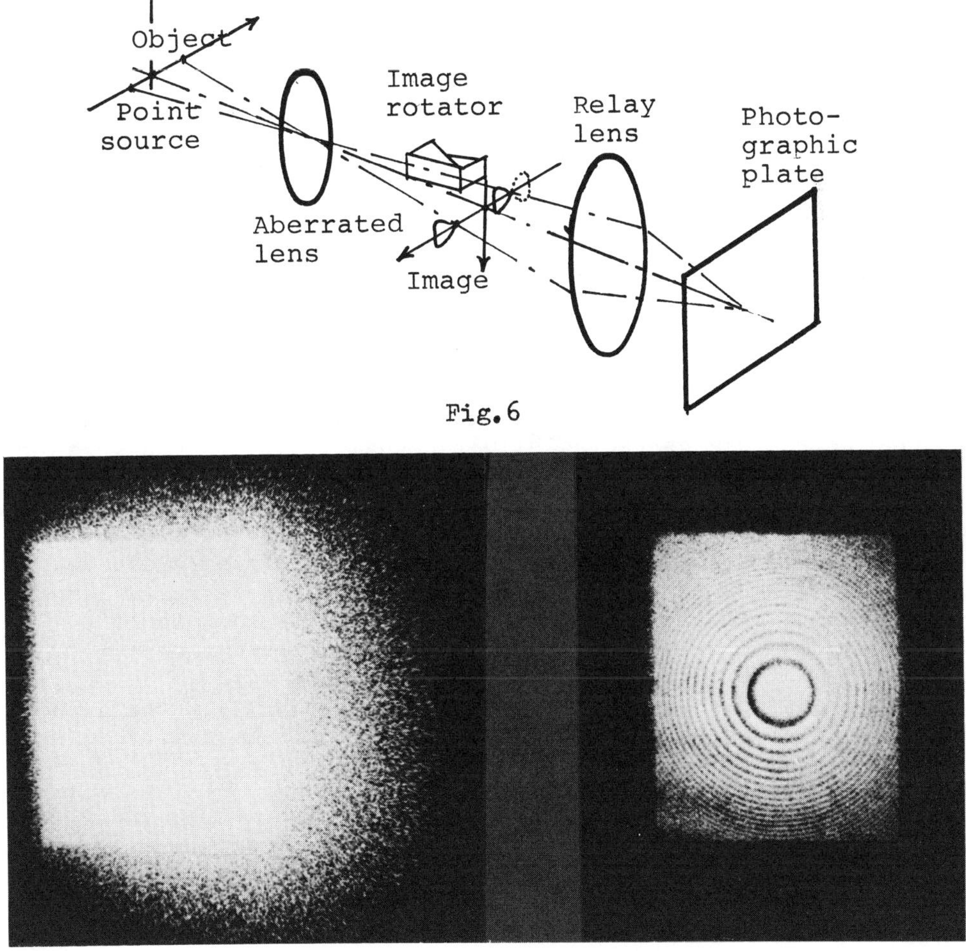

Fig.6

Fig.7 Correction of coma.

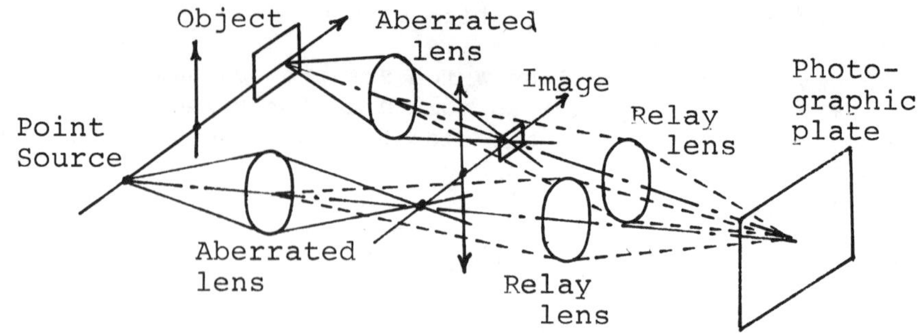

Fig.8

The result of an experiment using this method is shown in Fig.7. The lefthand photograph shows the blurred image formed by the lens with coma. The righthand photograph shows the corrected image. In this case coma can be corrected, but residual spherical aberration becomes twice as much as its initial values. Therefore we have a little blurred corrected image.

6. Correction Method with Two Lenses

The method mentioned above is very simple, but all aberrations cannot be corrected simultaneously. The method which uses two lenses with the same amount of aberrations has been employed. As is shown in Fig.8, an object and a reference point source make images independently by two lenses. The relay lenses in this figure form the images of pupils of these two lenses on the hologram. The aberrations of these relay lenses should be as small as possible. Fig.9 shows examples of the results obtained by this method. Lefthand photographs (A)(B) and (C) show blurred images made by the lens with spherical aberration, coma and astigmatism respectively. Righthand photographs show corrected images resulted by this method.

7. Application to the Correction of Turbulence Effect

The studies on the correction of turbulence effect on a wavefront of light using a hologram have been carried out by Goodman, Gaskill and the others as mentioned above. As is shown in Fig.10, the optical arrangement of Goodman's experiment is just the same arrangement for a lensless Fourier transform hologram. In this case, it is necessary that signal and reference

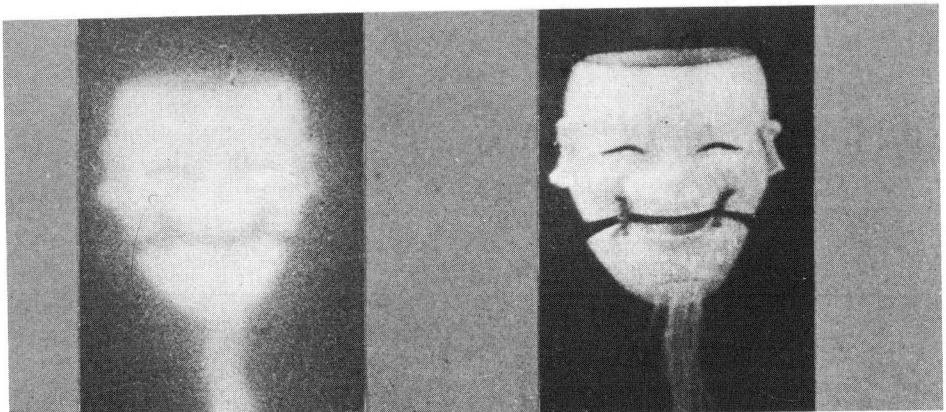

Fig.9(A) Correction of spherical aberration.

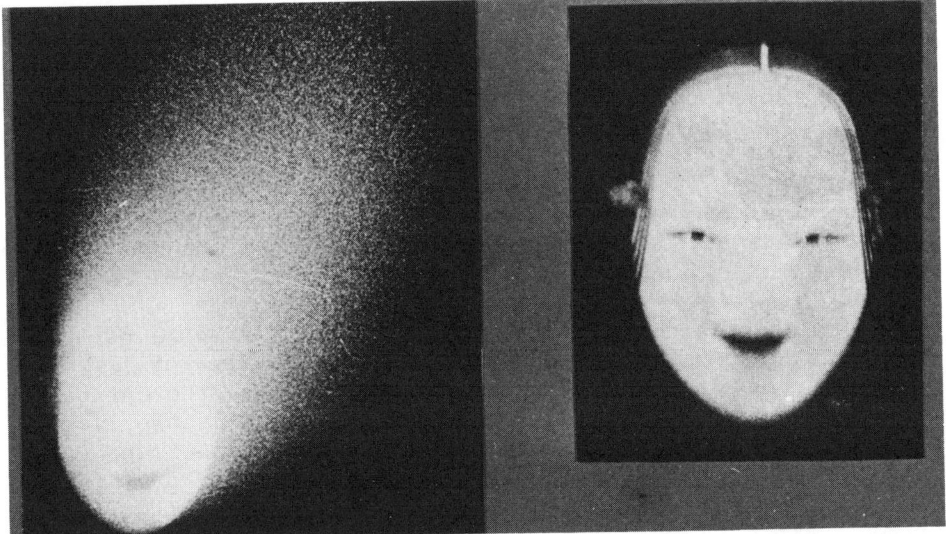

Fig.9(B) Correction of coma.

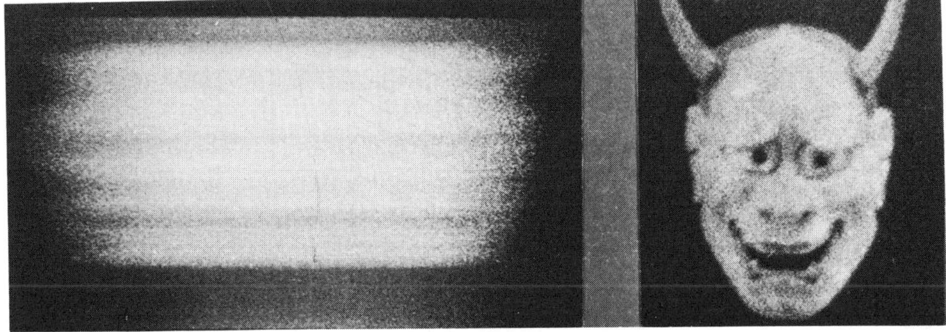

Fig.9(C) Correction of astigmatism.

beams pass through approximately the same portion of the perturbing medium with the same travelling direction. As Goodman has already pointed out, the photographic plate should be located close by the perturbing medium and also the reference source should be placed as near as possible to the object. In the case of correction of lens aberrations which is mentioned above, a relay lens is used in order to satisfy this condition. The authors tried an experiment using Wollaston prisms.

As is well known, when an unpolarized light beam passes through a Wollaston prism, it is divided into two polarized lights, that is, an ordinary light and an extraordinary light, and they propagate along two different directions which are defined by the construction of the prism. On the other hand when an ordinary and an extraordinary beam are incident on a Wollaston prism with specified incident angles, they propagate along the same direction after passing through Wollaston prism. Fig.11 shows the schematic diagram of the optical arrangement of this experiment. The reference and signal beams are polarized by the polarizers. P_1 and P_2, whose planes of polarization are perpendicular to each other. Two beams are combined in a beam by the first Wollaston prism W_1 so as to propagate in the perturbing medium along the same optical path. The light beam is again divided into two beams by the second Wollaston prism W_2, and we have the signal and reference beams. They are incident on a photographic plate P. Consequently, a lensless Fourier transform hologram is produced and its carrier frequency is defined with the construction of the Wollaston prisms.

Fig.12 shows an example of experiments in which a perturbing medium is a turbulent atmosphere caused by the flame of gas. (A) shows the image of original object formed by the optical arrangement itself without turbulence. (B) shows the image affected with turbulence. (C) shows the corrected image using this method. Righthand photographs show intensity profiles in

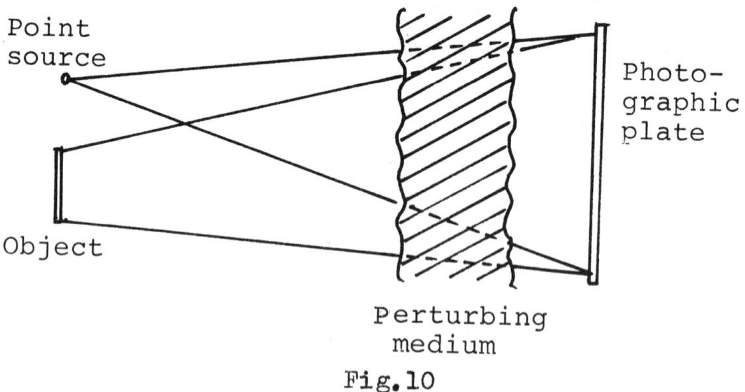

Fig.10

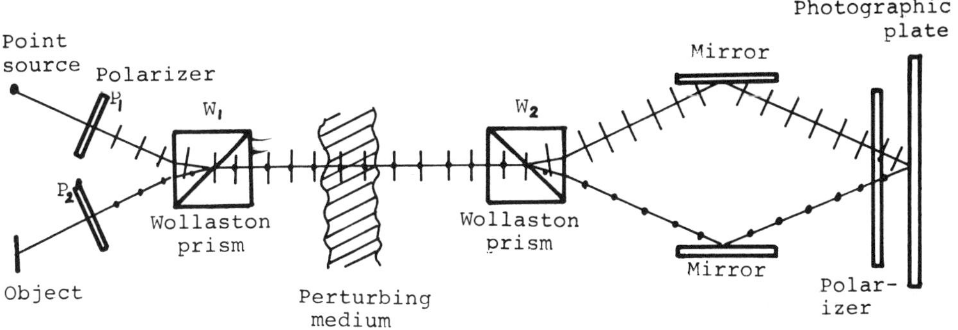

Fig.11

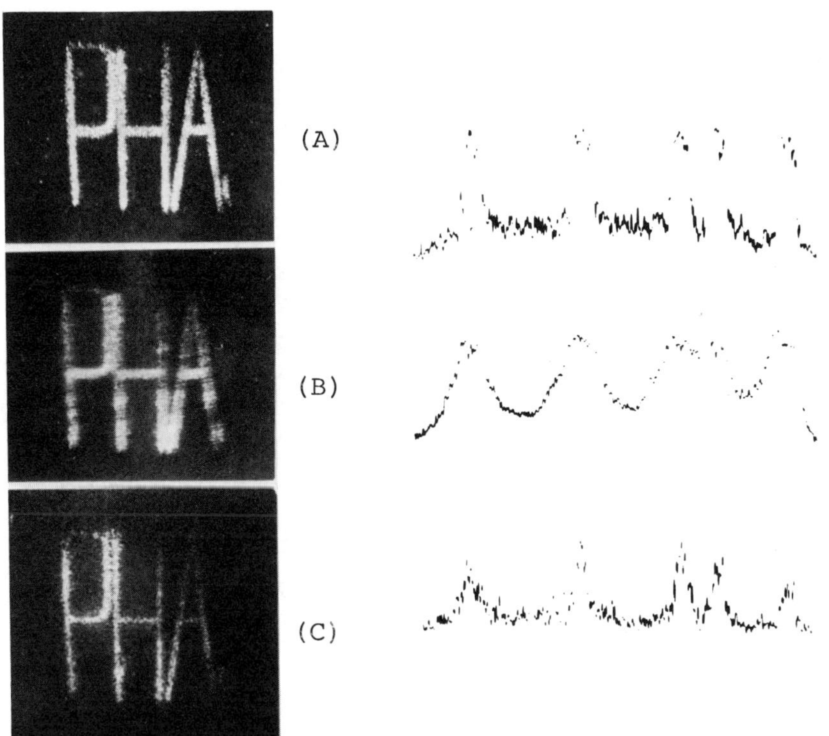

Fig.12 Correction of air turbulent effect. (A) original object. (B) Image affected with turbulence. (c) Corrected image.

lefthand photographs along the same direction. Comparing these profiles, it is noticed that this method is effective to improve degraded image quality caused by turbulence.

8. Conclusion

The method for the correction of lens aberration by holography have been described. From the point of view concerning image formation with a lens, it may seem unnecessary to try hard to correct lens aberration by holography, since many kinds of diffraction limited lenses are now produced by the recent progress of lens design and also by lens production techniques. On the other hand, from the point of view of image formation with a hologram, it is difficult to obtain a magnified image without aberrations, because an aberration-free image with a hologram can be reconstructed only when its magnification is unity. A combined optical system with lens and hologram should be taken into consideration in order to make it possible to expand the function of a conventional lens system as well as holography itself. In this system the lens forms a magnified image of an object and the hologram corrects the aberrations of the lens. This optical system has a disadvantage that it does not enable us the real-time observation of image. However, it gives us some benefits, such as a working distance which is longer than that of a conventional lens system because the defocusing aberration of the lens can be corrected with a hologram. Furthermore, the lens is quite simple in its construction, because almost all aberration can be corrected with the hologram.

The study presented here is a basic investigation with some preliminary experiment for the realization of such a combined optical system.

References

1. J.Upatnieks, A.Vander Lught and E.N.Leith : Appl. Opt., 5, 589 (1966).
2. T.Tsuruta and Y.Itoh : Appl. Opt., 7, 2139 (1968).
3. M.Noguchi and T.Ose : Seisan Kenkyu 20, 464 (1968).
4. H.Kogelnik : Bell syst. Tech. J.,44, 2451 (1965).
5. E.N.Leith and J.Upatnieks : J. Opt. Soc. Am., 56, 523 (1966).
6. H.Kogelnik and K.S.Pennington : J. Opt. Soc. Am., 58, 273 (1968).
7. J.W.Goodman, W.H.Huntley Jr., D.W.Jackson and M.Lehmann : Appl. Phys. Letters, 8, 311 (1966).
8. J.D.Gaskill : J. Opt. Soc. Am., 58, 600 (1968).
9. G.W.Stroke, R.Restrick, A.Funkhouser and D.Brumm : Phys. Letters, 18, 274 (1965).

OPTICAL TRANSFER FUNCTION MEASUREMENT BY HOLOGRAPHIC TECHNIQUES

Kazumi Murata and Hirofumi Fujiwara
Department of Applied Physics
Hokkaido University, Sapporo

1. INTRODUCTION

In recent years, a number of methods on the measurements of optical transfer function (called OTF hereafter) have been reported [1]. The OTF of an incoherent optical system is represented by

$$R(s,t) = |R(s,t)| \cdot \exp\{i\alpha(s,t)\} \tag{1}$$

which is a complex function of the spatial frequency s and t. OTF is equivalent to the autocorrelation of the pupil function f(u,v) of an optical system as

$$R(s,t) = \iint f(u,v)f^*(u-s,v-t)dudv \tag{2}$$

In addition, there is the Fourier transform relation between OTF and the point spread intensity h(x,y), that is

$$R(s,t) = \iint h(x,y)\exp\{-i(sx+ty)\}dxdy \tag{3}$$

The measuring method on the basis of eq. (2) is called as an autocorrelation (or shearing interferometric) method and the method of eq. (3) as a Fourier transform (or scanning) method.

Recently holographic technique has been applied to the measurement of OTF by Françon et al. [2], May [3], Ose et al. [4] and Matsumoto and Ose [5] who have all employed the autocorrelation method. In these methods, the wave front emerging from the pupil of a test lens is recorded on 2 identical holograms. In the reconstructing process, the autocorrelation output is obtained by measuring the flux emergent from the two holograms which are superposed and shifted relative to each other by s and t. The advantage of these methods may be due to the simple instrumentation since it is not necessary to use the conventional shearing interferometer. We will discuss the above method in Chapter 2.

The application of holography to the Fourier transform method for the measurement of OTF has been conducted by the present authors [6]. In this method, the point spread h(x,y) of a test lens is obtained by superposition of the holographically recorded point spread amplitude with the real point spread one by a coherent optical system. Then, this point spread h(x,y) is Fourier transformed by the successive optical system. It is of advantage in this method to observe the two dimensional distribution of OTF without the scanning of spatial frequency. Detailed discussions and some experiments are shown in Chapters 3 and 4, respectively.

2. AUTOCORRELATION METHOD BY HOLOGRAPHIC SHEARING INTERFEROMETRY [2,3,4,5]

The pupil function f(u,v) of an optical system to be tested is recorded on the two holograms placed in the Fresnel region [3] and contacted with their gelatine surfaces using a reference plane wave r(u,v) tilted by θ to the hologram plane as shown in Fig. 1. The two resulting identical holograms are developed under the condition of gamma -2. In reconstruction, these holograms, which are superposed and sheared by amount of s and t, are illuminated by the plane wave r(u,v). Then

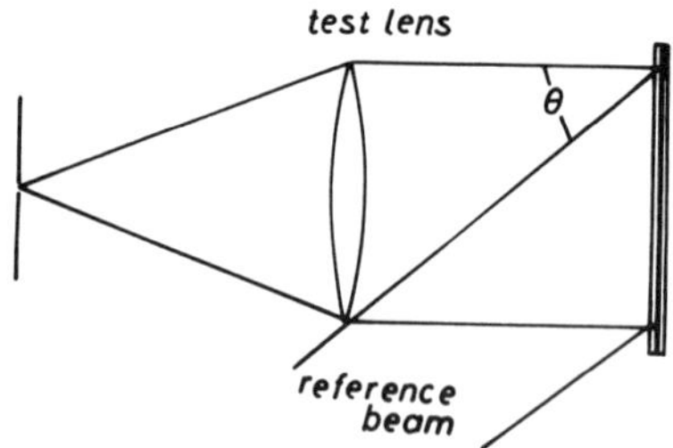

Fig. 1. Recording of pupil function on the two holograms (May [3])

the amplitude emerging from the holograms is

$$i(u,v) = r\{|f|^2+|r|^2+f^*r+fr^*\}\{|f_{s,t}|^2+|r_{s,t}|^2+f_{s,t}^*r_{s,t}+f_{s,t}r_{s,t}^*\} \quad (4)$$

where
$$r = \exp(i\beta u)\ ,\ \beta = 2\pi\sin\theta/\lambda$$
$$r_{s,t} = \exp\{i\beta(u-s)\} = r\exp(-i\beta s)$$
$$f = f(u,v)$$
$$f_{s,t} = f(u-s,v-t)\ ,$$

λ is the wave length of the light and * denotes the complex conjugate. If we put $f'_{s,t} = f(u-s,v-t)\exp(i\beta s)$ and $a = |f|^2+|r|^2 = |f_{s,t}|^2+|r_{s,t}|^2$, eq. (4) can be further expressed as follows;

$$\begin{aligned} i(u,v) &= a(f+f'_{s,t})+r(a^2+ff'^*_{s,t}+f^*f'_{s,t}) \\ &+r^2a(f^*+f'^*_{s,t})+r^3f^*f'^*_{s,t}+r^*ff'_{s,t} \end{aligned} \quad (5)$$

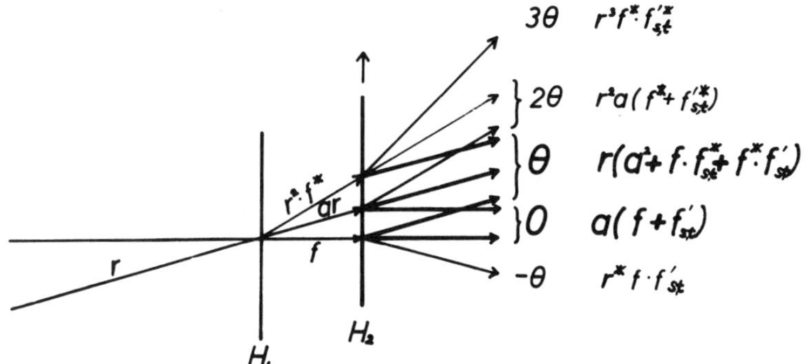

Fig. 2. Amplitude of transmitted light through the two holograms in various directions (Matsumoto and Ose [5])

The five terms in eq. (5) correspond to the reconstructing waves emerging from the set of holograms in the directions 0, +θ, +2θ, +3θ, and -θ, respectively, as shown in Fig. 2. The intensity of light in the +θ direction, equivalent to the direction of the illuminating beam, is obtained from the second term of eq. (5) and given by

$$\begin{aligned} I_{+\theta} &= \iint |a^2+ff'^*_{s,t}+f^*f'_{s,t}|^2\, dudv \\ &= 2a^2\left[C+\iint(ff'^*_{s,t}+f^*f'_{s,t})dudv.\right. \end{aligned} \quad (6)$$

In this equation the first term is not interesting for our purpose, while the second is the autocorrelation of the pupil function which is recorded on the hologram plane in the Fresnel region from the lens to be tested. If the aperture of the holograms is enough larger than that of the lens, the second term of eq. (6) represents a sinusoidal function of the frequency β modulated by OTF which is

$$I_{+\theta} = 2a^2\left[C+|R(s,t)|\cos\{\beta s-\alpha(s,t)\}\right]. \tag{7}$$

By detecting this second term photoelectrically under various shearing amounts of s and t between the holograms, both the modulus $|R(s,t)|$ and the phase $\alpha(s,t)$ can be obtained from the amplitude and phase of the oscillogram, respectively.

When the light flux in the O direction is detected, the output corresponding to the first term of eq. (5) is produced; it is given by

$$\begin{aligned} I_o &= a^2\iint|f+f'_{s,t}|^2\,dudv \\ &= a^2\iint\{a+(ff'^{*}_{s,t}+f^{*}f'_{s,t})\,dudv. \end{aligned} \tag{8}$$

The first term is a constant bias, while the second varies sinusoidally, which is the same as the second term of eq. (7).

These two cases based on the principle of the autocorrelation method using the interferometric holography were investigated by Françon et al.[2], May[3] and Ose and Matsumoto[4],[5]. The advantages of these methods compared with the usual autocorrelation ones [7] are that the devices for measurements are simple and that it is not necessary to modulate the light flux sinusoidally since the output has a carrier frequency automatically by shearing the holograms with constant velocity. With regard to the light flux to be detected, the waves propagating in the O direction are diffracted once by either of the two holograms, while the waves in the θ direction are diffracted twice by both holograms. Thus the photometric efficiency in the O direction is higher than that in the θ direction.

3. FOURIER TRANSFORM METHOD BY HOLOGRAM FILTER [6]

The complex conjugate of Fourier spectrum F of a two dimensional signal f can be recorded on the Fourier transform hologram, which is called an optical matched filter. By illuminating this hologram filter with the Fourier spectrum of the signal f, the power spectrum $|F|^2$ of the signal f is produced just behind the filter.

If the signal is the pupil function of an optical system to be tested, $|F|$ and $|F|^2$ correspond to the point spread amplitude $h_a(x,y)$ and intensity $h(x,y)$, respectively. By the use of the second Fourier transform lens, the amplitude distribution of the output image is the Fourier transform of $|F|^2$, namely, $R(s,t)$ as shown in eq. (3).

We now consider the hologram recording process illustrated in Fig. 3. Let the reference plane wave $r_1(x,y)$ be $\exp(i\beta x)$. Since the hologram may be produced by superposition of r_1 and h_a and developed under the proper condition, the amplitude transmittance of the hologram filter is represented by

$$H(x,y) = |r_1+h_a|^2 = 1+h+r_1h_a^*+r_1^*h_a \tag{9}$$

In the filtering process, the hologram filter is illuminated with $h_a(x,y)$, so that the amplitude of transmitted light emergent from it is equal to $H(x,y)\cdot h_a(x,y)$, which is then Fourier transformed by the second lens. The output corresponding to the third term of $H(x,y)$ is obtained as

$$\begin{aligned} i(s-\beta,t) &= \iint r_1|h_a|^2 \exp\{-i(sx+ty)\}\,dxdy \\ &= \iint h \exp[-i\{(s-\beta)x+ty\}]\,dxdy \\ &= R(s-\beta,t) \end{aligned} \tag{10}$$

which is distributed around the point ($s=\beta$, $t=0$) on the output image plane, where the square of the modulus of two dimensional OTF can be observed.

When a reference wave is introduced onto the image plane, the modulus $|R|$ and phase α can be observed simultaneously in interference patterns. Let the reference plane wave r_2 be

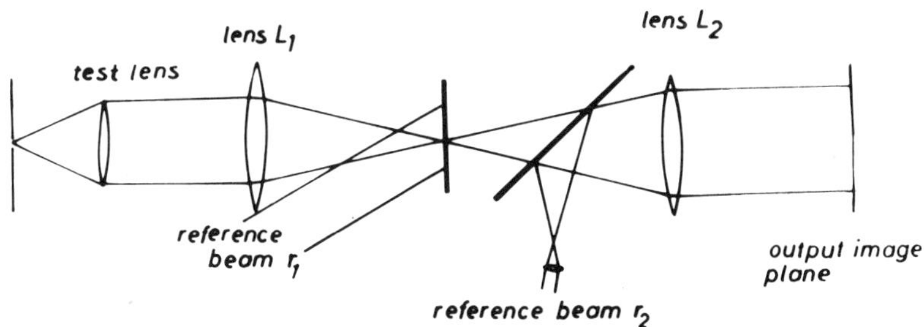

Fig. 3. Principle of Fourier transform method by hologram filter

$\exp(i\gamma s)$. The intensity $I(s-\beta,t)$ on this plane is represented as

$$I(s-\beta,t) = |\exp(i\gamma s)+R(s-\beta,t)|^2$$
$$= 1+|R(s-\beta,t)|^2+2|R(s-\beta,t)|\cos\{\gamma s-\alpha(s-\beta,t)\}. \quad (11)$$

If the reference plane wave is incident along the optical axis, i.e. $\gamma = 0$, eq. (11) is rewritten in the form,

$$I(s-\beta,t) = 1+|R(s-\beta,t)|^2+2|R(s-\beta,t)|\cos\{\alpha(s-\beta,t)\} \quad (12)$$

In eq. (11) the interference pattern introduced as a carrier is phase-modulated by the phase α of OTF, but in eq. (12) this pattern follows the variation of the phase α.

Now, let us take an example in order to explain the principle described above. The intensity of the output image is assumed to be given by a dotted curve shown in Fig. 4(a). Fig. 4(b) and 4(c) show the modulus $|R|$ and phase α of OTF which are obtained from Fig. 4(a). When one usually needs to measure only the modulus $|R|$, it can be obtained without the use of reference wave as explained in eq. (10).

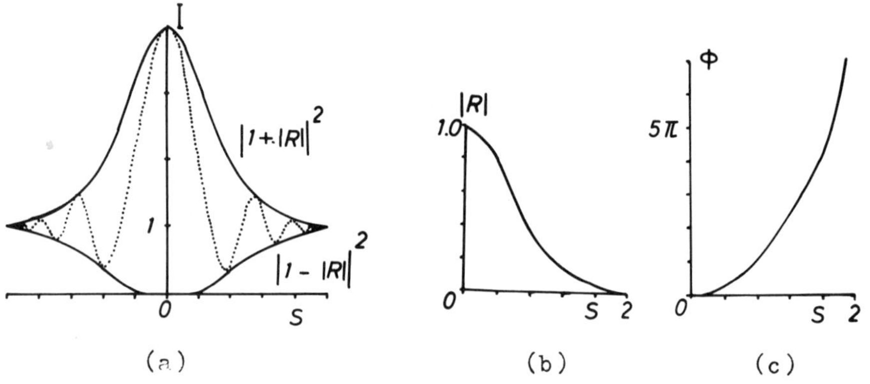

Fig. 4. An illustrative example of the intensity of the output image (a), the modulus (b) and the phase (c) of OTF.

4. EXPERIMENTS

In order to measure the OTF of a test lens, the experimental set-up shown in Fig. 5 is used. In the process of making a hologram filter, the reference point source is placed on the same plane as a test lens, and in the filtering process, the hologram filter is illuminated only by the Fourier transform of the wave front of the pupil. The output image around the point $(s=\beta, t=0)$ can be observed by a microscope or can be photographed. If it is necessary to measure the phase, the second reference wave passed through two mirrors M_2, M_3 and a half mirror HM_2 illuminates the output plane. In the first experiment, a singlet lens is tested on its optical axis. Fig. 6 shows a two dimensional OTF pattern (a) with its microdensitometer trace (b).

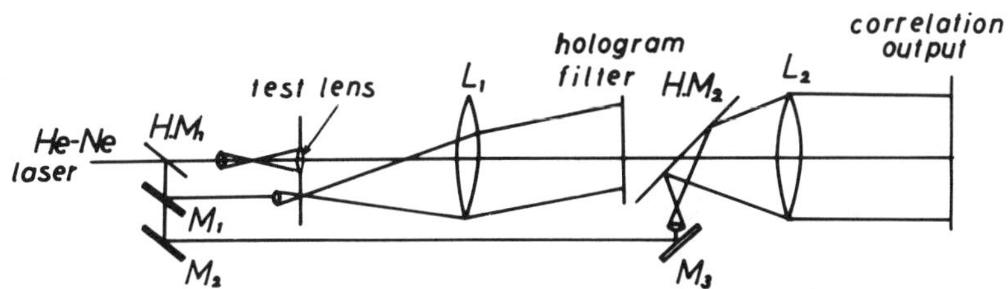

Fig. 5. Experimental set-up

In the second experiment, a phase plate which is shown in Fig. 7(a), is coaxially attached to an aberration-free lens and is tested. The inner part of the plate has a phase retardation π compared with the outer part. A photograph of the output image with its microdensitometer trace is shown in Fig. 7(c) and (d). Moreover, the square of the modulus

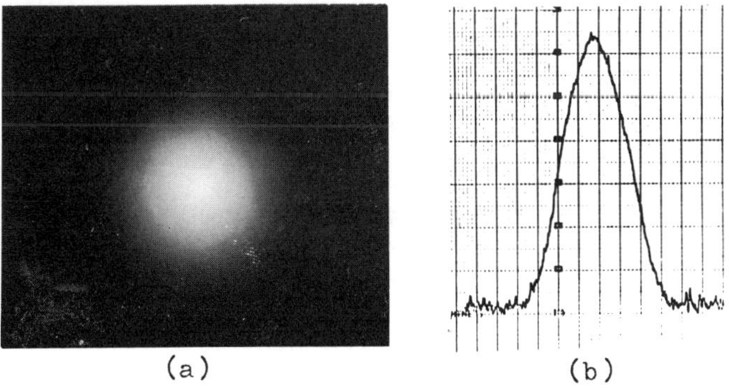

(a) (b)

Fig. 6. Two dimensional OTF pattern of the singlet lens (a) and its microdensitometer trace (b).

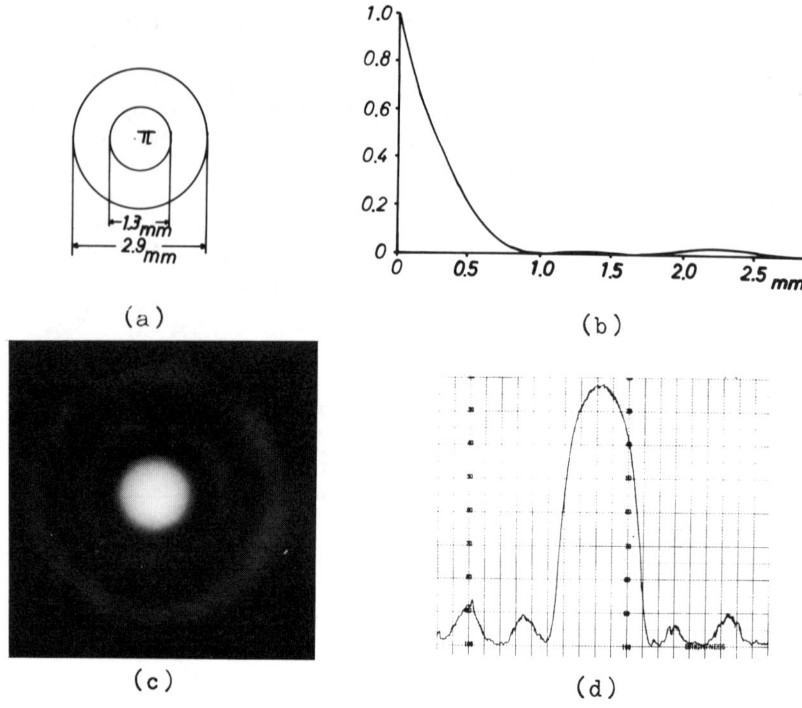

Fig. 7. Annular aperture with phase plate (a), the square of the modulus of its OTF calculated (b), a photograph of the two dimensional OTF pattern of it (c) and its microdensitometer trace (d).

$|R|$ of OTF is theoretically calculated and shown in Fig. 7(b). The microdensitometer trace shows that the experimental result almost agrees with the theoretical one.

5. CONCLUSION

In recent years, the image evaluation of optical systems by means of OTF has come into wide use and various practical instruments for measuring of OTF have been used at institutes and industries. These instruments are generally complicated and delicate with optical, mechanical and electrical parts. Compared with those instruments, the holographic devices for measuring of OTF are simple in construction and operation.

The autocorrelation methods by holographic interferometry described in Chapter 2 need no complicated interferometer but only a hologram scanner and a photometer as an usual optical autocorrelater. When one of the holograms is sheared with constant velocity, the amplitude and the phase of the oscillating output of the photometer give the modulus and the phase of OTF, respectively.

The Fourier transform method by holographic filtering described in Chapters 3 and 4 is very simple since neither mechanical scanner nor electronic device is needed. The two dimensional OTF $R(s,t)$, having not only the modulus $R(s,t)$ but also the phase $\alpha(s,t)$, can be observed in the output plane. This is very convenient for testing the lenses with non-symmetrical point spread. It may be possible to apply this holographic method to testing a lot of lenses for industry. For this purpose, there are two proposals; firstly by using the photochromic material to make a hologram filter, OTF can be measured in a short time without the alignment of this filter and secondly by using the computerized hologram filter of the ideal lens, OTF can be inspected in a real time.

The authors wish to thank Dr. M. May for kind discussions on her papers.

REFERENCES

[1] K. Murata: Progress in Optics (North-Holland, Amsterdam, 1966) Vol.5, ed. E. Wolf, p.119.

[2] M. Françon, S. Lowenthal, M. May and R. Prat: C. R. Acad. Sc. (Paris) 263, (1966) 237.

[3] M. May: Symposium on Applications of Coherent Light, (Florence, Sept., 1968); Optica Acta 16, (1969) 569.

[4] T. Ose, K. Matsumoto and I. Yamaguchi: U.S.-Japan Seminar on Holography (Tokyo, Oct., 1967).

[5] K. Matsumoto and T. Ose: Japan. Jour. Appl. Phys., 7, (1968) 621.

[6] K. Murata and H. Fujiwara: to be published.

[7] H. H. Hopkins: Optica Acta 2, (1955) 23; L. R. Baker: Proc. Phys. Soc. B68, (1955) 871, and others.

APPLICATIONS OF CLASSICAL THEORY OF INTERFEROMETRY TO HOLOGRAPHY

Tadao TSURUTA

Research Laboratory, Nippon Kogaku K.K.

Shinagawaku, Tokyo, Japan

The elaborate theory and practice of interferometry have been established principally to meet the situation that the available light sources have finite size (finite spatial coherence) and finite coherence length (finite temporal coherence). The theory of localization of fringes provides means to reduce the effect of the source size in a plane to be considered and give rise to a system of bright and well-defined fringes exclusively in this plane. The symmetric arrangement of two arms of the interferometer is convenient to match the path lengths to a desired accuracy. The advent of lasers seriously changed the situation. If a single mode emission from a He-Ne laser is used in the laboratory experiment, an arrangement of two arms is sufficient to produce clear fringes and no considerations are necessary to compensate both the source size and the coherence length. Once two waves originating from a laser intersect together, there appears a system of non-localized fringes, however complex in shape the waves may be. This ensures the astonishingly well-defined reconstruction of a three-dimensional and diffuse object from a hologram which is nothing but a record of interference fringes between a simple plane or spherical wave (reference wave) and a rather complex wave issuing from the object. So far, the coherence theory of fringe formation has little to do with the holography.

However, there still remain several important subjects, in which the theory plays essential part. Among them, we will discuss the fringe formation and localization in the holographic interferometry, and the modulation transfer evaluation in the incoherent holography in which a laser is replaced by a conventional thermal source.

FORMATION AND LOCALIZATION OF HOLOGRAPHICALLY PRODUCED INTERFERENCE FRINGES*

Introduction

The holographic interferometry is now very widely employed to measure a change in shape of a diffusely reflecting surface. Interference fringes produced on the surface indicate the contour of equal height variations of change took place over the surface between the production of and the reconstruction from a hologram. But, if the object undergoes a minute translation and rotation between the application of the change, fringes come to focus at a position away from the object and lose their contrast. This situation is quite different from that of the conventional interferometry, which uses an extended thermal source and is composed of flats and spheres of optical quality. In the conventional optical interferometry, we have a means of calculation of visibility and localization of fringes. It is derived from the van Cittert-Zernike theorem and has a very simple and straightforward physical meaning. But, our situation is quite different from this. The source is now point forming and the object is diffusely reflecting. The purpose of this study is to derive explicit expressions for the fringe formation and localization in the presence of translation between the identical objects whose surfaces are diffusely reflecting, and to give a clear physical meaning of the phenomena.

Theory

By using real-time or double-exposure holographic interferometry, we can realize two interfering waves of complex shape, which are exactly identical in shape not only in its broad structure but in its detailed ones due to the surface finish and minute machining marks. Therefore, we may assume that there are two surfaces which have a common profile $\varphi(x,y)$ as shown in Fig.1(a). By this profile, we postulate that it has a microscopic variation of height which exceeds the wavelength of light and which has no correlation across the surface, in order to ensure the diffuse reflection of the surface. In Fig.1(a), one surface is rotated by 2ϕ in the plane of the page and translated by 2X and 2Y with

---

* T. Tsuruta, N. Shiotake and Y. Itoh: Published in Optica Acta 16(1969)723-733. A part of this work was presented at the Symposium on Application of Coherent Light, Florence, September, 1968. A preliminary study was published in ref.1-2. Several articles independently dealing with this phenomenon have been published in these two years (ref.3-6).

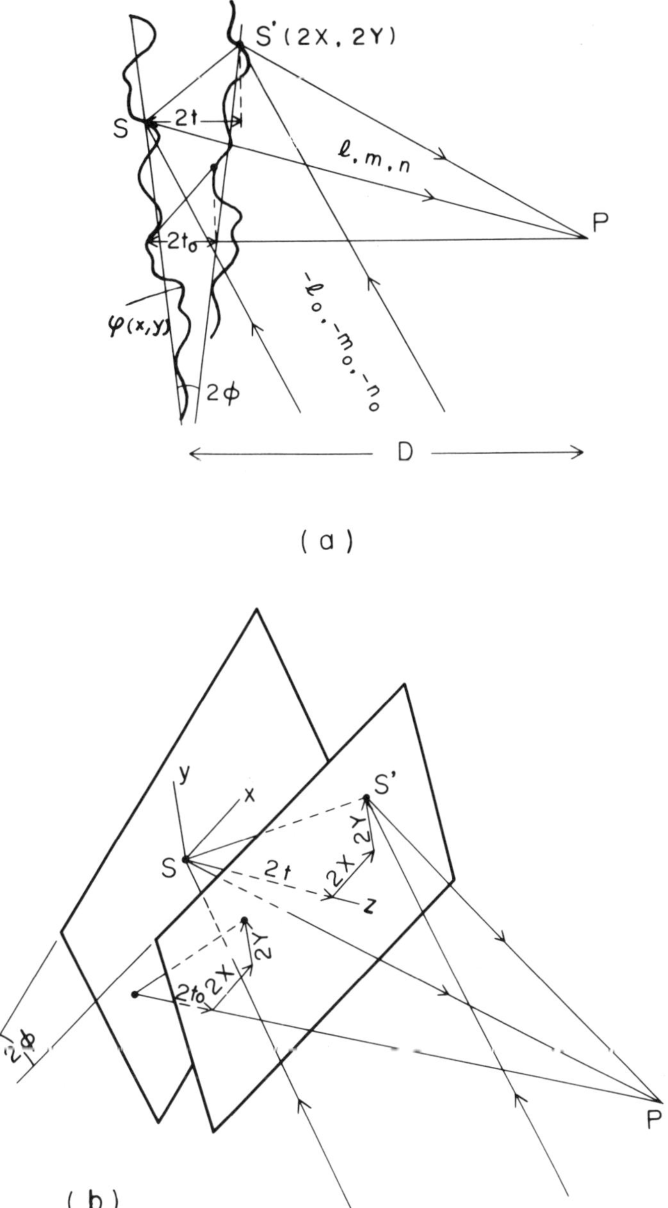

Fig.1 Geometry relating to the tilt of a diffusely reflecting surface and the localization of fringes.

respect to the other surface. Here, x axis is taken in the plane of the page and y axis perpendicular to this plane. When these two surfaces are supposed to be illuminated **simultaneously by a** collimated beam from a laser, the disturbance at an arbitrary taken point P is given by the sum of the waves diffracted by the two identical surfaces which are displaced to each other. This is:

$$D(P)=D_1(P)+D_2(P)$$
$$=\int \exp ik(R_1(x,y)+\alpha\phi x+\alpha\varphi(x,y))dxdy + \int \exp ik(R_2(x,y)-\alpha\phi x+\alpha\varphi(x-2X,y-2Y))dxdy. \tag{1}$$

where R_1 and R_2 are the optical path lengths from points on the planes of the wedge to P, $k=2\pi/\lambda$ is a wave propagation number and α is an obliquity factor and remains nearly constant over the surface. The intensity at P is simply given by multiplying the complex conjugate to this disturbance as given by:

$$I(P)=D(P).D(P)^*$$
$$=|D_1(P)|^2+|D_2(P)|^2+\mathrm{Re}\iint \exp ik\Delta.dx_1dx_2dy_1dy_2. \tag{2}$$

where

$$\Delta=R_1(x_1,y_1)-R_2(x_2,y_2)+\phi(x_1+x_2)+\varphi(x_1,y_1)-\varphi(x_2-2X,y_2-2Y). \tag{3}$$

The first two terms of eq.2 are the so-called bias term of the interference pattern and the third term represents the interference effect. Now, we assume that the microscopic height variation $\varphi(x,y)$ is a perfectly random function and its profile has spatially white noise spectrum to sufficiently high spatial frequencies. Then, its auto-correlation function is represented by the Dirac's delta function and the double integral of eq.2 is reduced to a single one as the following:

$$I(P)=|D_1(P)|^2+|D_2(P)|^2+2\mathrm{Re}\int \exp ik(R_1(x,y)-R_2(x+2X,y+2Y) +2\alpha\phi x)dxdy. \tag{4}$$

This equation shows that we may only consider pairs of corresponding points S and S' as shown in Fig.1 for the analysis of the fringe formation. The corresponding points are separated to each other by the rotation and translation of the identical surface. This equation also shows that the microscopic height profile of the surface has no effect on the formation of the fringes, while the relative displacement of the corresponding points S and S' determines the fringes, i.e., their shape, location and visibility.

We next calculate an actual optical path length as indicated in eq.4. A cartesian coordinate system is so chosen as the z axis perpendicular to the bisector of the two surfaces, the x axis in this plane and perpendicular to the wedge junction and the y axis is parallel to the junction. Then, x, y and z components of the displacement from S to S' are 2X, 2Y and 2t, respectively as shown in Fig.1. Let us assume that the direction cosines of the parallel beam of light illuminating the surface are $-l_o$, $-m_o$ and $-n_2$, and those of the scattered ray from the pair S and S' is very small, say, several multiples of a wavelength, then the optical path difference Δ between the rays emanating from S and S' is given by,

$$\Delta = d\{L(l+l_0)+M(m+m_0)+N(n+n_0)\}, \tag{5}$$

where $d=2(t^2+X^2+Y^2)^{1/2}$ is the distance between S and S' and L, M and N are the direction cosines of this displacement which are variable according to the position of S over the surface. Interference fringes are most distinct where the variation of Δ for changes in l and m are as small as possible*. This is the meaning of the localization of fringes of diffusely reflecting surface. This requirement may be expressed mathematically by differentiating Δ with respect to l and m to find the distance D from the wedge at which $\partial\Delta/\partial l=\partial\Delta/\partial m=0$. For this purpose, the path difference given in eq.5 must be rewritten in the form in which all parameters except l and m are constant over the surface. This is readily obtained as follows:

$$\begin{aligned}\Delta &= 2t\{X(l+l_0)/t+Y(m+m_0)/t+(n+n_0)\} \\ &= 2X(l+l_0)+2Y(m+m_0)+2(t_0+\tan\phi\,.\,lD/n)\,(n+n_0),\end{aligned} \tag{6}$$

where D is the distance between P and the surface, $2t_o$ is the z component of the distance between the corresponding points at the projection of P to the wedge and

$$t=t_0+\tan\phi\,.\,lD/n. \tag{7}$$

By differentiating the above equation, we obtain:

$$\begin{aligned}\frac{\partial\Delta}{\partial l} &= 2X-2(t_0+\tan\phi\,.\,lD/n)l/n \\ &\quad +2\tan\phi\,.\,\frac{1-m^2}{n^3}\,(n+n_0)D \\ &= 2t\left(\frac{L}{N}-\frac{l}{n}\right)+2\tan\phi\,.\,\frac{1-m^2}{n^3}\,(n+n_0)=0,\end{aligned} \tag{8}$$

*Michelson first introduced a similar analysis in his paper describing his Michelson interferometer(ref.7) where $X=Lt/N$. Similarly, we obtain:

$$\frac{\partial\Delta}{\partial m}=2t\left(\frac{M}{N}-\frac{m}{n}\right)+2\tan\phi\,.\,\frac{lm}{n^3}\,D(n+n_0)=0. \tag{9}$$

Several important properties of fringes are readily obtained from these equations.

(1) The simplest solution is obtained by letting $Ln=lN$ and $Mn=mN$. In order that these requirements are fulfilled over the entire surface for variable L and M and non-zero n and N, $L=0$ and $M=0$ must always hold so that $l=m=0$ follows. Therefore, fringes are found to be localized on the surface $D=0$, when translations along x and y axes are zero and they are observed perpendicular to the surface.

(2) A solution for non-zero translation can be obtained by letting $m=0$ and $M=Y/d=0$ in eq.9. Then the localization position is obtained from eq.8,

$$D=-\frac{X-l\,.\,t/n}{\tan\phi\,.\,(n+n_0)/n^3}=-\frac{2n(X-l\,.\,t/n)}{2\tan\phi\,.\,(n+n_0)/n^2}. \tag{10}$$

This means that fringes are localized away from the surface when the displacement is made simply along the x axis and the observation is made in a plane perpendicular to the wedge junction. The numerator represents an apparent displacement of the corresponding points S and S' in the case of oblique observation $l\neq 0$, and the denominator is approximately an effective wedge angle for n nearly equal to unity.

(3) When translation along the y axis is present, the value of D satisfying both of eqs.8 and 9 does not exist. This means that there is no priviledged position where the visibility of the fringes has the maximum value. In other words, there is no localization in this case.

When the interfering surfaces are placed parallel to each other, that is $\phi=0$, fringes appear at infinity. Then, the expression for the path difference is given by:

$$\Delta=2(l+l_0)X+2(m+m_0)Y+2t_0(n+n_0). \tag{11}$$

The first two terms represent the Brewster fringes and the last one the Haidinger rings.

To obtain a satisfactory interpretation of the theoretical results given above, let us suppose that there is a minute grating with its normal coinciding with that of the surface concerned and its grooves of arbitrary spacing and orientation (Fig.2). By a proper choice of grating, an incident wave $(-l_o,-m_o,-n_o)$ can be

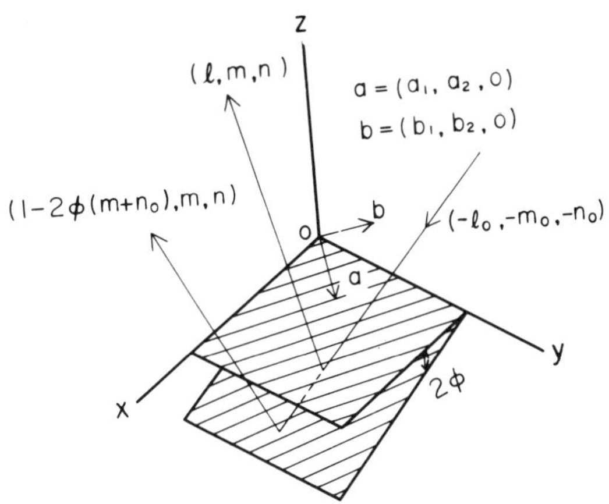

Fig.2. Diffraction from two identical gratings tilted by 2∅ with respect to each other.

diffracted to the direction (l,m,n) which was previously considered as that of the scattered wave. Let $(a_1,a_2,0)$ to be the direction cosines of a unit vector, perpendicular to the grooves of the grating, which lies in the (x,y) plane, and $(b_1,b_2,0)$ be those of a unit vector parallel to the grooves. Then the grating equations are immediately obtained:

$$a_1(l+l_0)+a_2(m+m_0)=p\lambda/d, \tag{12}$$
$$b_1(l+l_0)+b_2(m+m_0)=0, \tag{13}$$
$$a_1/a_2=-b_2/b_1, \tag{14}$$

where d is a grating spacing constant, λ is the wavelength and p is any integer. When the grating is rotated by 2∅ around the y axis, the above equations are modified as follows:

$$a_1(l'+l_0)+a_2(m' \mid m_0) \mid 2\phi a_1(n'+n_0)-p\lambda/d, \tag{15}$$
$$b_1(l'+l_0)+n_2(m'+m_0)+2\phi b_1(n'+n_0)=0, \tag{16}$$

where (l',m',n') are the direction cosines of the diffracted wave from the tilted grating. By subtracting eq.15 from eq.12 and eq.16 from eq.13, we obtain:

$$a_1(l'+l)+a_2(m'-m)+2\phi a_1(n'+n)=0, \tag{17}$$
$$b_1(l'-l)+a_1b_1/a_2\,.\,(m'-m)+2\phi b_1\,(n'+n)=0. \tag{18}$$

These two equations should hold simultaneously, so that the

relation between the diffracted rays from different setting is:

$$l' = l - 2\phi(n + n_0), \tag{19}$$

$$m' = m. \tag{20}$$

A ray diffracted from a minute grating will intersect that from an identical grating, tilted by 2ϕ and translated by $(2X, 2Y, 2t)$, only if the following requirements are fulfilled:

$$2X = (\mu + \nu)l - 2\phi\nu(n + n_0), \tag{21}$$

$$2Y = (\mu + \nu)m, \tag{22}$$

$$2t = \mu n + \nu n', \tag{23}$$

where μ and ν are any numbers and $n' = (n^2 + 4l\phi(n + n_0))^{1/2}$. For a wedge made by tilting identical plane surfaces, the transalations X and Y are the same over the surface, while t varies along the wedge. For the given direction cosines of the observation (l,m,n,), X and Y completely determine μ and ν, so that t is a single value, corresponding to a particular line on the object. For a general configuration therefore, there exists no region where the diffracted rays intersect on the object. For Y=m=0, two diffracted rays necessarily meet in the (x,z) plane, and the position of intersection which concides exactly with eq.10 is readily obtained.

We may therefore regard a diffusely reflecting surface as being composed of minute gratings with their normals coinciding

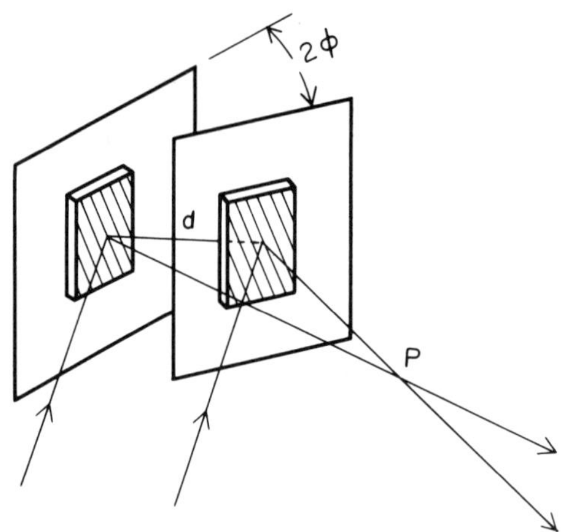

Fig.3. Ray tracing method to determine the localization plane.

with that of the surface, and having arbitrary spacings and orientations of the grooves. Two rays falling onto a pair of identical gratings are diffracted to meet at a point, if it exists, which gives the localization position of fringes (Fig.3). Because the spacing and orientation of the grooves are arbitrary, we can take them to be such that the diffracted rays be in the desired direction of observation. When the interfering surfaces are placed parallel to each other, the diffracted rays from any pair of gratings emanate parallel and form interference fringes at infinity. This grating model of interference has thus completely explained the theoretical results, revealed the physical nature of the fringe formation in holographic interferometry, and presented a practical ray tracing method similar to that of conventional interferometry, in which the region of localization includes those points, if they exist, which lie at the intersection of the two rays derived from a single incident ray from a broad source.

For the metrological use of interferometry, the quantity to be determined is the depth of the wedge t, so that the effect of the variation in l and m on the path difference Δ has to be eliminated over the object. Figure 4 shows an example of a hologram interferometer satisfying this requirement. An object is placed perpendicular to the incident parallel beam from a laser. A beam of light scattered by the object and incident upon the camera lens placed at the focus of the collimator is limited by a diaphragm, whose angular aperture in the object space is given by $\alpha = a/f$, where a is a radius of the diaphragm and f is the focal length of the collimator lens. If the translation is confined to the line parallel to the wedge, that is $X \neq 0$ and $Y=0$, and the camera is focused at the localization plane, the optical path length Δ responsible for

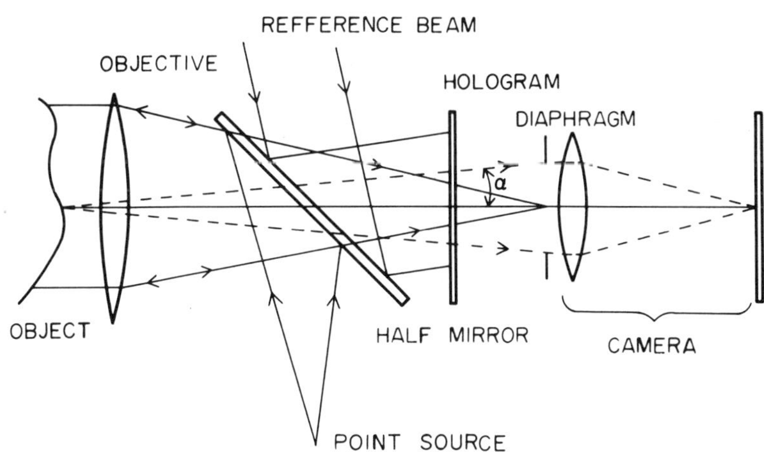

Fig.4 A hologram interferometer suitable for metrological applications.

the fringe contours is given by:

$$\Delta = 4t_0 + 2t_0 (l^2 + m^2). \tag{24}$$

The reduction in visibility is unappreciable provided the thickness t_o is small and $2t_o(a/f)^2$ does not exceed about a quarter of a wavelength. When a translation Y along the junction is present together with X, and the camera is focussed at the localization plane for X, the optical path difference Δ is given by the following expression neglecting powers of l and m higher than the first:

$$\Delta = 2Ym + 4t_0. \tag{25}$$

As Δ varies with m, the visibility γ will drop. From coherence theory this is expressed by

$$\gamma = 2J_1\left(\frac{4\pi}{\lambda} Y.\alpha\right) \Big/ \frac{4\pi}{\lambda} \gamma.\alpha, \tag{26}$$

where J_1 is the Bessel function of the first order. When the plane of the observation is on the object, in spite of the departure of the fringe localization plane due to the translation, the visibility is given by:

$$\gamma = 2J_1\left(\frac{4\pi}{\lambda} (X^2+Y^2)^{1/2}.\alpha\right) \Big/ \frac{4\pi}{\lambda} (X^2+Y^2)^{1/2}.\alpha. \tag{27}$$

If the requirement $\gamma > 0.8$ is fulfilled, the light emanating from the laterally separated identical points will spread by diffraction and effectively overlap at the image plane to produce clear interference fringes.

Deformations, translations and rotations taking place on the actual object, however complex they may be, are the aggregate of tilts and translations of various amounts and orientations over the surface. Therefore, the discussions described so far will serve as a governing principle for analyzing the interference fringes obtained by using a hologram interferometer with a diffusely reflecting surface as an object.

Experiments

An experiment was first performed to demonstrate some of the interesting properties of interference fringes discussed in the previous section. A flat disc of aluminium of 5mm in thickness and 70mm in diameter was mounted on an air-tight chamber, and the air was pumped out to cause the disc to sag by a few hundredths of a millimeter. The surface of the disc was sand-blasted to scatter the incident light from a He-Ne laser. The chamber was fastened to a table which could be translated by an adjustable screw. The disc was placed normal to the incident light as shown in Fig.4, and a hologram was made by exposing a Kodak 649F plate twice, before and after the application of the deformation. By illuminating the hologram by the reference beam, concentric fringes representing contours of the deformation were observed over the reconstructed image of the disc. When a small lateral translation as well as the deformation occurs between the exposures, fringes come to focus away from the disc, and the reduction of the fringe visibility also follows. Photographs shown in Fig.5 were obtained from a hologram which received a translation of 15 microns between the exposures. The first two photographs were made by focusing the camera at the reconstructed image of the disc with an angular aperture of 1/75 for (A) and 1/25 for (B). In (A) fringes are clearly seen, while in (B) it is rather difficult to recognize their existance. By calculation, using eq.27, we obtain $\gamma=0.75$ for (A) and effectively zero for (B), in agreement with the experiment. Other photographs in Fig.5 were made for the imaging position away from the disc. The angular aperture of the camera was taken 1/25. For every photograph of different distance D from the disc, there exist fringes of the most distinct contrast, whose spatial frequency s satisfies the following relation:

$$s=\lambda D/2X. \qquad (28)$$

As denoted in Fig.5, the measured spatial frequencies s_{meas} agree well with the calculated ones. The direction of the translation is parallel to that of the wedge along the horizontal line, so that the requirement of the localization of fringes is fulfilled exclusively along this line. Along an azimuth of angle θ to the horizontal line, the visibility decreases according to eq.26, in which the translation along the junction of the wedge is given by $2X\sin\theta$. An azimuth along which the fringes disappear is readily calculated as $\theta=36°$. The region of distinct fringes over the disc shown in the photographs agrees qualitatively with this result.

The second experiment was concerned with a quantitative verification of the theory, expressed by eq.10. We employed a double-exposure hologram interferometer using two collimated reference beams of different incident angles to the photographic plate. The details of this interferometer described in ref.2 is schematically shown in Fig.6. The first exposure using one of the reference beams was made with the object in position and the second after the object was tilted. After the photographic processing, the hologram was replaced to its original position and illuminated simultaneously by the two reference beams. Then the two reconstructed images exactly coincide in space and clear Fizeau fringes are produced on the images. The distance between the successive fringes is inversely proportional to the amount of tilt given between the exposures. The lateral translation between the reconstructed

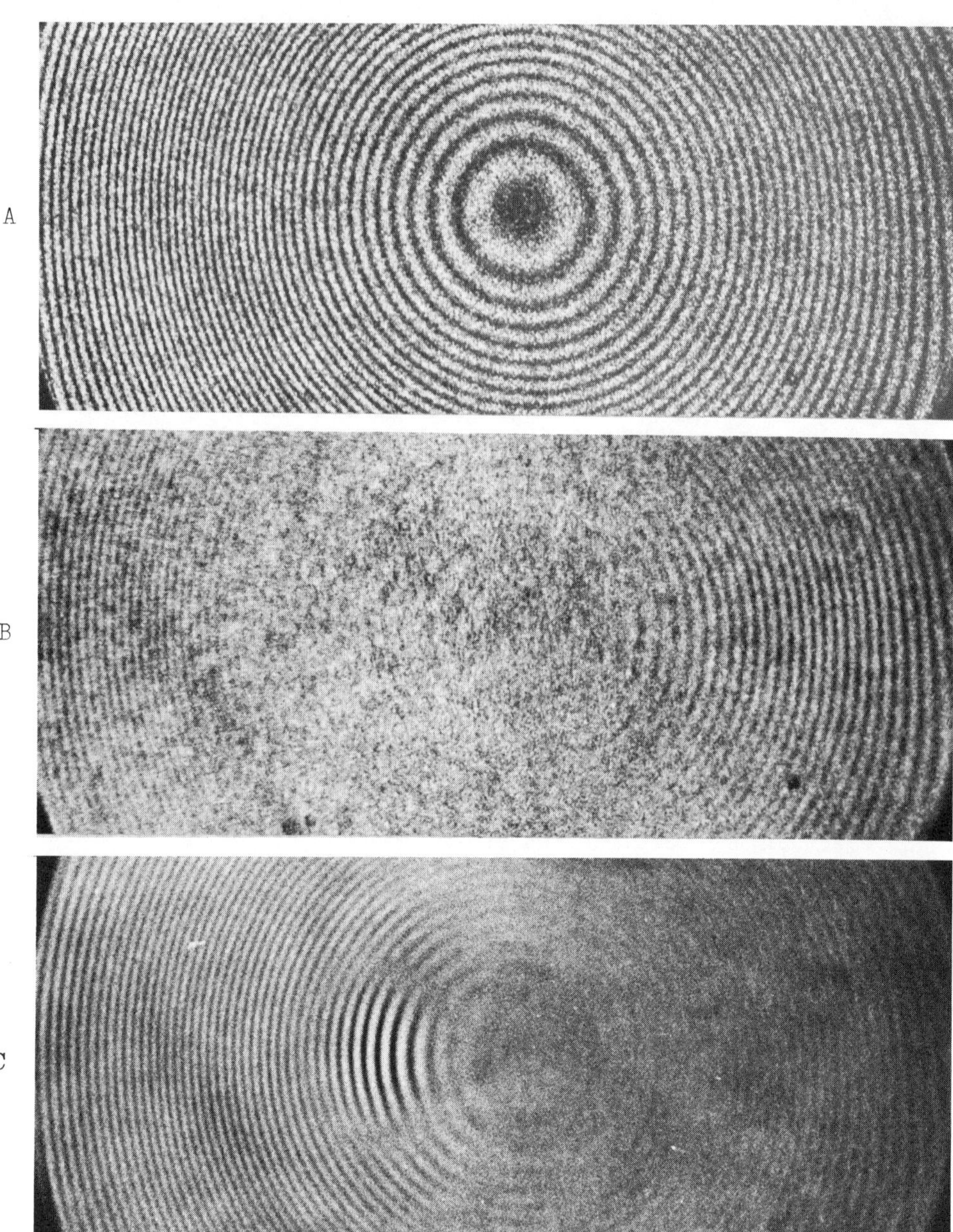

Fig.5. Interference fringes produced from a double-exposure hologram. A disc of alminium undergoes a lateral translation od 15 microns as well as the longitudinal deformation between the exposures. (A) focused at the image of the disc with an angular aperture $\alpha = 1/75$. (B) focused at the disc with an angular aperture $\alpha = 1/25$. (C) at 35mm behind the disc with $\alpha = 1/25$. $s_{cal} = 0.69$ lines per mm and $s_{meas} = 0.70$ lines per mm.

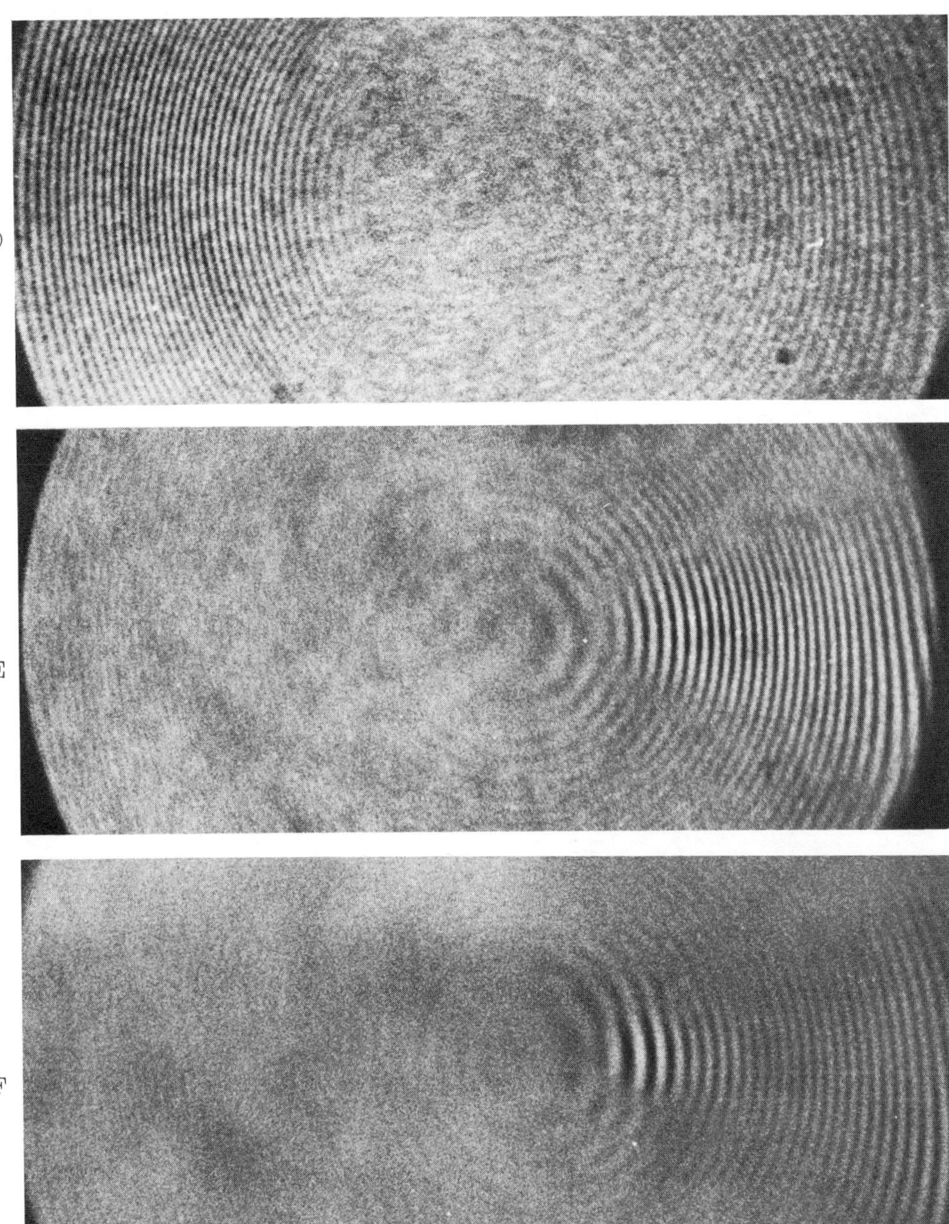

Fig.5. (continued) (D) at 15mm behind the disc with $\alpha = 1/25$. $s_{cal} = 1.6$ lines per mm and $s_{meas} = 1.5$ lines per mm. (E) at 25mm in front of the disc with $\alpha = 1/25$. $s_{cal} = 1.0$ lines per mm. (F) at 40mm in front of the disc with $\alpha = 1/25$. $s_{cal} = 0.61$ lines per mm and $s_{meas} = 0.6$ lines per mm.

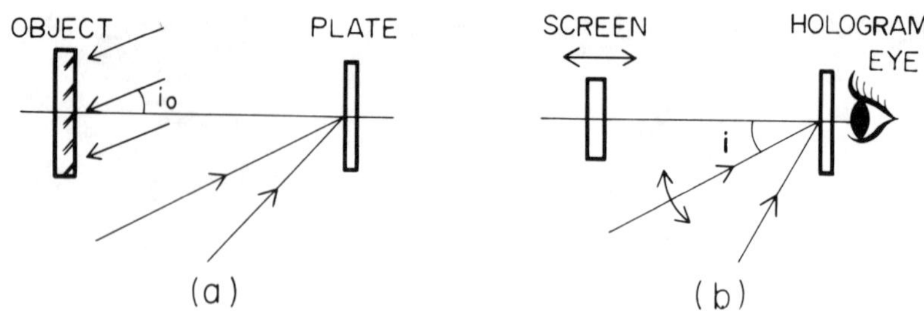

Fig.6. Double-exposure hologram interferometer using two reference beams. (a) Production of a holcgram. (b) Measurement of the localization position.

images were realized by a slight rotation of one of the hologram illuminating beams. The rotation of the beam, however, was accompanied by an additional tilt between the images so that eq.10 should be modified to:

$$D = -\frac{2X}{2(n_0+1)\phi - 2X/P}, \tag{29}$$

$$2X = P\cos i \, . \, \Delta i. \tag{30}$$

where P is the distance between the object and the hologram, 2ϕ is an angle of tilt of the object between the exposures, i is an angle of inclination of the hologram illuminating beam to the normal of the hologram and Δ i is an angle of rotation of the beam to give various amounts of translation in the reconstruction process.

The specimen was a glass plate of 60mm in diameter which was deposited with MgO on its surface to make it perfectly diffusely reflecting. It was illuminated by a parallel beam at 30° to the normal, giving $n_o = \cos i_o = 0.87$. Slight tilts of the specimen were given by 0", 11", 50" and 65" of arc, and these angles were accurately measured by an autocollimator. Exact coincidence between the reconstructed images, that is X=Y=0, was obtained by adjusting one of the hologram illuminating beams to the position where no parallax took place between the reconstructed image and the fringes. A screen with lines parallel to the fringes was then put in position instead of the object and moved along the line of sight, and the plane of the fringe localization was determined by measuring a position of no parallax between the screen and the fringes for a given angle of rotation of an adjusting beam. Figure 7 shows the experimental results. This good agreement between the theory and the experiment is sufficient to confirm the discussions we have presented.

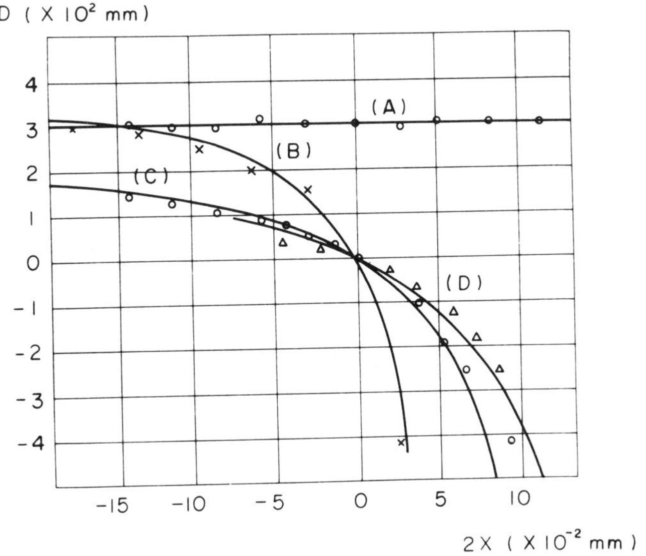

Fig.7. Variations in the localization plane with the translation 2X for various values of tilt 2ϕ. (A) P=300mm, 2ϕ=0, t_o=0. (B) P=400mm, 2ϕ=13", t_o=0.2mm. (C) P=300mm, 2ϕ=50", t_o=0. (D) P=300mm, 2ϕ=65", t_o=0. Solid curves are the theoretical results from eq.29.

References

1) J. Tsujiuchi and T. Tsuruta, OYO BUTURI 36, 232(1967)
2) T. Tsuruta, N. Shiotakc and Y. Itoh, Jap. J. appl. Phys., 7, 1092(1968).
3) K.A. Stetson, Presented at the Symposium on the Engineering Uses of Holography, Glasgow(1968).
4) S. Walles, Institute of Optical Research, Stockholm, Technical Report TR 30.6.1968.
5) C. Froehly, J. Monneret, J. Pasteur et J.ch. Viénot, Opt. Act., 16, 343(1969).
6) W.T. Welford, Optics Communications, 1, 123(1969)
7) A.A. Michelson, Lond. Edinb. Dubl. Phil. Mag., 13, 236(1882).

HOLOGRAPHY USING AN EXTENDED SPATIALLY INCOHERENT SOURCE*

Introduction

When a light source other than a laser is employed for making a hologram, a considerable reduction of the image quality results, owing to the small spatial and time coherence of the source. In 1967, Leith and Upatnieks made a hologram of very excellent quality with their achromatic fringe system (ref.1). They placed a photographic plate at the image plane of a grating and recorded a hologram between two beams of different orders of diffraction, in either of which was inserted the transparent object to be recorded. They introduced the time coherence requirement which made possible the application of a high-pressure mercury arc, but did not deal with the spatial coherence. Kato and Suzuki later extended their method to the spatial coherence problem and obtained some qualitative results which related the source size to the resolution limit of the reconstructed image(ref.2). These studies, however, were concerned mainly with their particular arrangements and no attempts were made to derive a general theory of the image formation for a hologram made by using a thermal source.

According to the author's view point, the analysis of the image formation of this type of holography should be based upon the theory of fringe formation that is well known in optical interferometry and described in the literature(refs.3,4). Generally, interference fringes formed by different parts of the extended source are displaced with respect to each other and this results in the reduction of the fringe visibility. The reduction is great for closely spaced fringes, which is the case in side-band holography recording, consisting of very close fringes formed by the object and the reference beams. Clear fringes will be recorded in a photographic plate only when the phase difference between pairs of beams, one diffracted from the object and the other striking the plate directly, are the same or nearly the same, no matter from what part of the extended source the light originates. The author presents a theory of the image formation of this type of holography with the aid of the theory of the fringe formation of two-beam interferometry and shows some interesting properties of the reconstructed image.

Fringe Formation by an Extended Monochromatic Source

Figure 1 shows a schematic arrangement of an interferometer with two arms. A Michelson interferometer may be considered as an

* Published in the January issue of J. Opt. Soc. Am.(1970).

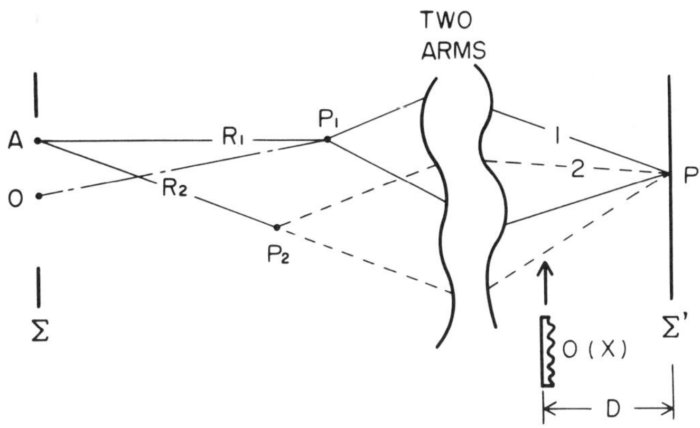

Fig.1 Schematic arrangement of an interferometer with two arms.

example. An extended source of uniform radiance is situated at the source plane Σ and the fringes are recorded in the fringe plane Σ'. Suppose that P_1 and P_2 are conjugate to P with respect to the optical systems of paths 1 and 2 of the interferometer. In the Michelson interferometer, P_1 and P_2 are the virtual images of P formed by the two end mirrors, respectively. Then, the visibility of the fringes at P in the plane of Σ' is related to the modulus of the degree of coherence Γ_{12} between P_1 and P_2 by:

$$\gamma=[2(I_1\cdot I_2)^{\frac{1}{2}}/I_1+I_2]\Gamma_{12}, \tag{1}$$

where I_1 and I_2 are the transmissions of the separate arms. According to the van Cittert-Zernike theorem, which relates the coherence between any pair of points in an illuminated plane to the distribution on the illuminating source, the degree of coherence Γ_{12} is given by:

$$\Gamma_{12}=\int_\Sigma \frac{\exp ik(R_1-R_2)}{R_1\cdot R_2}d\sigma \Big/ \int_\Sigma \frac{1}{R_1\cdot R_2}d\sigma, \tag{2}$$

in which $k=2\pi/\lambda$, where λ is the wavelength. R_1 and R_2 are the optical path lengths from a point A upon the source to P_1 and P_2, respectively, and the integration is over the region of the source surface. Because R_1 and R_2 vary very little over the region of integration, we may regard them as constant in the denominator, though not in the exponential. The above equation is reduced to:

$$\Gamma_{12}=\int_\Sigma \exp ik(R_1-R_2)d\sigma \Big/ \int_\Sigma d\sigma. \tag{3}$$

This indicates that the degree of coherence is given by the complex amplitude at P_2 in the aberration-free diffraction pattern that is associated with an aperture identical in shape with the source and with P_1 to which all rays converge in phase. If P_2 is located in a plane perpendicular to OP_1 and crossing P_1, where O is the center of the circular source, the degree of coherence is given by:

$$\Gamma_{12}=2J_1(p)/p, \tag{4}$$

in which $p=k\alpha\langle P_1P_2\rangle$, where $\langle P_1P_2\rangle$ is the distance between P_1 and P_2, α is the semi-angular aperture subtended by the circular source at P_1, and J_1 denotes the Bessel function of the order one. On the other hand, when P_2 lies along OP_1, the degree of coherence is given by:

$$\Gamma_{12}=\frac{\sin q/4}{q/4}, \tag{5}$$

where $q=k\alpha^2\langle P_1P_2\rangle$. Interference fringes will be most distinct at any point P whose conjugate images P_1 and P_2 coincide, to give $\Gamma_{12}=1$. But this requirement is too severe for most interferometers, so that we may set $\Gamma_{12}>0.8$, which provides a region of practically sufficient fringe visibility. We calculate next this region for two types of interferometers, the Michelson and triangular interferometers.

The Michelson interferometer can be reduced to a system of two mirrors, M_1 and M_2, that form an air wedge as shown in Fig.2(a). The system is illuminated perpendicularly to the inner bisector of the wedge. The wedge angle is 2ϕ. The distance between the surfaces of the mirrors through a point O on the bisector is designated b and a point P at which the fringes should be recorded is at a distance a from the inner bisector of the wedge and crossing O. According to the definition given in the first paragraph of this section, P_1 and P_2 are the virtual images of P formed by the mirrors M_1 and M_2, respectively. Then, the lateral and longitudinal projections lat and lon of $\langle P_1P_2\rangle$ to a plane perpendicular to OP_1 and OP_1 itself are given, respectively, by:

$$\text{lat}=4\phi a, \tag{6}$$

$$\text{lon}=(1+\cos 2\phi)b. \tag{7}$$

Suppose that a conical beam from a circular source of a semi-angular aperture of α falls upon the wedge with its central ray coinciding with OP. The lateral splitting given by eq.6 becomes zero for a=0. The depth of the region $\pm\Delta a$ around a=0 that gives $\Gamma_{12}>0.8$ is given from eq.4 by:

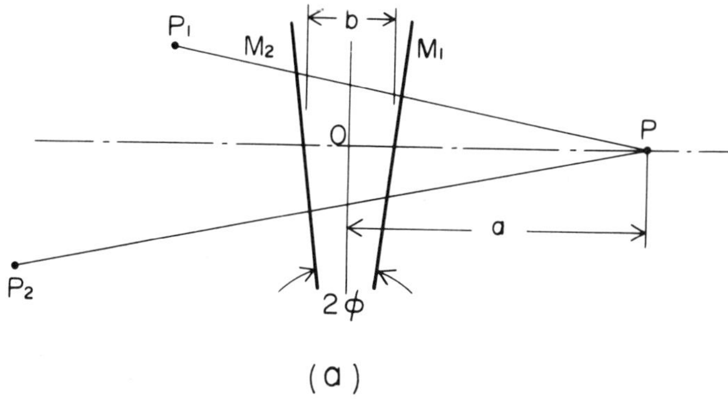

(a)

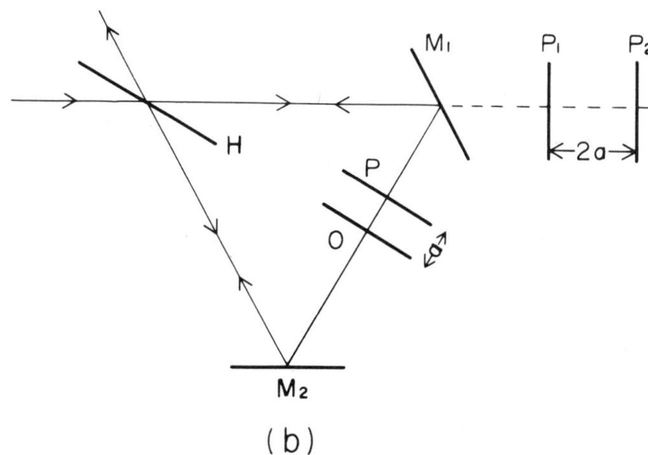

(b)

Fig.2. Fringe formation in (a) a Michelson interferometer reduced to an air wedge of an angle 2ϕ and (b) a triangular interferometer. In (a) P is a point in a plane in which fringes are recorded; M_1 and M_2 are plane mirrors; P_1 and P_2 are mirror images of P by M_1 and M_2. In (b) H is a beam splitter; M_1 and M_2 are plane mirrors; P_1 is a mirror image of P by M_1; P_2 is an image of P by successive reflections by M_2 and H; O is midpoint of the triangular path, at which the path difference between the beams falling from opposite directions is nearly the same.

$$\Delta a < 1.3\lambda/8\pi\alpha\phi. \tag{8}$$

This relation holds approximately only if the decrease of Γ_{12} due to the longitudinal separation lon in eq.7 is ignored. This condition is given by setting $\Gamma_{12} > 0.8$ in eq.5:

$$\pi\alpha^2 b/\lambda < 1.1. \tag{9}$$

This requirement is the same as that necessary to eliminate Haidinger rings produced by parallel mirrors at a distance b apart, down to a diameter within which the path difference does not exceed a quarter of a wavelength. As a numerical example, we consider a fringe system of 100 lines per mm over a region of 60mm with the path difference between the interfering beam coinciding at the center of the field. Then, with the source $\alpha = 1/200$, we obtain the depth of focus ± 0.41mm and the left side of eq.9 is 0.12, much smaller than 1.1.

A triangular interferometer (ref.5) is shown in Fig.2(b). A light beam falls upon a beam splitter H as indicated by an arrow. The solid line indicates the central ray coming from a circular source at infinity. The beam is divided by H into two beams that are adjusted so as to traverse triangular paths along the same path in opposite directions before they are recombined. If a photographic plate is placed in a triangular path, a system of standing waves is recorded with the optical separation of half wavelength, provided that a monochromatic radiation enters the interferometer. In a plane located at O at which the path lengths from the beam-splitter, arriving from opposite sides, are exactly the same, a system of white light fringes may be recorded, just as in Lippmann photography. Suppose that a point P in a plane Σ' in which fringes are to be recorded is located at a distance of a from O. The conjugate images P_1 and P_2 of P are the virtual images produced by the reflection M_1 and reflections by M_2 and H, respectively. Because they are along an extended line of the central ray at a distance of 2a, eq.8 is automatically satisfied and the only requirement to restore sufficient visibility is eq.9. From the localization point of view, in which fringes are defined as localized at the position where the lateral splitting of the images P_1 and P_2 is zero, the fringes of this type are localized everywhere, and provide a very thick region of sufficient visibility for an extended source of monochromatic radiation.

Transfer Function of the Hologram

In this section, we consider a holographic arrangement in which a photographic plate is placed in exact coincidence with the plane of the most distinct fringe visibility and then a transparent object is inserted in either of the interferometer arms, say arm 2 as shown in Fig.1, at a distance D from the plate. As shown in Fig.3, the complex amplitude distribution O(X) of the object illuminated by a plane wave is related to its spectrum $o(\theta)$ by the Fourier transform:

$$O(X)=\frac{1}{2\pi}\int o(\theta)\exp(ik\theta X)d\theta \tag{10}$$

$$o(\theta)=\frac{1}{2\pi}\int O(X)\exp(-ik\theta X)dX, \tag{11}$$

where X denotes a two-dimensional cartesian coordinate at the object plane and θ is an angle of diffraction from the object. These equations show that the object is to be considered as an aggregate of a beam deflector, which deviates the incident light by θ , with a corresponding amplitude transmission proportional to the Fourier component $o(\theta)$ of the object, where θ is related to the spatial frequency l of the object by:

$$\theta=\lambda l. \tag{12}$$

We now consider the visibility of the interference fringes at the plane of the plate between a beam deflected by the object and the beam from the arm 1. A ray emanating from P and passing through a beam deflector placed at the position of the object is deviated as if it were at a position away from P by $D\theta = \lambda Dl$. In Fig.4 are shown three possible arrangements: Leith and Upatnieks type(a), the triangular interferometer(b), and Gabor's in-line arrangement(c). Prisms drawn in Fig.4 denote deflectors corresponding to the particular component of spatial frequency l of the transparent object. In Leith and Upatnieks's arrangement, as shown in (a), a photographic plate is placed in the image plane of a grating. A ray incident upon the grating is diffracted into several orders: they meet at a point in the image plane of the grating. Two beams of different orders are made to pass and fall upon the plate and the object is placed in one beam, the other is the reference beam. This system

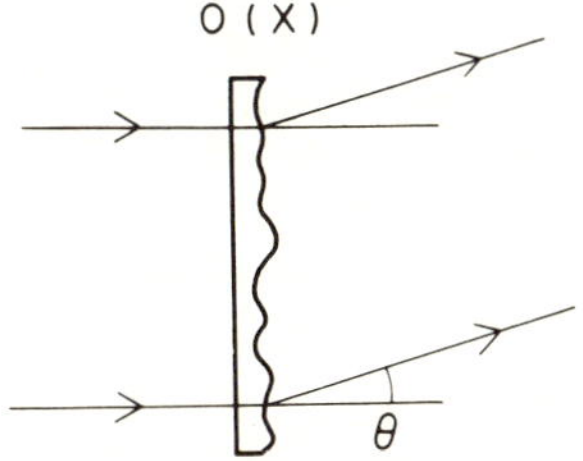

Fig.3 Illustrating an object illuminated by a plane wave and the resultant diffracted wave.

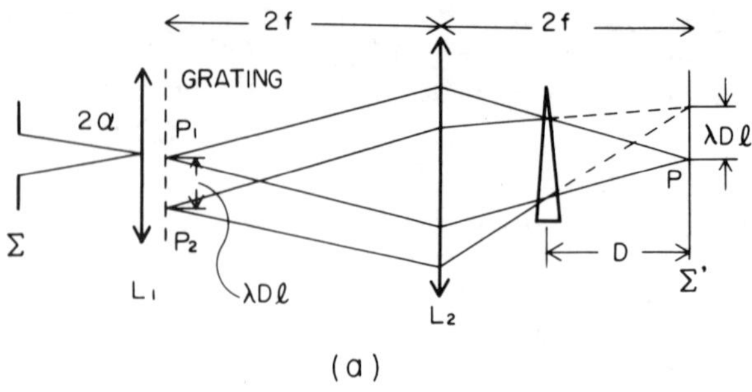

(a)

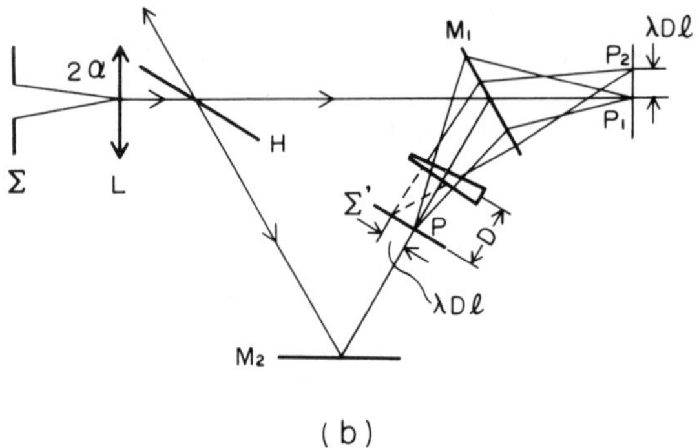

(b)

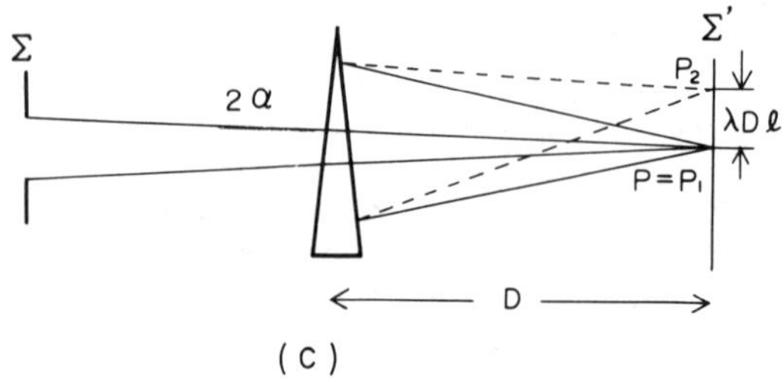

(c)

Fig.4. Source-size compensated arrangements. (a) Leith and Upatnieks's achromatic fringe system; L_1 is a collimator to locate the source at infinity; 2α is an angular aperture of the source; L_2 is a lens to image a grating onto a hologram plate at Σ'; f is the focal length of L_2. A prism in the object plane at a distance D from the hologram indicates a beam deflector corresponding to a particular spatial-frequency component of the object. (b) triangular interferometer; H is a beam splitter; M_1 and M_2 are plane mirrors; Σ' is the hologram plate; (c) Gabor's arrangement.

functions nearly the same with the two mirror system of Fig.2(a), except that it gives lat=lon=0, i.e., $<P_1P_2>=0$ over the whole hologram plane between the reference beam and the undiffracted beam from the object. It should also be noted that the path difference between them is always zero over the whole hologram plane. This is the meaning of the achromatic arrangement. In the triangular arrangement(b), a plate is placed at the mid-point 0 as shown in Fig.2(b), and an object is placed at a distance D from it. A beam falling from the opposite direction serves as the reference beam. Then, we also have $<P_1P_2>=0$ between the reference beam and the undiffracted (directly transmitting the object) beam. In Gabor's apparatus(c), the undiffracted beam is nothing but the reference beam itself, so that $<P_1P_2>=0$. Therefore, we indicate the light path of the undiffracted beam in Fig.4 in order to illustrate P_1 instead of the reference beam so as to simplify the drawing. Common to the three arrangements, the translation of P_2 with respect to P_1 for the diffracted beam from the object of the spatial frequency l is equal to the distance $D\theta=\lambda Dl$ taken in the plane of the hologram as mentioned above. We now have:

$$\text{lat}=\lambda Dl, \tag{13}$$

$$\text{lon}=0. \tag{14}$$

The visibility of the fringes corresponding to the Fourier spectrum $o(\lambda l)\lambda\Delta l$ of the object and recorded in a hologram plate is obtained from eq.4 by:

$$\gamma=\frac{2I_1^{\frac{1}{2}}|o(\lambda l)\lambda\Delta l|}{I_1+|o(\lambda l)\lambda\Delta l|^2}\cdot\frac{2J_1(2\pi\alpha Dl)}{2\pi\alpha Dl} \tag{15}$$
$$\doteqdot 2|o(\lambda l)\lambda\Delta l|/I_1^{\frac{1}{2}}\cdot 2J_1(2\pi\alpha Dl)/2\pi\alpha Dl,$$

where $|o(\lambda l)\lambda\Delta l|^2$ is supposed to be sufficiently small compared to I_1, the transmittance of the arm 1, as is the case in the ordinary holography arrangement with the arm 1 used as the reference beam. For ordinary holography arrangements of side-band as well as Lippmann types, the image is reconstructed by light of a single diffraction order from the hologram. So, when the developed plate or the hologram is illuminated by a laser, a wave diffracted by the hologram to form a reconstructed image differs by

$$H(l)=2J_1(2\pi\alpha Dl)/2\pi\alpha Dl \tag{16}$$

from the original wave coming from the object, for each spatial component, so that we may regard this quantity as the optical transfer function of the hologram for both amplitude and phase distribution of the object. For Gabor's in-line holography, development of the hologram to a gamma of two is necessary for the above discussion to be valid. The source size that gives the cut-off spatial frequency l_c is deduced from eq.16 as:

$$\alpha = 3.8/2\pi D l_c. \tag{17}$$

Baez, using a different approach (ref.6), obtained a similar result for Gabor's in-line holography. The three typical arrangements are thus shown to have the same source size requirement, if they are properly arranged.

By Fourier transformation of eq.16, the spread function h(X) of the imaging is obtained as:

$$\begin{aligned} h(X) &= \text{unity} \quad \text{for} \quad |X| < \alpha D, \\ &= 0 \quad \text{for} \quad |X| > \alpha D. \end{aligned} \tag{18}$$

This is a step function with diameter $2\alpha D$. This can be interpreted as a geometrical shadow of a pin hole illuminated by the source of semi-angular aperture α falling upon a photographic plate placed at a distance D from it. The quantities α and D can readily be evaluated from the configuration of an experimental arrangement. It should be noted that the image reconstructed from a hologram made by use of a source of finite size is similar to the defocused image in ordinary photography. For a given source size, a better image will be obtained as this distance is made smaller. By use of the triangular interferometer, it is possible to locate a transparent object very close to the hologram plate without obstructing the reference beam falling from the reverse side of the plate, so that this interferometer may provide a good means for obtaining a high resolution hologram by using a spatially incoherent light source. In the above discussions, and especially in the derivation of eq.16, we have neglected the effect of diffraction caused by the finite aperture of the hologram. In practical arrangements, however, the resolution limit given by eq.17 is far below that of diffraction, which is determined by the angular aperture of the hologram seen from the object plane, so that the analysis may be applied to almost all of the arrangements that use a spatially incoherent light from a thermal source for the production of a hologram.

When the object is illuminated diffusely by a diffuser close behind the transparent object, the object may be considered to have spatial Fourier components continuously ranging up to very high spatial frequencies, depending on the light scattering property of the diffuser. The light that is scattered beyond an angle λl_c does not contribute to the fringe formation, but falls upon the hologram plate as an incoherent background which reduces the signal to noise ratio of the reconstructed image. This was dealt with in detail by Hamasaki(ref.7). A reduction of image quality results from this effect. This explains explicitly the difficulty of making a hologram of a diffusely reflecting object by use of an extended source.

Time Coherence

In hologram production, there are two causes of reduction of fringe visibility that depend on the finite time coherence of the light source. The first originates from the path difference between the interfering beams, the reference beam and the undiffracted beam from the object. If this is compensated over the hologram plate, a light source of quite short coherence length can be used for the production of the hologram. Leith and Upatnieks' achromatic fringe system is a typical example. Gabor's in-line holography also belongs to this class, because the undiffracted beam serves as a reference beam. This may be accomplished also by use of the triangular interferometer, in which path lengths of two beams falling upon the plate from opposite directions can be made to coincide. The second cause of reduction of fringe visibility is diffraction from the object. The diffracted wave from the object falls upon the plate at an angle θ to the undiffracted wave, where $\theta = \lambda l$, as mentioned in eq.12. Then, an additional path difference Δ is introduced in spite of the exact equality of the path lengths of the reference and undiffracted object waves, where:

$$\Delta = D\lambda^2 l^2/2. \tag{19}$$

If this path difference is equal to the coherence length L of the source, the highest spatial frequency that can be reconstructed is given by:

$$l = (\lambda)^{-1}(2L/D)^{\frac{1}{2}}. \tag{20}$$

Summary

We have applied the theory of two-beam interferometry to the analysis of the reconstructed image from a hologram produced by using a spatially incoherent light source and have drawn attention to several important physical requirements for this type of holography. The hologram should be placed in a plane where the reference beam and the undiffracted beam from the object form the most distinct fringes. An optical transfer function has been derived for this configuration and is found to have the same form for three different systems: Leith and Upatnieks' , triangular interferometer, and Gabor's arrangements. They are compensated for the path difference between the reference beam and the undiffracted beam from the object in the plane of the hologram, and are convinient for use with thermal light, which has small time coherence as well as spatially incoherence.

References

1) E.N. Leith and J. Upatnieks, J. Opt. Soc. Am. 57, 975(1967).
2) M. Kato and T. Suzuki, J. Opt. Soc. Am. 59, 303(1969).
3) M. Born and E. Wolf, Principles of Optics (Pergamon Press, Inc., New York, 1959).
4) H.H. Hopkins, J. Opt. Soc Am. 47, 508(1957).
5) P. Hariharan and D. Sen, J. Sci. Instr. 38, 428(1961).
6) A.V. Baez, J. Opt. Soc. Am. 42, 756(1952).
7) J. Hamasaki, Appl. Opt. 7, 1613(1968)

APPLICATION OF HOLOGRAPHIC INTERFEROMETRY TO MECHANICAL EXPERIMENTS

Hiroyoshi Saito, Ichirou Yamaguchi and Toshinori Nakajima

The Institute of Physical and Chemical Research
Yamato-machi, Saitama Pref.,Japan

1. Introduction

One of the merits of holographic interferometry (1,2) is that distinct interference fringes appear even on a rough surface. The use of interferometry is not restricted only to an object with specularly reflecting surfaces. Another merit is the capability of time shearing interferometry. A wave front from an object can be compared not with any reference wave front but with that of the object itself in different states. Such advantages make the interferometric method applicable extensively to engineering problems. Moreover, the interferometry enables us to measure essentially the longitudinal displacement more easily than the lateral one. In the present paper, some new attempts are made mainly for bending experiments where the measurement of deflection is a fundamental task. The following three experiments ----- the first two are static and the third dynamic ----- are carried out;

1) Determination of Poisson's ratio of plate;
2) Analysis of moments in bending slab;
3) Flexural vibration of plate.

Further the following experiment is supplemented;

4) Analysis of moving object.

2. Determination of Poisson's Ratio of Plate (3,4)

When a prismatic bar with a rectangular cross section is subjected to pure bending in one of its principal planes, the surface of the bar deforms to a saddle-like shape. The solution of this problem is well known in the three-dimensional theory of elasticity (5). Taking the Cartesian coordinates in such a way that the origin is at the centroid of the cross section and the xz-plane is the principal plane of bending as shown in Fig. 1, the contour lines for this anticlastic surface become hyperbolas represented by the following equation:

$$x^2 - \nu y^2 = constant. \tag{1}$$

The smaller angle 2α between the asymptotes of hyperbolas, which are represented by making the constant in Eq.(1) equal to 0, is related to the Poisson's ratio ν as follows:

$$\nu = \tan^2 \alpha \ . \tag{2}$$

Therefore, by taking the contour lines by the interferometric method, the Poisson's ratio of the bar material is easily determined.

The experimental arrangement used is shown in Fig. 2. A test plate supported by a pair of steel rods is loaded through another pair on the back surface. A collimated beam of a He-Ne laser is divided into two beams by a beam splitter. The reflected beam illuminates normally the test surface and the scattered light of reflection from the surface falls on a hologram plate placed at the rear focal plane of the field lens in the normal direction of the test plate. The beam transmitted through the beam splitter is reflected by a mirror and brought to the hologram plate as a reference beam. The angle of incidence of reference beam to the hologram plate is about 15° of arc.

The double exposure method was tried in the first stage of this experiment but later the real time method was preferred because it was difficult by the former to bring out a symmetrical fringe pattern owing to the small tilt of bending apparatus and the asymptotes did not always appear over the interferogram. In the latter case, as the expected form of fringe pattern is known, it is easy to replace and adjust the hologram position. Asymmetry of fringe pattern reduced the accuracy of the angle measurement considerably.

In the stage of reconstruction, a diaphragm of an adequate size was placed close to the hologram. The direction of view and the angular spread of available light were uniquely determined by the position and the size of the hologram aperture, respectively.

Figure 3 shows an example of interferograms. In the lower graph the fringe orders along the x-axis are plotted on a logarithmic scale. The slopes of both lines are exactly equal to 2 as expected from Eq.(1). Four test pieces were prepared from the same steel plate. Their dimensions were 125mm in length, 3 mm in thickness and 30, 20, 15 and 10 mm in width. The interval of inner supports and that of outer supports of the plate were 80 and 110 mm respectively.

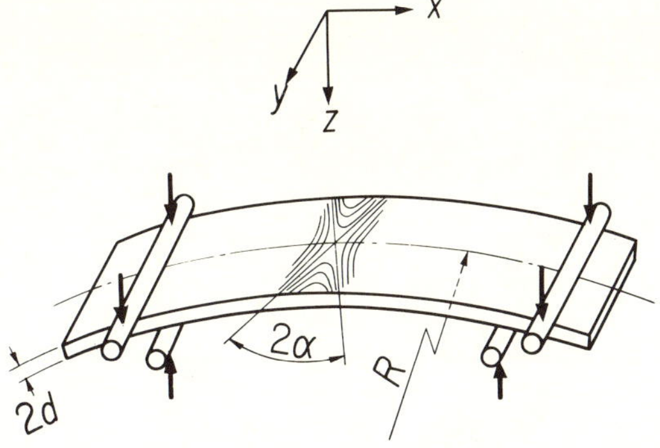

Fig. 1. Deformation of a plate by pure bending.

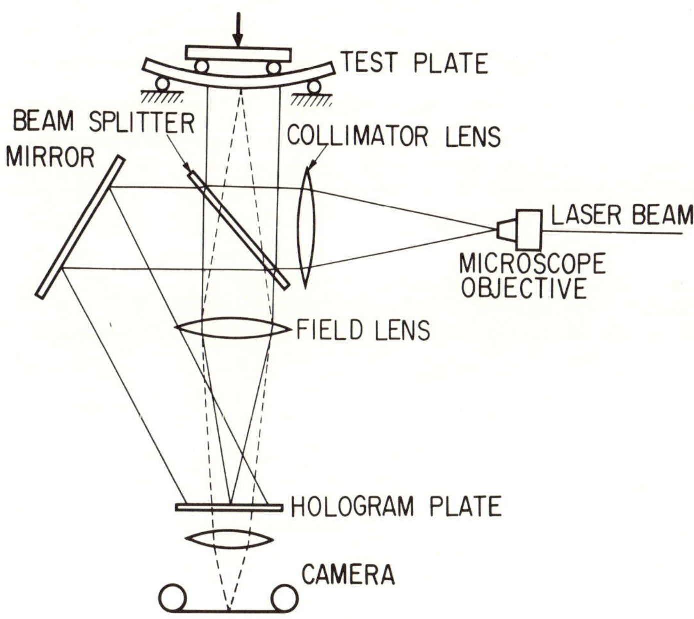

Fig. 2. Schematic diagram of experimental arrangement used in holographic interferometry.

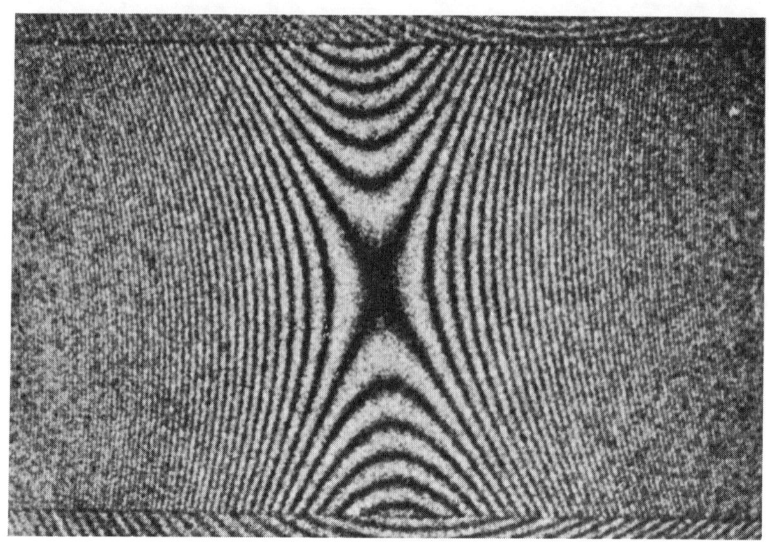

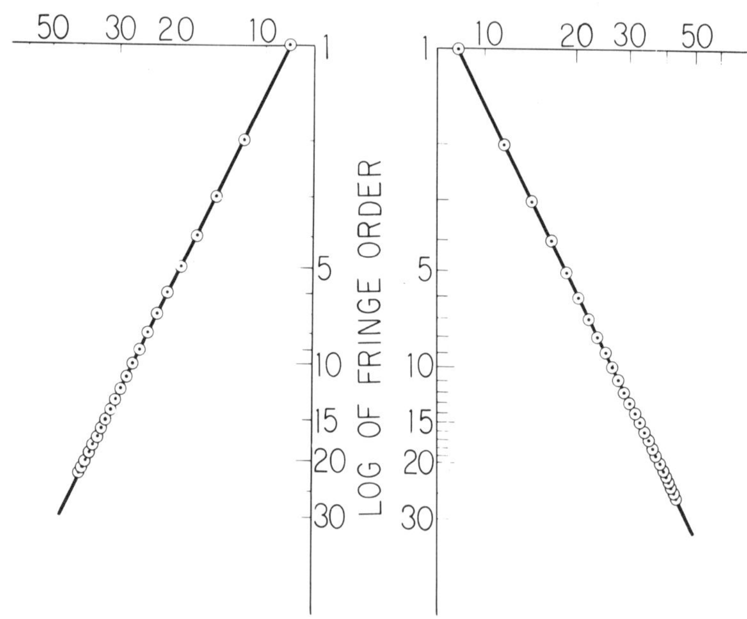

Fig. 3. Example of interferograms for a steel plate in pure bending. Lower graph shows the fringe orders along x-axis.

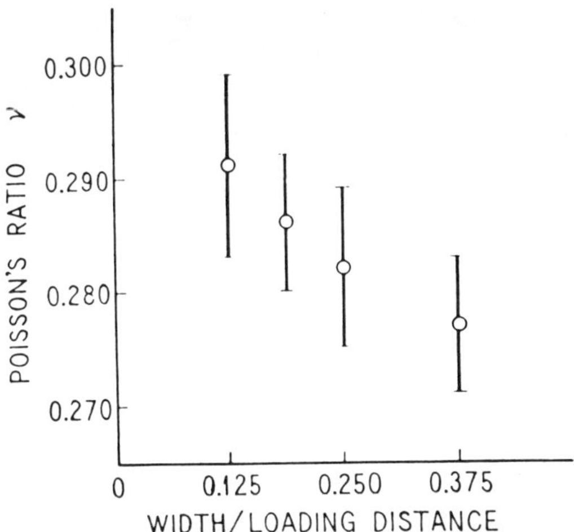

Fig. 4. Experimental results of Poisson's ratio for steel plates of different widths.

About ten interferograms corresponding to different amounts of deflection were taken for each test piece. The dispersion of the measured angle was within ± 0.5° for each piece. The resultant values of ν vs. the ratios of the plate width to the distance between the loads are plotted in Fig. 4. It may be noted that the assumption of pure bending, i.e., loading lines are sufficiently far from the middle of the plate, has not been fully satisfied in the experiment. A smaller value of ν is obtained from a wider plate, whose loading and supporting positions are relatively close to the middle. However, the reproducibility of the measured value for each specimen is fairly good. Accuracy of the measurement depends mainly on that of the angle measurement, which is limited by the width of fringes, and the accuracy attained in this experiment was about 3 %.

The method is most effectively applied to the materials whose surface cannot be polished to be optically flat. Similar experiments were performed for plates of cement and wood. Fig. 5 shows the fringes in the case of cement plate and the Poisson's ratio is measured as 0.18 from this interferogram. Fig. 6 shows examples of interferogram for wood plate along and across the grains of wood and fine irregularities of fringes due to the grains are observed. It may be interesting to study the bending problem of orthotropic plate like fiber reinforced plastics.

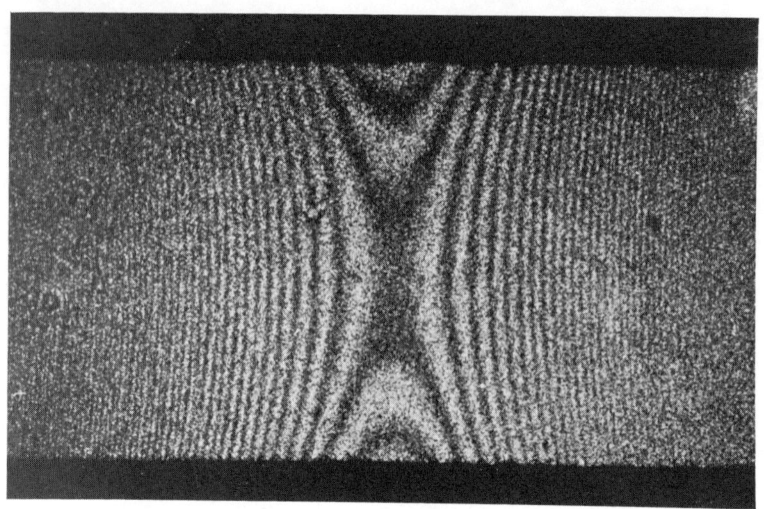

Fig. 5. Interferogram for a cement plate in pure bending.

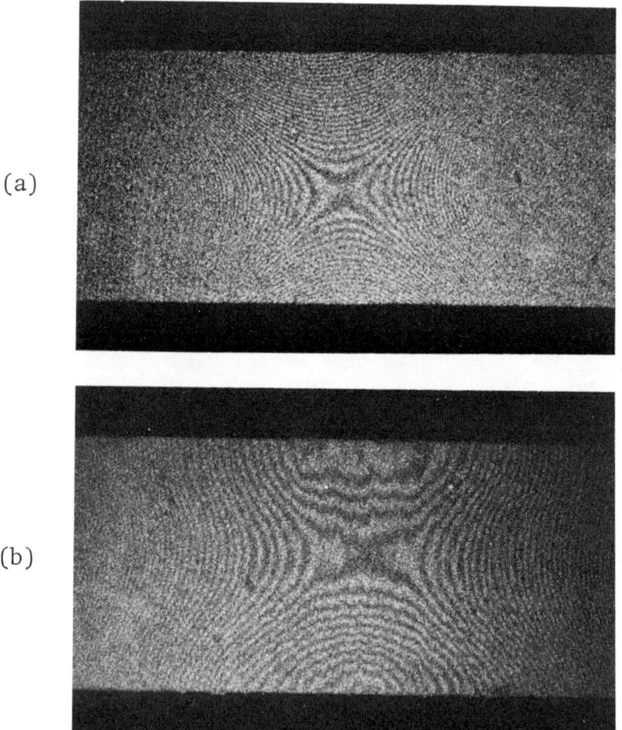

Fig. 6. Interferograms for wood plates in bending. (a) and (b) correspond to the plates along and across the grains of wood respectively.

3. Analysis of Moments in Slab Model

The calculation of the stress distribution in bending of a plate of arbitrary shape has always been a difficult matter. Some experimental methods have been devised such as, for instance, the moiré method and Salet-Ikeda's method (6). Here holographic interferometry is applied for the purpose of deducing the distribution of moments.

Now consider the case where a plate is bent by lateral loading (Fig. 7). As is well known from the theory of elasticity (7), there exist between the moments per unit length m_x, m_y and m_{xy} and deflection w the following relations:

$$\begin{aligned} m_x &= -D\left(\frac{\partial^2 w}{\partial x^2} + \nu \frac{\partial^2 w}{\partial y^2}\right) \\ m_y &= -D\left(\frac{\partial^2 w}{\partial y^2} + \nu \frac{\partial^2 w}{\partial x^2}\right) \\ m_{xy} &= -m_{yx} = D(1-\nu)\frac{\partial^2 w}{\partial x \partial y} \quad , \end{aligned} \tag{3}$$

where $D = Eh^3/12(1-\nu^2)$ is the modulus of plate rigidity, E is Young's modulus and h the thickness of plate. The fiber stresses of a plate are given by the following equations:

$$\sigma_x = 6m_x/h^2 \quad , \quad \sigma_y = 6m_y/h^2 \quad , \quad \tau_{xy} = 6m_{xy}/h^2 . \tag{4}$$

Therefore, if the second derivatives of w at every point are obtained, the distribution of stresses in the plate can be known from Eqs. (3) and (4). It is clear that an interferogram between the surface of the plate before and after bending indicates the topography of w, because this topography is nearly the same as that of middle plane for a thin plate.

Two experimental methods are applied (Fig. 8), one is for an optically smooth surface and the other for a rough surface. When the bending surface is smooth, the reflected wave front corresponding to the deflection w(x,y) without and with a lateral shift

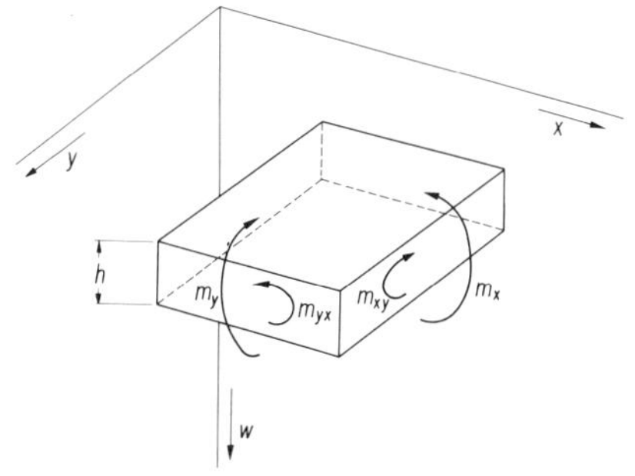

Fig. 7. Notations used for a bending slab.

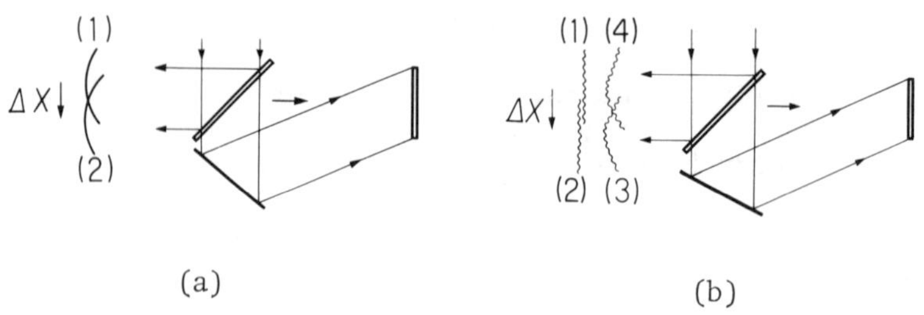

Fig. 8. Schematic diagram for explanation of the two methods in obtaining the first derivative fringe of deflection. (a) is for a smooth surface and (b) for a rough surface.

of position, i.e., corresponding to $w(x-\Delta x,y)$, are recorded in a hologram successively. Then the two wave fronts reconstructed from this hologram cause shearing interference (8) and the intensity of interference is proportional to $1+\cos\frac{4\pi}{\lambda}\frac{\partial w}{\partial x}\Delta x$. Thus, the fringes represent the first derivative of w.

The second method is for a rough surface. In the stage of recording, a hologram plate is exposed four times, i.e.,

1) in the state of surface before bending,
2) in the same state as 1) and with slight lateral shift,
3) in the same position as 2) and with bending of plate,
4) in the same bending as 3) and with repositioning to the first place.

Then, in the reconstruction from this hologram, four wave fronts interfere with each other. The intensity of the interferogram is given by $1+\cos\frac{2\pi}{\lambda}\frac{\partial w}{\partial x}\Delta x\cdot\cos\frac{4\pi}{\lambda}w$, excluding a proportionality factor. In this case the interferogram which shows the contour of deflected surface is modulated by its first derivative. The experimental arrangement was the same as already mentioned except that the bending apparatus was mounted on a carriage capable of a slight lateral shift.

Two bending problems were treated, i.e., one was the anticlastic bending with four point loads on square plate and the other the bending of circular disk supported along the circumference and loaded in the center. Fig. 9 is an example of shearing interferogram for the square plate obtained by the first method. Only the first derivative fringes appear as parallel, straight fringes at equal intervals. From the theory of elasticity, the contour of $\partial w/\partial x$ must be given by straight lines at equal intervals along the y-direction. The result is in good agreement with the theory.

Fig.10 shows examples of interferograms for the circular disk with central loading. The left figure represents the contours of

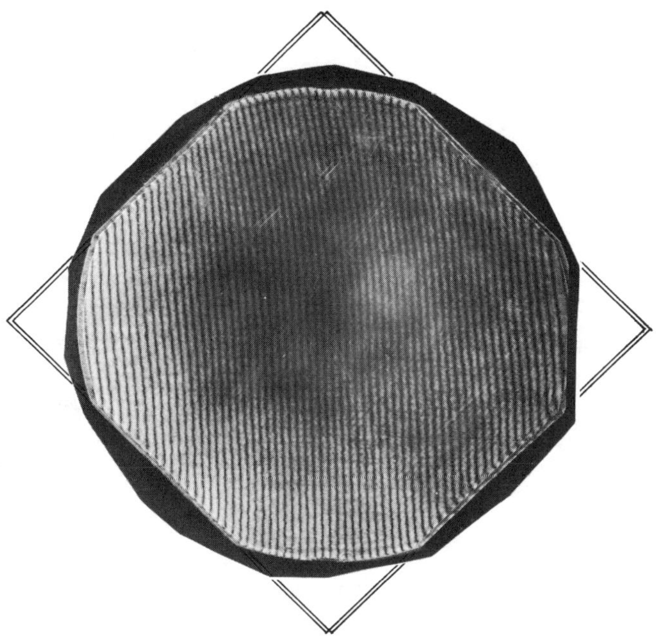

Fig. 9. Shearing interferogram which indicates the contours of $\partial w/\partial x$. The specimen is a square plate in anticlastic bending.

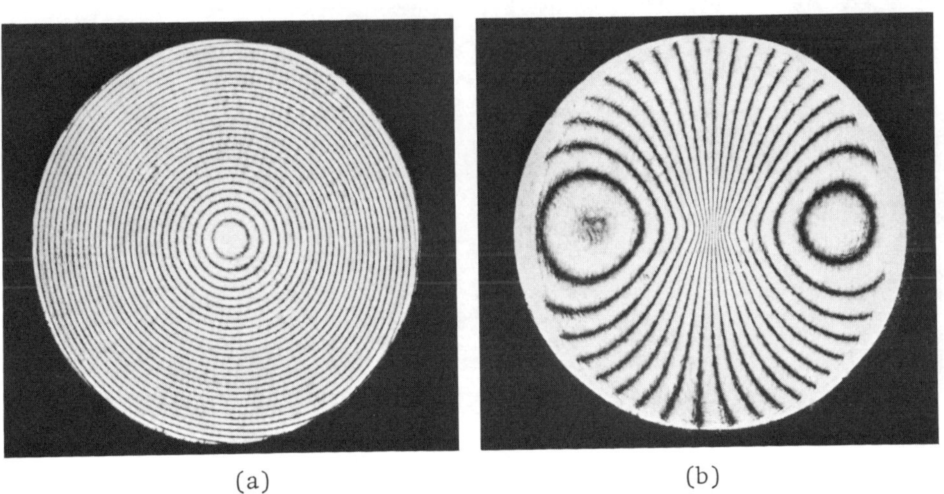

(a) (b)

Fig.10. Interferograms for the circular disk in bending with central load. (a) represents contours of deflection w and (b) those of $\partial w/\partial x$.

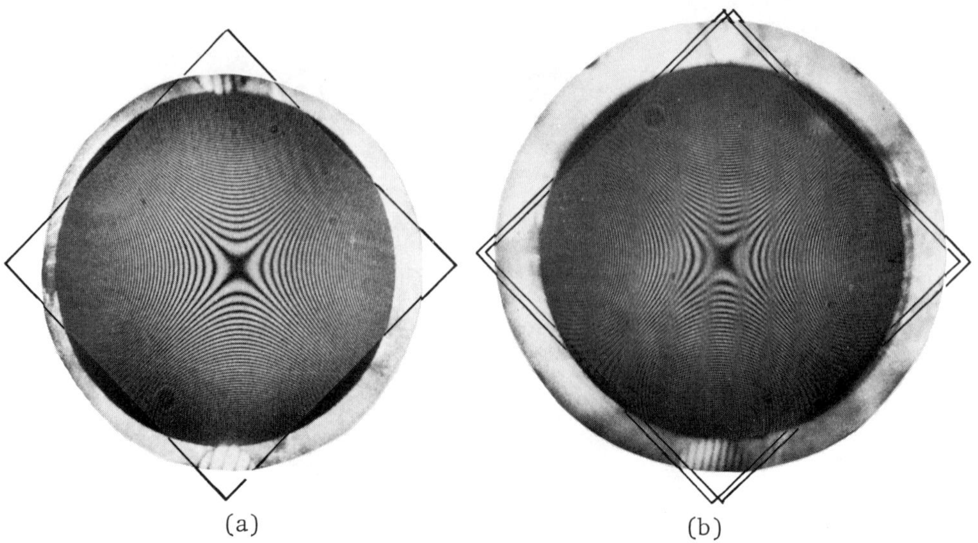

Fig.11. Interferograms for the case of a square plate in anticlastic bending with four point loads. (a) represents contours of deflection w and (b) those of $\partial w/\partial x$.

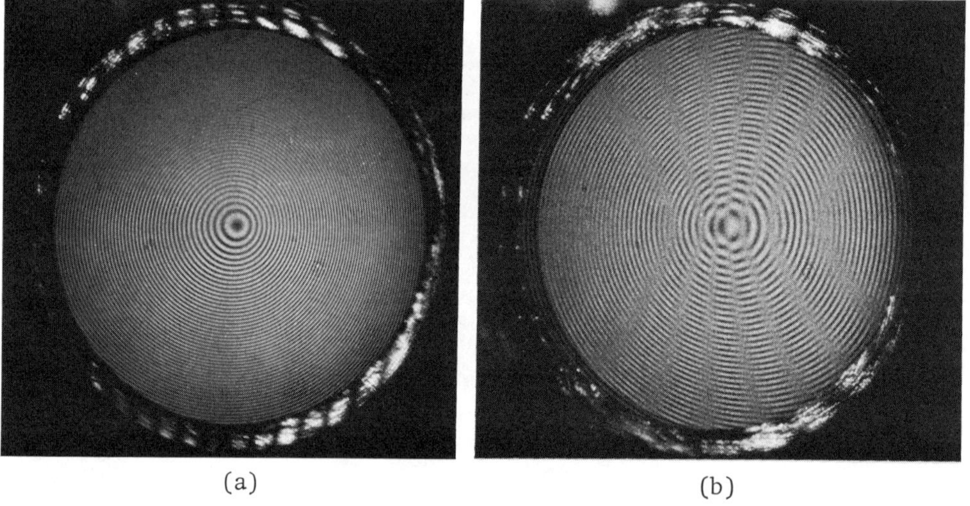

Fig.12. Interferograms for a circular disk in bending with central load. (a) represents contours of deflection w and (b) those of $\partial w/\partial x$.

deflection w and the right one the contours of its first derivative $\partial w/\partial x$ obtained by shearing interference. Both fringes have shapes as expected from the theory of elasticity.

Interferograms for rough surfaces obtained by the second method, or the four time exposure method, are shown in Fig.11. The surface of specimen was covered with white soot of MgO so as to diffuse the reflected light. Fig.11 (a) shows the contours of deflection w, i.e., the symmetrical hyperbolas. Fig.11 (b) shows the contours of the first derivative of w with respect to x. This interferogram shows an interesting feature of modulation; and parallel, straight half-tone fringes that represent the first derivative of w appear at equal intervals.

Another example of the circular disk in bending is shown in Fig.12. The left figure corresponds to the deflection and the right to its first derivative. Similar features are seen in these fringes. The interferograms obtained by the above experiments also have a good agreement with the expected patterns from the theory of elasticity.

However, the first derivative fringes do not have enough visibility in principle; thus, it is difficult to acquire the second derivative successively by any optical technique. On the other hand, the first method gives distinct interference fringes for the first derivative and successive differentiation may be possible optically. Since the initial slight deformation and poor surface finish cannot be compensated, the first method is good only for optically flat surface and is thus restricted to few applications.

4. Experiments on Flexural Vibration of Plate

Two techniques, a time-averaged technique (9) and a stroboscopic one (10), have been developed for the analysis of surface vibration by holography. In the time-averaged technique,a hologram plate is exposed sufficiently for a longer time than the period of vibration. Then the intensity of fringe pattern is proportional to the square of the zero order Bessel function of the first kind. Thus the contours of equal amplitude of vibration can be obtained and dark fringes correspond to the roots of the Bessel function. The fringes are not at equal intervals in the amplitude of vibration and their intensity decreases rapidly to zero.

In the stroboscopic technique, on the other hand, the light of illumination is chopped synchronously with the motion of the plate and two states of the plate in vibration are recorded in a hologram. This is a natural version of two beam interferometry. The fringe order is proportional to the amplitude of vibration and the

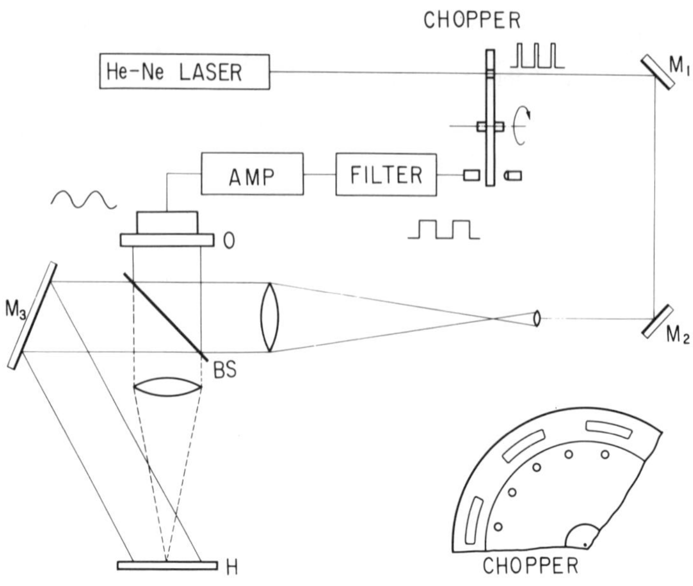

Fig.13. Experimental arrangement of the stroboscopic technique for analysis of vibration of plates.

intensity of fringes does not decrease in the higher orders.

The vibration of a metal plate is demonstrated by the latter technique. The experimental arrangement is shown in Fig.13. It is the same as a usual holographic interferometer except that chopping means are provided immediately behind the output of the laser. The inset in the figure shows details of a part of the chopper disk. A metal disk with many holes equally spaced along a circle is used as a light chopper and rotated by a synchronous motor at the rate of 3000 rpm. It also plays a role as the pickup of signals which drive the plate. The light signal which passes through the other series of holes in parallel to the illumination holes and has a nearly rectangular shape of pulse is converted electronically into a sinusoidal signal. The plate is excited by this signal synchronously with the light of illumination. It is also possible to change the phase relation between the vibration and the illumination by shifting the position of the pickup system.

A circular aluminum plate which had a diameter of 8 cm and a thickness of 0.1 mm was supported with its circumference on a frame of a usual radio speaker. The plate was driven by a central

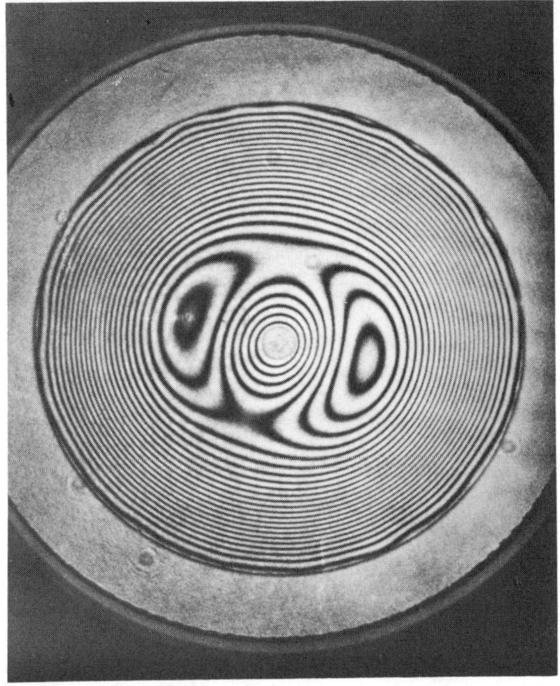

Fig.14. Interferogram of vibrating disk excited in center.

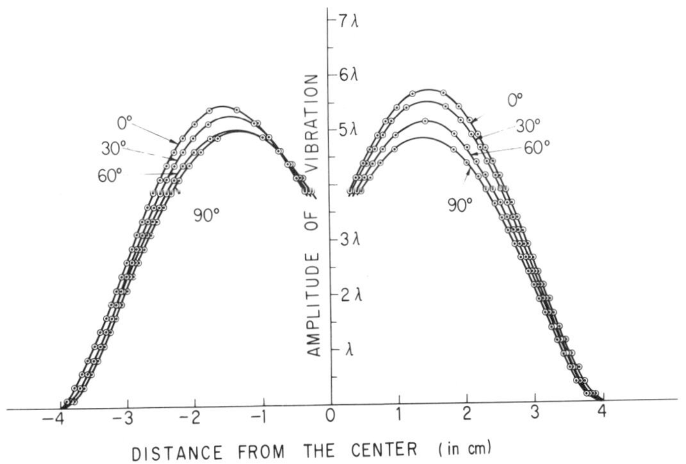

Fig.15. Amplitude distributions along diameters.

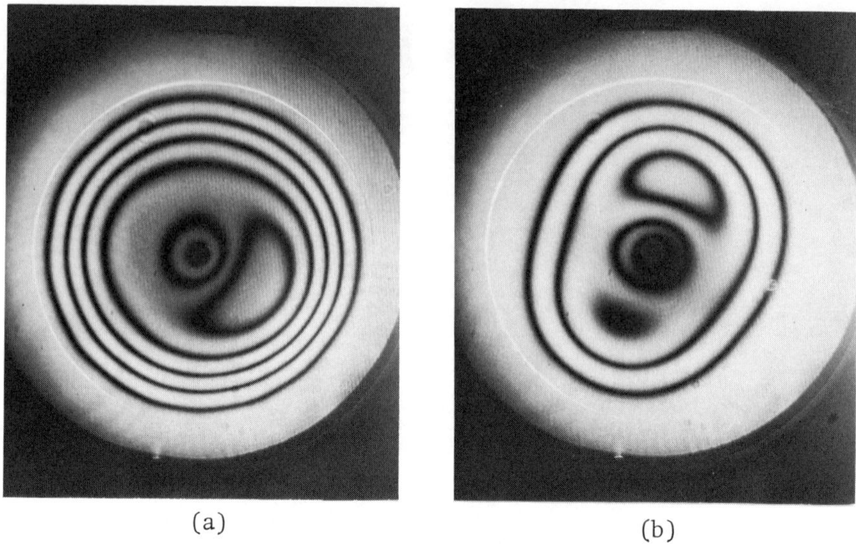

(a) (b)

Fig.16. Interferograms of a vibrating disk excited in center with frequencies of (a) 400 and (b) 800 c/s.

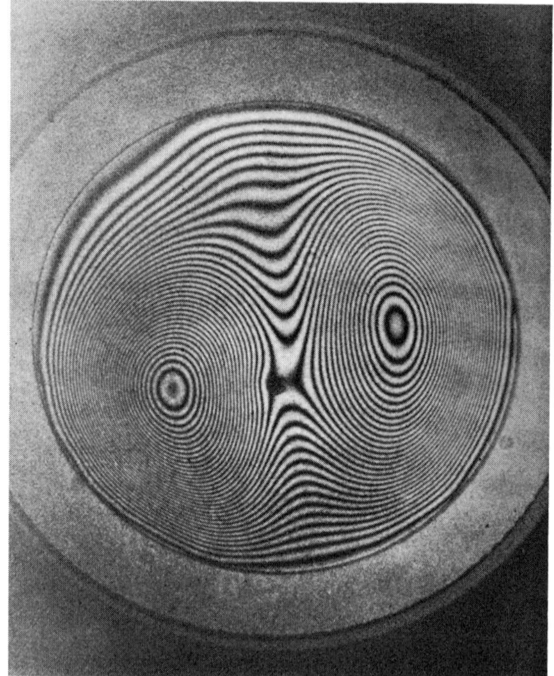

Fig.17. Interferogram of a vibrating disk with eccentric excitation.

or sometimes eccentric excitation. The deformed states in one and the other peaks of vibration were exposed to a hologram plate by chopping the illumination. Fig.14 is an example of the interference patterns when the plate is subjected to vibration by central excitation. Since the directions of illumination and observation are normal to the vibrating plate, the intervals of fringes correspond to an amplitude of about 0.16 μ. The frequencies of vibration and illumination are set in the desired range between 200 and 1600 c/s by changing the rate of rotation of the disk. In the example of Fig.14, the frequency of vibration was 200 c/s and the pulse width of illumination was about 1/20 of a period of vibration T_0. From this interferogram, the amplitude distribution of the vibrating plate is obtained as shown in Fig.15.

Figs.16 (a) and (b) are other examples, which correspond to the cases excited in the center at 400 and 800 c/s, respectively. Another example of the case of eccentric excitation is shown in Fig.17. The exciting point is indicated by a black point below the center. Triple beam interference can also be achieved. For instance, two wave fronts corresponding to the peaks of vibration and to the stopped state interfere with each other and the resultant interferogram is shown in Fig.18.

Through this experiment an interesting phenomenon was found that a part of an image of object became dark when the amplitude of vibration was increased. Fig.19 shows this phenomenon. The amplitude of vibration is successively increased and the central portion of the disk becomes dark. This may be expected from the fact that the movement of a plate during a pulse duration of exposure brings about the degradation of the contrast of carrier fringes in a hologram. The calculation of the effect of pulse width on the reconstructed image was carried out according to the Goodman's theory (11) in which the hologram for a moving object performs a temporal filtering process. The results show that the intensity of reconstructed image is modulated by an attenuation function A(a,T) which is expressed by:

$$A(a,T)=\left\{J_0\left(\frac{4\pi}{\lambda}a\right)+2\sum_{n=1}^{\infty}(i)^n \operatorname{sinc}\left(\frac{n\omega T}{2}\right)J_n\left(\frac{4\pi}{\lambda}a\right)\right\}^2, \quad (5)$$

where T is the pulse width of illumination and a and ω the amplitude and the angular frequency of vibration respectively. It is assumed in the calculation that there exists a linear relation between the amplitude transmittance of a hologram and the light intensity incident on the hologram.

It is known that in the case of time-averaged method the exposure time T is considerably longer than the period of vibration and the intensity of reconstructed image depends only on the first term of the attenuation function, i.e., $\left\{J_0\left(\frac{4\pi}{\lambda}a\right)\right\}^2$. However, the stroboscopic method relates to the case where the pulse exposure T

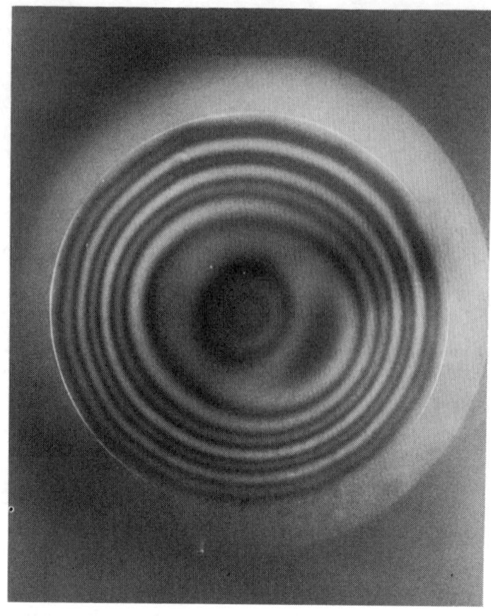

Fig.18. Interferogram obtained by three wave front interference.

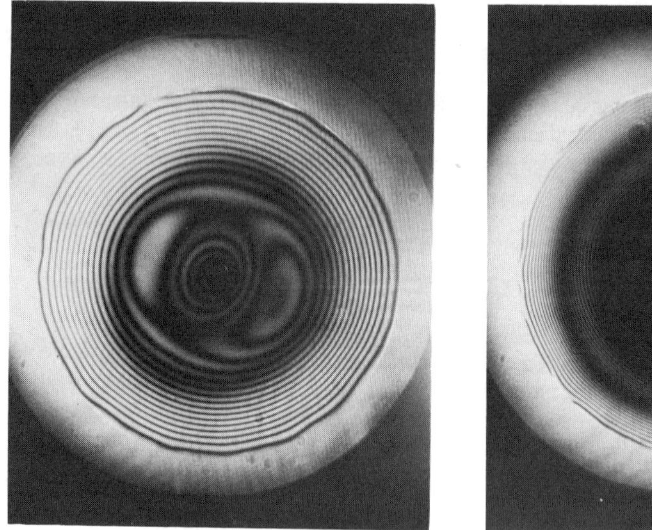

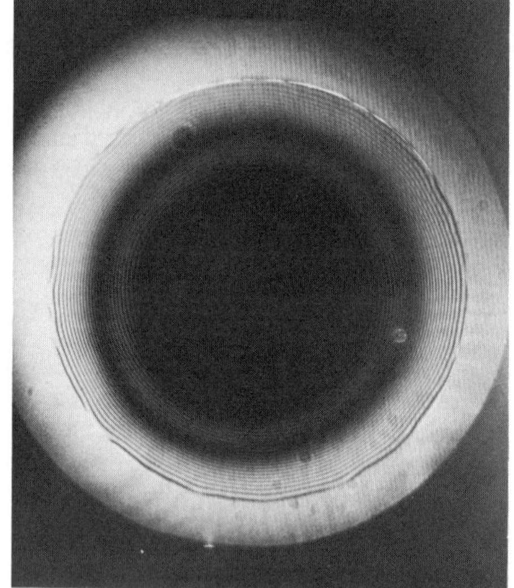

Fig.19. Interferograms which show the decrease of fringe visibility with increase of vibration amplitude.

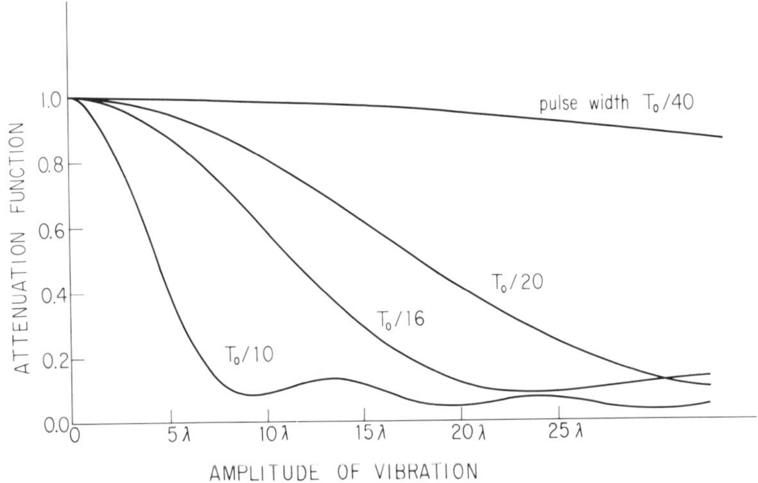

Fig.20. Attenuation function due to the effect of pulse width of exposure on the intensity of reconstructed image.

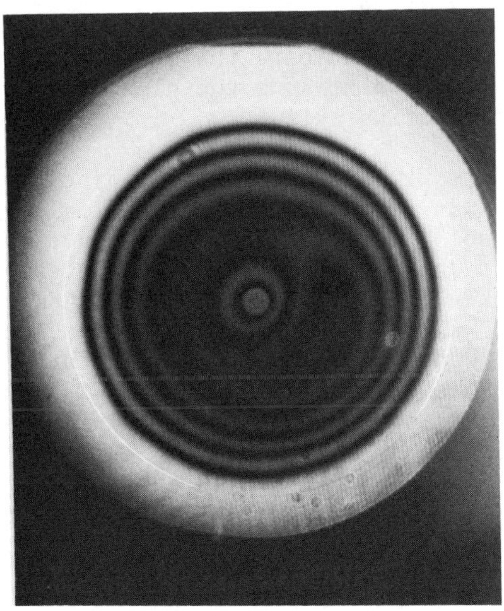

Fig.21. Interferogram obtained by the time-averaged method for comparison with that by the stroboscopic method, Fig.16 (a).

is fairely shorter than the period of vibration. Fig.20 shows the results of calculation of A(a,T) for T being equal to 1/10, 1/16, 1/20 and 1/40 of T_0 respectively. From this figure it is known that, for instance, the intensity of reconstructed image degrades to 80 % when the amplitude of vibration is 2.6 λ for a pulse width of illumination of $T_0/10$. The calculated variation of intensity is in agreement with the experiment. It is noted that the exposure pulse width of illumination should be short in the stroboscopic technique in order to bring about distinct interference fringes. The intensity of reconstructed image by the stroboscopic method decreases more slowly than that by the time-averaged method as shown in Fig.21.

5. Analysis of Object Motion

J.W.Goodman (11) and J.D.Redman (12,13) suggested that the time-averaged method in holographic interferometry can be used for the analysis of an object motion. A few such experiments were tried. Fig.22 (a) shows a picture of rotating plate taken by the time-averaged technique. The plate was rotated by the usual hour hand of an electric table clock. The linear velocity of an edge of the plate in the viewing direction was about 15 μ/sec. The intensity of the reconstructed image decreases according to the square of sinc function of velocity and several fringes that indicate the contours of velocity can be seen at equal intervals. The double exposure technique was also applied to the above-mentioned case. The fringes shown in Fig.22 (b) also indicate the contours of velocity because the velocity of each point on the plate varies linearly with distance from the rotating axis.

Another experiment of analysing the object motion was carried out. A small box was filled with some material of high polymer which can flow rheologically. After the surface of material became flat, the box was raised up and then the contents began to flow. Fig.23 (a) was taken by the time-averaged technique with exposure times of 1/10, 1/5 and 1/2 sec. Contours of velocity along the viewing direction can be seen as fringes in this figure. In this case the usual double exposure technique of holographic interferometry can also be applied and the fringes indicate the increment of displacement during the interval between exposures. Therefore the motion of objects can be analysed by the successive double exposure holograms. Fig.23 (b) shows an example of interferograms. These techniques may be useful for the application to the experiments of creep and flow phenomena.

6. Conclusion

In summary, the holographic interferometry which has many advantages over the conventional one can be applied in many fields

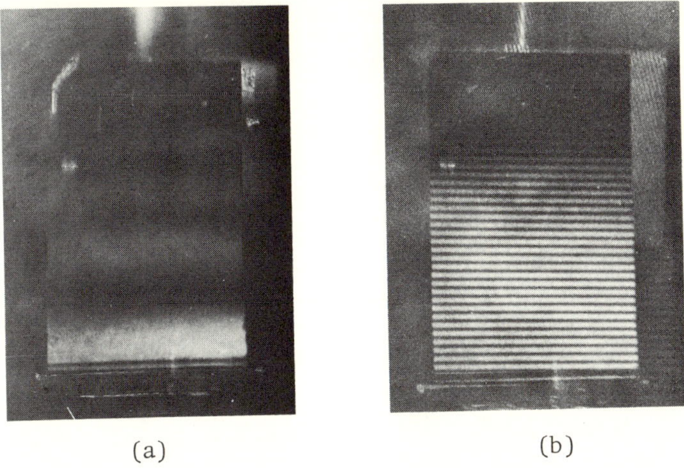

(a) (b)

Fig.22. Contours of velocity on a slowly rotating rectangular plate. (a) is obtained by the time-averaged method and (b) by the double exposure method.

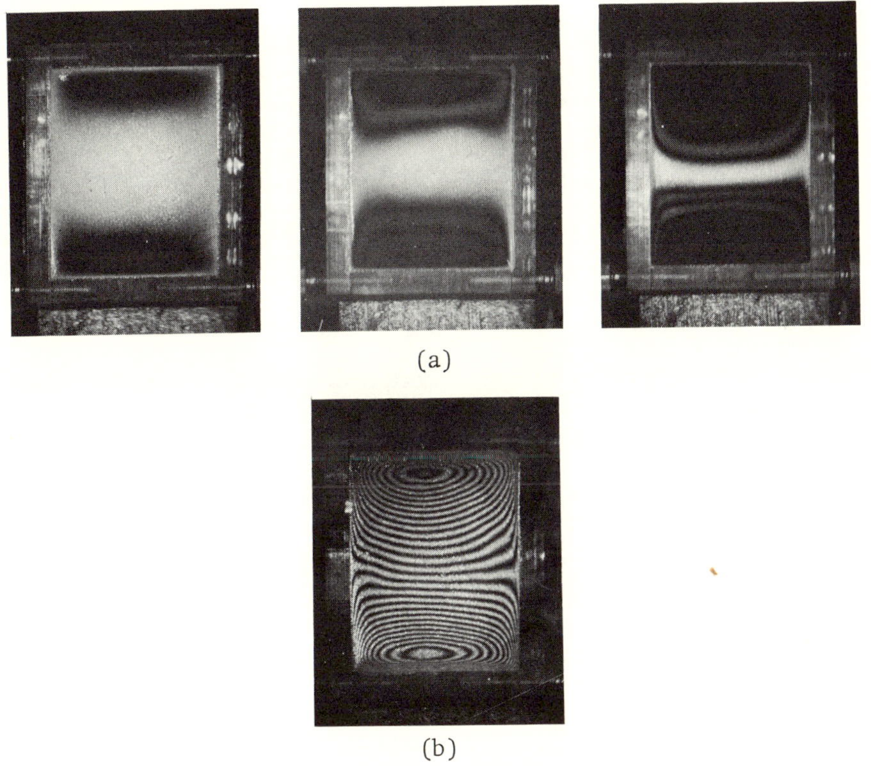

(a)

(b)

Fig.23. Contours of velocity in rheological flow of high polymer. (a) is obtained by the time-averaged method (b) by the double exposure method.

of engineering. Here the features of holographic interferometry -- the appearance of fringes on a rough surface, the multiplex recording of a state of objects, self-compensation of initial conditions, and so on -- are utilized in the mechanical experiments. The new character of the technique produces a new method. Experiments in three dimensions may be promising in engineering by using the reproducibility of three-dimensional information, which is also one of the distinctive features of holography.

Acknowledgment. The authors wish to thank Kiyoshi Hachimine and Nobuo Hiyama for their help in performing the experiments.

Appendix. Effects of the Pulse Width of Exposure Time on the Image Reconstructed from Holograms

According to the theory of temporal filtering of holograms by Goodman, the intensity of reconstructed direct images is considered.

The amplitudes at a point of (x,y) on a hologram plate are denoted by $U_o(x,y;t)$ and $U_r(x,y)$, the former representing the amplitude of wave coming from the object and the latter that of the reference wave respectively. If there exists a linear relation between the light intensity recorded on the hologram and the amplitude transmittance of the hologram, the amplitude of a reconstructed direct image $E'(x,y)$ is expressed except for a proportionality factor as follows:

$$E'(x,y) = U_r^*(x,y)\int_{-T/2}^{T/2} U_o(x,y;t)\,dt \quad , \tag{A1}$$

where T is the exposure time and the asterisk means the complex conjugate.

Equation (A1) can be rewritten in the form

$$E'(x,y) = U_r^*(x,y)\int_{-\infty}^{\infty} rect\left(\frac{t}{T}\right) U_o(x,y;t)\,dt \quad , \tag{A2}$$

where

$$\begin{aligned} rect\left(\tfrac{t}{T}\right) &= 1 \qquad -\frac{T}{2} \le t \le \frac{T}{2} \quad , \\ &= 0 \qquad t < -\frac{T}{2} \;, \quad t > \frac{T}{2} \;. \end{aligned} \tag{A3}$$

Applying the Parseval theorem,

$$E'(x,y) = U_r^*(x,y)\,T\int_{-\infty}^{\infty} sinc(\pi T\nu)\,\tilde{U}_o(x,y;\nu)\,d\nu, \tag{A4}$$

where $sinc(\pi T\nu) = sin(\pi T\nu)/\pi T\nu$ (A5)

and $\tilde{U}_o(x,y;\nu)$ denotes the Fourier transform of $U_o(x,y;t)$. Assuming that the surface of an object oscillates sinusoidally with an amplitude a and an angular frequency ω,

$$U_o(x,y;t) = \exp\left[i\left(\frac{4\pi}{\lambda}a\right)\cos(\omega t+\mu)\right] , \tag{A6}$$

where μ is the phase of oscillation. Therefore,

$$\begin{aligned}\tilde{U}_o(x,y;\nu) &= \int_{-\infty}^{\infty} U_o(x,y;t)\,e^{-i2\pi\nu t}\,dt \\ &= \sum_{n=-\infty}^{\infty}(i)^n J_n\left(\frac{4\pi}{\lambda}a\right)\int_{-\infty}^{\infty} e^{in(\omega t+\mu)}e^{-i2\pi\nu t}dt ,\end{aligned} \tag{A7}$$

and since

$$e^{iz\cos\theta} = \sum_{n=-\infty}^{\infty}(i)^n e^{in\theta} J_n(z) . \tag{A8}$$

Eq.(A7) is rewritten in the following form:

$$\tilde{U}_o(x,y;\nu) = \sum_{n=-\infty}^{\infty}(i)^n e^{in\mu} J_n\left(\frac{4\pi}{\lambda}a\right)\delta(2\pi\nu - n\omega) . \tag{A9}$$

Substituting Eq.(A9) into Eq.(A4),

$$\begin{aligned}E'(x,y) &= U_r^*(x,y)T\int_{-\infty}^{\infty} \mathrm{sinc}(\pi T\nu) \\ &\quad\times\sum_{n=-\infty}^{\infty}(i)^n e^{in\mu} J_n\left(\frac{4\pi}{\lambda}a\right)\delta(2\pi\nu - n\omega)\,d\nu \\ &= U_r^*(x,y)T\sum_{n=-\infty}^{\infty}(i)^n e^{in\mu}\,\mathrm{sinc}(n\omega T/2)J_n\left(\frac{4\pi}{\lambda}a\right) .\end{aligned} \tag{A10}$$

In the case of exposure at the time of peak amplitude of oscillation, i.e., $\mu = 0$,

$$\begin{aligned}E'(x,y) &= U_r^*(x,y)T\sum_{n=-\infty}^{\infty}(i)^n\,\mathrm{sinc}(n\omega T/2)J_n\left(\frac{4\pi}{\lambda}a\right) \\ &= U_r^*(x,y)T\left\{J_0\left(\frac{4\pi}{\lambda}a\right) + 2\sum_{n=1}^{\infty}(i)^n\,\mathrm{sinc}(n\omega T/2)J_n\left(\frac{4\pi}{\lambda}a\right)\right\} \\ &= U_r^*(x,y)T\cdot\{A(a,T)\}^{1/2} .\end{aligned} \tag{A11}$$

The calculation of $A(a,T)$, attenuation function as it is called, is made for $T = T_o/10$, $T_o/16$, $T_o/20$ and $T_o/40$, where $T_o = 2\pi/\omega$, a period of the oscillation. The results are shown in Fig.20 and the intensity of reconstructed image is modulated by the attenuation function $A(a,T)$ depending on the amplitude of oscillation.

References

1) R.J.Collier, E.T.Doherty and K.S.Pennington, Appl.Phys.Lett., 7, 223 (1965).
2) K.A.Haines and B.P.Hildebrand, Appl. Optics, 5, 595 (1966).
3) I.Yamaguchi and H.Saito, Japanese J. Appl. Phys., 8, 768 (1969).
4) W.G.Gottenberg, Experimental Mechanics, 8, 405 (1968).
5) S.Timoshenko and J.N.Goodier, Theory of Elasticity, 2nd ed. (McGraw-Hill Inc.,1951), p.250.
6) J.P.Duncun and C.J.E.Brown, Experimental Mechanics, Proc. 1st International Congress in New York, (Pergamon Press,1961), p.149.
7) S.Timoshenko, Theory of Plates and Shells, (McGraw-Hill,New York, 1940), p.81.
8) O.Bryngdahl, J.Opt.Soc. Amer., 58, 865 (1968).
9) R.L.Powell and K.A.Stetson, J.Opt.Soc.Amer., 55, 1593 (1965).
10) E.Archbold and A.E.Ennos, Nature, 217, 942 (1968).
11) J.W.Goodman, Appl. Optics 6, 857 (1967).
12) J.D.Redman, J.Sci.Instrum., 44, 1032 (1967).
13) J.D.Redman, J.Sci.Instrum., Ser.2, 1, 821 (1968).

EXPERIMENTAL ASPECTS OF HOLOGRAPHIC INTERFEROMETRY †

Ralph F. Wuerker

TRW Systems Group
Redondo Beach, California

Over the last years, the author has been concerned with the industrial applications of holography and holographic interferometry; particularly the recording of high speed elusive phenomena, aerodynamic visualization, non-destructive testing, and contouring. All represent problems and applications which can be uniquely solved by holography with state-of-the-art laser illuminators and some new optical systems insensitive to the incoherence of these illuminators.

The laser which has been used in the work at TRW is the pulsed Q-switched ruby laser. For holography, this system is still unique. It emits in the visible region of the spectrum within the range of sensitivity of photographic emulsions. It emits one pulse with enough energy ($\sim$ 1 calorie) to expose high resolution photographic plates. Its pulse duration is of the order of 50 billionths of a second. The short pulse duration and high energy means that it can make holograms free and independent of massive vibration isolation equipment, as well as make holograms which are impossible with gas lasers.*(1) The only problem with the ruby laser is that it is not very coherent, particularly in its more conventional forms, such as a large $\sim$ 1 cm diameter room temperature rod, electro-optical Q-switch, no intercavity aperture, nor other mode-determing devices. A later paper at the Conference described techniques for increasing the coherence of the ruby laser.** These necessarily increase the size and inflexibility of the illuminator.

---

*An illustrative example was that of holograms of people shown at this Conference (Reference 1)**Ibid.

†Work reported on in this paper was supported in part by the TRW Independent Research Program, NASA/Ames, Air Force Avionics Laboratory, Air Force Rocket Propulsion Laboratory and NASA/JPL.

This paper will emphasize holography with the more conventional Q-switched ruby laser, and using optical arrangements to compensate for the spatial and temporal incoherence of these smaller, more mundane illuminators. The approach was pioneered at TRW.(2)

Elusive Phenomena

The lack of lenses in holography means that the new technique is able to overcome photography's age-old depth-of-focus problem. Teamed with the 50 billionth of a second ruby laser, holography can study small ballistic phenomena, ballistic impacts, spray patterns, and any uncertain distributed phenomena. Illustrative of this unique capability is the example shown in Figure 1, which is two different photographs from the reconstruction of the same hologram. In one picture, one mosquito is in good focus. In the other, another is in good focus. Without belaboring a point, the author claims that this feat is impossible photographically! Holographically, it is simple. One records the hologram with a 50 nanosecond, pulsed ruby laser. The developed plate is then reconstructed and the insect of choice is photographed with a conventional camera. One chooses whichever insect is to be recorded by changing the focus of the copy camera.

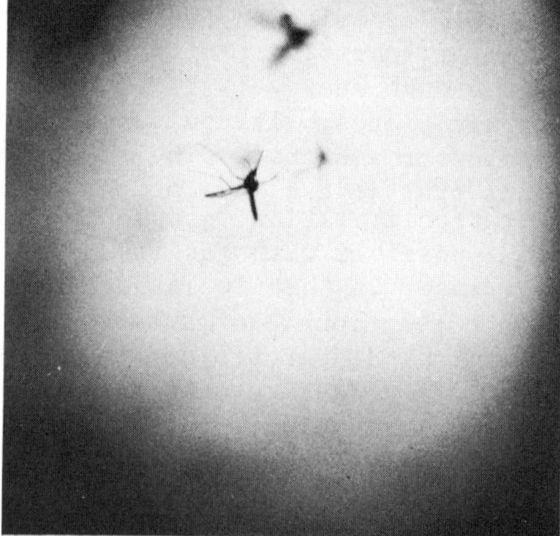

Figure 1. Photographs of the reconstruction of a Q-switched ruby laser hologram of a flight of mosquitoes. The photographs differ by the focus of the copy camera.

Possibly of more practical value, and certainly more difficult is the recording of hypergolic fuel combustion. Here holography is used to record phenomena in the presence of luminosity. The extreme intensity of laser light enables one to make the 3-D records free of the effects of flame light. The work was sponsored independently by NASA/JPL and the U.S. Air Force Rocket Propulsion Laboratory.(3,4) Since hypergolic liquid rocket engines are dangerous, equipment had to be developed which would permit the recording of holograms at an out-of-doors rocket test stand. The initial work was done at NASA Edwards Test site on the California Mojave Desert. Climatic conditions vary from 21° F. to 110° F. The mean solar flux is also a maximum (1000 watts/meter^2). In addition, the test rocket engines themselves produce extreme sonic and thermal conditions.

The ruby laser holographic apparatus which was developed for the rocket work is shown in Figure 2. A schematic is given in Figure 3. This latter diagram illustrates the guiding principle behind all the holographic devices first developed at TRW for recording holograms with ruby lasers of low or unknown spatial and temporal coherence. Reference to the schematic (Figure 3) shows the laser on top in a protecting cannister. The beam from the

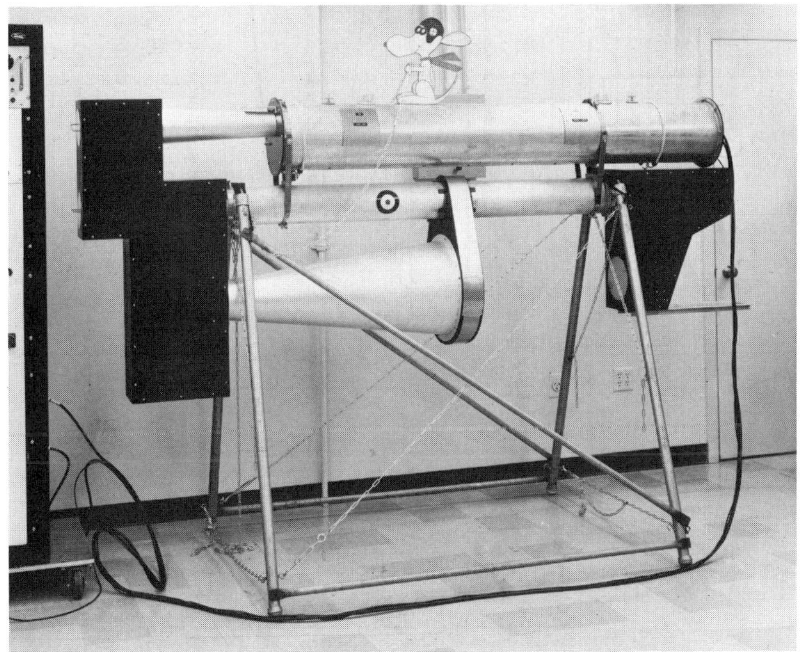

Figure 2. Photograph of the holocamera developed and used to record hypergolic flame combustion, both open flame and confined within experimental liquid rocket engines with transparent walls. The electronics for operation are in the cabinet on the left. Courtesy NASA/JPL

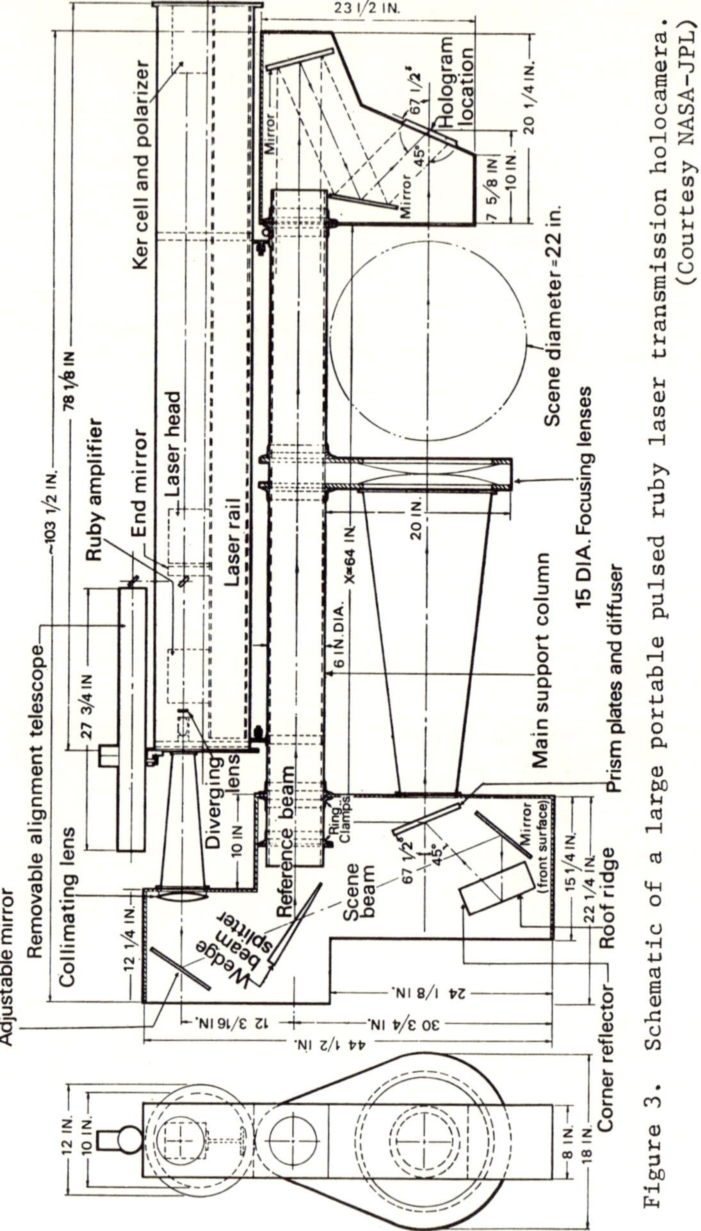

Figure 3. Schematic of a large portable pulsed ruby laser transmission holocamera. (Courtesy NASA-JPL)

laser is expanded with a Galilean telescope and then directed via a mirror into the holocamera proper. A large glass wedge divides the collimated beam into scene and reference beam components. The reference beam is directed up the interior of the main support column, after which it is reflected off of two front surface mirrors and onto the hologram in the far end of the camera. The stronger scene beam passes through the wedge, is reflected off a front surface mirror, then from a pair of mirrors arranged as a roof reflector. The beam is then incident on an element called a prism plate. It is a piece of lucite machined with prismatic steps which refracts the light beam through 45 degrees. The prism plate is followed by several pieces of ground glass. These serve the purpose of uniformly scattering the scene beam light.

Halfway between the prism plate and hologram is a pair of large diameter condensing lenses, each of focal length equal to half the prism plate-hologram separation. These lenses gather the light scattered by the diffusing glass and focus it onto the hologram. The basic principle behind this type of transmission holocamera is that each ray, divided at the wedge beam splitter, recombines with itself again at the hologram. That is to say, the large condensing lenses spatially match the scene and reference beams. In addition, the beam splitter, mirrors, and prism plate in the holocamera are located so that each pair of rays travel the same optical path distance. As a result, the scene and reference beams are also temporally matched all across the hologram.

When the holocamera is correctly aligned, holograms recorded in it are as bright upon reconstruction as holograms recorded on granite tables, with highly coherent gas lasers. The holocamera concept makes possible recording of high quality holograms with ruby lasers of such limited coherence that they might be considered worthless for holography.

The reconstructions of holograms recorded in the Figure 2-3 apparatus appear three-dimensional, as though the background were illuminated from behind by a uniform diffuse source.

The scene or event is placed between the lens set and the hologram. Scene depth is determined by the hologram-lens distance, not by the temporal coherence of the illuminator. Figure 1 is an example of a photograph of a reconstruction of a hologram made in this type of transmission holocamera. The viewing angle is determined only by the diameter of the focusing lenses. Using even larger quadruple condensing lenses, transmission holocameras of 13 inches scene diameter and 60 degrees viewing angle have been constructed.[(5)]

These transmission holocameras have the further property that the recording is independent of the velocity of the event.

Extremely high speed events produce images which are smeared, no different than in conventional photography. Unlike reflected light holography, the scene and reference beams are stationary. The event moves through the scene beam, blocking or eclipsing it. Image smear is thus just the product of the velocity component perpendicular to the scene beam direction times exposure time. For a 50 nanosecond ruby laser and a one kilometer per second particle, image smear in the reconstruction would be 50 microns.

Outside of the problem of loss of resolution due to image smear, an equally important question is the resolution of holograms of static events made in this apparatus, particularly holograms recorded on standard commercial plates and developed by standard dark room procedures.* A summary of present state-of-the-art work is presented in Figure 4.(6) The pictures were recorded with the Figure 2-3 holocamera via a camera mounted directly behind the hologram location.** A resolution chart (Type USAF 1951) was placed in the scene volume 45 centimeters away from the hologram location. The picture on the left is a photograph made with an incandescent lamp placed behind the prism plate and diffuser of the holocamera. This picture only tests the resolution of the copy camera. Inspection of the negative with a microscope showed that it could be read to 80 line pairs per millimeter (6th column, 3rd row).

The middle picture in Figure 4 is a reproduction of a photograph made with ruby laser illumination. For this picture, the reference beam was blocked, and the ruby laser fired. The copy camera photographed the image the same as with the companion white light photograph. This picture measures the resolution of the camera for ruby laser light illumination. It also directly compares, under identical conditions, white light photography with laser light photography. An obvious difference is the speckle pattern which is a result of the spatial coherence of the laser light. Inspection of the middle negative with a microscope showed that it could be read to 40 line pairs per millimeter (5th column, 3rd row). The factor of two difference between a scene illuminated in white light and one illuminated in coherent light seems to be a rule of thumb which one experiences whenever one changes from incoherent to coherent light.

The right picture in Figure 4 is the photograph of the reconstruction of a hologram made under identical conditions. The hologram was reconstructed in the Figure 2-3 apparatus, using the same

---

* TRW uses exclusively Agfa 8E75 plates (Product of Agfa-Gevaert, Antwerp, Belgium), developed in 1:4 solution of Eastman HRP developer, rinsed, and fixed in a 1:4 solution of Eastman Rapid Fix.

** The camera had a Schneider Kreuznach 1:5.6/300 mm lens. The pictures were recorded on Eastman S0243 film.

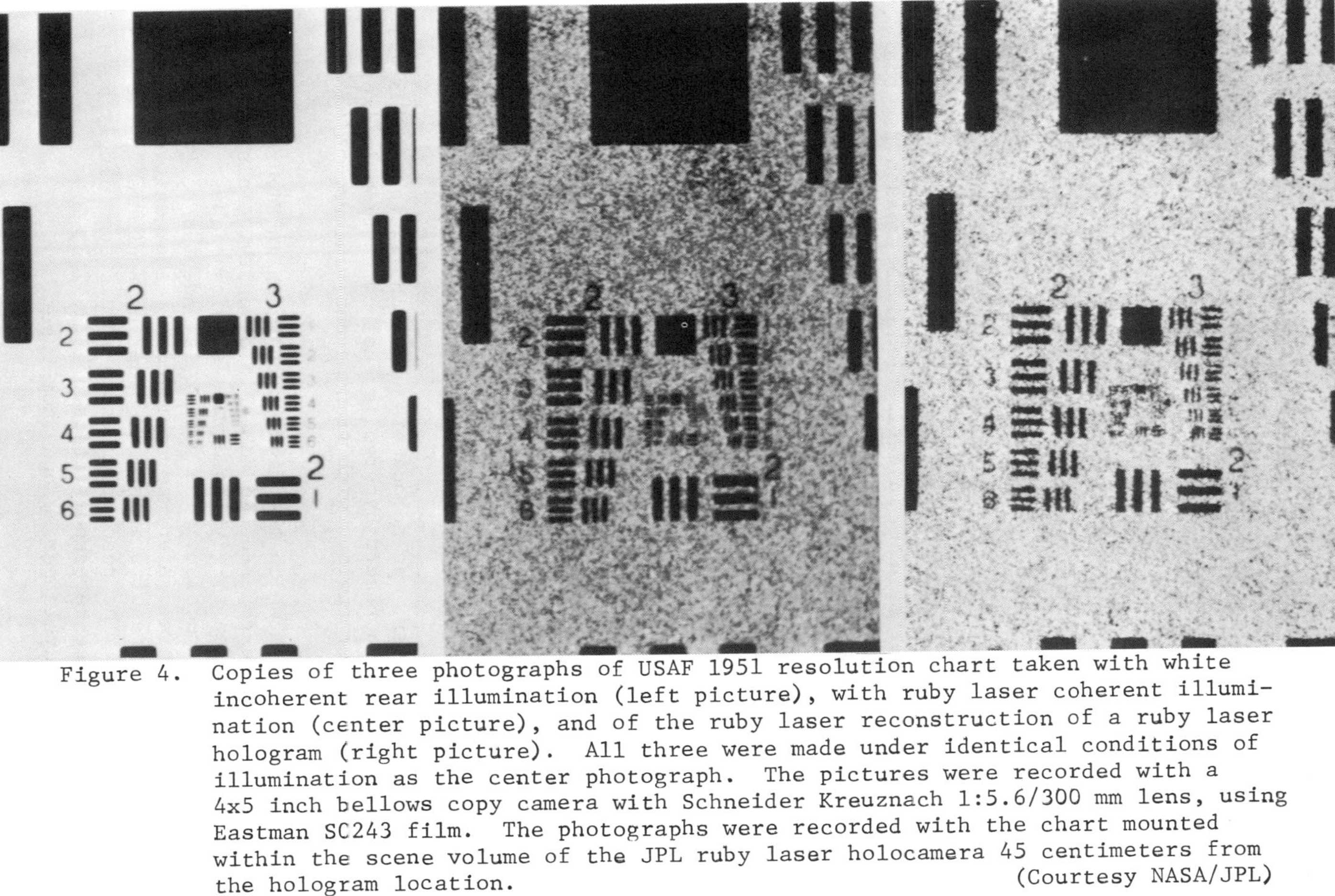

Figure 4. Copies of three photographs of USAF 1951 resolution chart taken with white incoherent rear illumination (left picture), with ruby laser coherent illumination (center picture), and of the ruby laser reconstruction of a ruby laser hologram (right picture). All three were made under identical conditions of illumination as the center photograph. The pictures were recorded with a 4x5 inch bellows copy camera with Schneider Kreuznach 1:5.6/300 mm lens, using Eastman SC243 film. The photographs were recorded with the chart mounted within the scene volume of the JPL ruby laser holocamera 45 centimeters from the hologram location. (Courtesy NASA/JPL)

wavelength and illuminator as was used to record it. The photograph was made in the following manner: first an 8E75 plate was placed in the hologram location and illuminated with scene and reference beams. Then the plate was removed, developed, fixed, etc. The developed hologram was replaced in the hologram apparatus, before the lens of the copy camera (which had not been changed). The scene beam was blocked. The ruby laser was fired, now the reference beam passed through the hologram. The reconstruction of the hologram was generated, which was photographed by the copy camera.

Examination of the negative with a microscope shows a resolution of 20 line pairs per millimeter, a factor of two less than the resolution of the direct laser photograph (center picture). Although not evident in the figure, the speckle pattern of the hologram is coarser than that of the direct photograph. The result, depending on one's point of view, is either remarkable or indicates the need for further careful investigation. The reason for the factor of two difference between the hologram reconstruction and the direct laser photograph is not known.*

Holograms were also recorded with the resolution chart mounted at 94 centimeters and 25 centimeters from the hologram location. Reconstruction of the former showed a resolution of 10 line pairs per millimeter. Reconstruction of the 25 centimeter distance hologram showed resolution of 20 line pairs per millimeter, no better than the hologram at 45 centimeters range. These holograms were also reconstructed with a helium-neon laser and with a "continuous wave" ruby laser.** They were examined by projecting a real image. Both gave the same resolution; however, when the helium-neon laser was used, the plate had to be skewed to compensate for the astigmatism introduced by the 10% change in wavelength. Using a helium-neon laser as reconstructor, one plane at a time could be brought into sharp focus; other planes, however, were astigmatic.

As noted earlier, the holocamera shown in Figure 2 and 3 was successfully used to record hypergolic fuel combustion in experimental liquid rocket engines with transparent windows.[2,3,4]

---

* The hologram, from which the right photograph in Figure 4 was made, had a scene-reference beam ratio of unity. This was empirically chosen to give holograms showing highest contrast ratio in reconstruction. The author is indebted to Professor Gabor, who at this conference reminded all that unit scene-reference ratio results in higher order noise terms in the reconstruction. The resolution tests should accordingly be repeated at scene-reference ratios of 1:2, 1:3, 1:4.

** Siemens IPS 60 cycle per second ruby laser.

Double Exposure Holographic Interferometry

The simple change from a single-exposed hologram to a double-exposed hologram converts the Figure 2-3 holocamera into a transmission holographic interferometer.(7) The first exposure records the empty scene, the wavefront of the diffuser pattern. The second exposure records the scene with the superimposed phase changes due to the event being recorded. The double-exposed hologram after development reconstructs the separately recorded wavefronts at the same time. The two wavefronts interfere with one another, producing fringes or dark bands at regions of destructive interference between the two separately recorded wavefronts. Bright fringes result from regions of constructive interference. For transmission holograms, the fringes map regions of constant optical path length between the two exposures. Optical path is the integral of the product of the index of refraction along the physical path.

An example of a photograph of the reconstruction of a double-exposed holographic interferogram made in a transmission holocamera of the type shown in Figure 2 and 3, is reproduced in Figure 5. The subject in this case was a high speed (3500 ft/second) 22 caliber rifle bullet fired through the scene volume of a transmission holocamera like the one shown in Figure 2. The example emphasizes the applicability of holographic interferometry to the study of aerodynamic phenomena. In gas flow problems, the local index of refraction is proportional to the gas density.(8) In simple regions of axial symmetric flow, theory and experiment have been closely matched.(7,8)

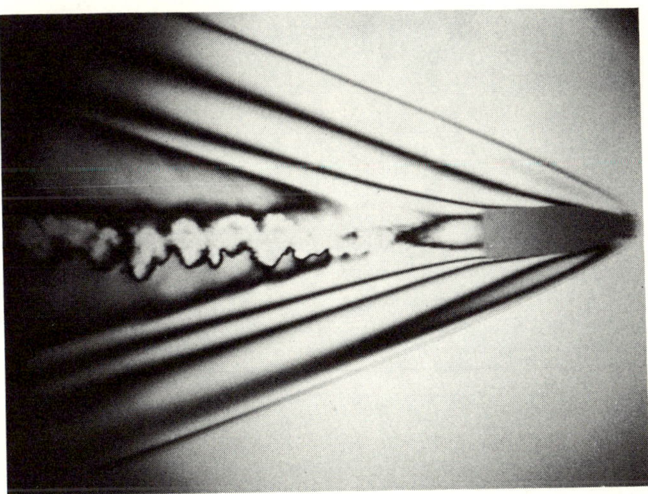

Figure 5. Photograph of the reconstruction of a double-exposed ruby laser hologram made in air.

The example does not illustrate the three-dimensional nature of these holographically derived interferograms. Suffice it to say that fringe pattern is a function of the viewing angle.

Also implicit in the Figure 5 example is the timing and firing of the laser relative to the location of a high speed bullet in the scene volume of the holocamera. A Kerr cell Q-switch in the laser oscillator and electronic delay circuitry synchronized the emission of the laser to the event. Electronic Q switches, however, tend to reduce temporal coherence. The holocamera corrects for the lack of single frequency emission, making possible high quality interferometric recordings of high speed phenomena without concern for multi-frequency emission of the laser. Dye cell Q switches can be used to increase temporal coherence of a ruby laser, but their use negates precise timing.(1)

Holography makes the recording of interference phenomena easier than with a classical interferometer. Precision optics are no longer required; in fact, the interferogram shown in Figure 5 was recorded with an "interferometer" with a ground glass diffuser in one leg. A more obvious example is shown in Figure 6. It is the interferometric record of the burning of an acetylene-air mixture inside a transparent lucite pipe. The lucite pipe was standard finish material. Holographic interferograms record only phase differences. They are not sensitive to complexity of the phase of the scene beam.

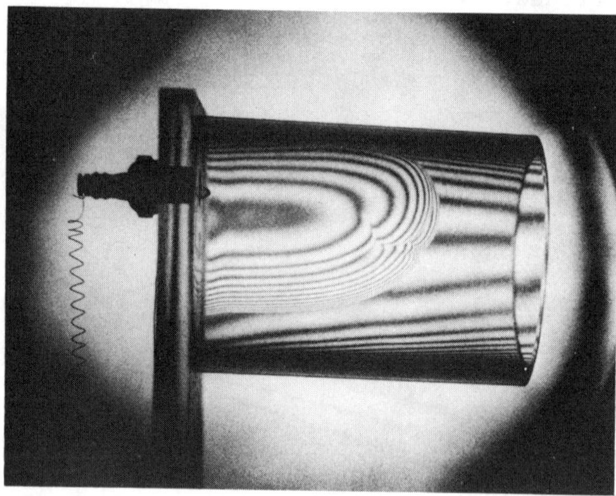

Figure 6. Photograph of the reconstruction of a double-exposed ruby laser hologram of acetylene-air burning within a transparent lucite cylinder. The example emphasizes the insensitivity of holographic interferometry to the shape of optical surfaces.

Reflected Light Holography

Reflected light holograms also may be recorded with Q-switched ruby lasers of low coherence using optical configurations based on the same principle as the transmission holocamera shown in Figures 2 and 3.(9) With electronically-timed Q-switched ruby lasers, a reflection holocamera was used to study the transient deformation of metal plates, etc.(10) In the study of vibration phenomena of plane surfaces, little temporal coherence is necessary, because of the constant path length, providing scene-reference path lengths are matched. Reflected light holocameras provide the spatial match of the scene and reference beams.

Recently, the author has succeeded in recording reflected light holograms without a pair of intermediate lenses for spatially matching the scene and reference beams at the hologram.(4) This was a result of recently available ruby laser rods of greater homogeneity.

Contouring

The broad bandwidth of the lasing transition in ruby can result in the simultaneous emission of may modes. This type of emission severely limits the temporal coherence of the emitted beam. The modes, however, can be restricted by the use of resonant reflectors. These elements, along with the adjustment of laser rod temperature and of total cavity length, can result in a ruby laser which emits basically two very close modes of different frequency. Reflected light two-beam holograms recorded with such an illuminator, show on reconstruction a scene with repeating light and dark zones. The zones follow one another at greater and greater depth. Each zone turns out to be an equal range contour. The contour spacing is inversely proportional to the wavelength difference $\Delta\lambda$ between the two lasing lines; namely,

$$\lambda^2/2\Delta\lambda \cos\theta/2 \quad ,$$

where θ is the angle between the direction of viewing and the direction of the incident light for each point of the scene.(10) An example of one such holographically-produced contour map is shown in Figure 7. The hologram in this case was recorded with a ruby laser with a pair of quartz resonant reflectors, each of 23 millimeter optical thickness. The contour spacing in the reconstruction has the same value. By changing to a 1/8 inch thick sapphire resonant reflector, holograms of 7.7 mm contour spacing were recorded.(11)

Figure 7. Photograph of the reconstruction of a hologram of a reflected light scene recorded with a single pulse from a ruby laser emitting basically two frequencies, separated from one another by approximately 1/8 Å. This type of contouring makes possible recording of re-entrant structures.

The significance of the two-frequency ruby laser holographic contouring technique is that the contour spacing is within the range of everyday engineering measurement; in addition, re-entrant structures can be contoured. Since the ruby laser is of high intensity and short duration, contouring can be done free of mounting on granite tables, etc. Deployable spacecraft dish antennas are an interesting future possibility for ruby laser holographic contouring.

Summary

Becuase of its high energy, short emission time, narrow band width, and emission in the visible region of the spectrum; the ruby laser holds a unique position in the application of holography. Small relatively incoherent ruby laser illuminators can produce high quality holograms and holographic interferograms when used with special coherence insensitive optical arrangements ("holo-cameras"). Ruby lasers have recently been constructed which emit two frequencies separated by $\lesssim$ 1/4 Å. These more special sources have been used to make holographic contour maps of complex scenes.

REFERENCES

1. D. Ansley, "Pulsed Laser Holography," U.S.-Japan Seminar on Information Processing by Holography, Washington, D.C., October 17, 1969 (this conference).

2. R. E. Brooks, L. O. Heflinger, and R. F. Wuerker, "Pulsed Laser Holograms," IEEE J. of Quantum Electronics, Vol. QE-2, 275-279, August, 1966.

3. R. F. Wuerker, B. J. Matthews, and R. A. Briones, "Producing Holograms of Reacting Sprays in Liquid Propellant Rocket Engines," Final Report, JPL Contract #952023, NAS7-100, TRW Report No. 68-4712.2-024, July 31, 1968.

4. R. F. Wuerker and B. J. Matthews, "Laser Holocamera Droplet Measuring Device," Technical Report AFRPL-TR-69-204, November, 1969.

5. B. J. Matthews and R. F. Wuerker, "The Investigation of Liquid Rocket Combustion Using Pulsed Laser Holography," AIAA 5th Propulsion Joint Specialist Conference, June 9-13, 1969.

6. R. F. Wuerker, B. J. Matthews, and B. J. Heckert, "Analysis of Holograms of Reacting Sprays," Final Report, JPL Contract #952357, TRW Report # 12299-6001-R0-00, January 1970.

7. L. O. Heflinger, R. F. Wuerker, and R. E. Brooks, "Holographic Interferometry," J. Appl. Phys., Vol. 37, 642-649, February, 1966.

8. A. B. Witte and R. F. Wuerker, "Laser Holographic Interferometry Study of High Speed Flow Fields," AIAA 4th Aerodynamic Testing Conference, Paper #69-347, Cincinnati, Ohio, April April 28-30, 1969.

9. R. F. Wuerker and L. O. Heflinger, "Pulsed Laser Holography," 1968 Symposium on the Engineering Uses of Holography, September 17, 1968, now being published in Harvey and Robertson, Engineering Uses of Holography, Cambridge University Press.

10. R. F. Wuerker, "Holographic Interferometry of Transient Vibrations with a Pulsed Ruby Laser," U. S. Navy Underwater Sound Laboratory, Holographic Interferometry Seminar, August, 1968.

11. L. O. Heflinger and R. F. Wuerker, "Holographic Contouring via Multifrequency Lasers," Appl. Phys. Ltrs., 15, 28-30, July, 1969.

PULSED LASER HOLOGRAPHY WITH TEM_{00} MODE RUBY LASERS

David A. Ansley

Conductron Corporation

Ann Arbor, Michigan

ABSTRACT

The development of stable pulsed ruby lasers operating in a single transverse and axial mode (TEM_{00}) has significantly extended the domain of holographic applications. Both transmission and reflection holography have been utilized for recording holograms of mechanical systems, flow patterns, and human beings ... analyzing stress and vibration ... and in biophysical and medical research.

INTRODUCTION

Since the development by our laboratory of a single-mode pulsed ruby laser[1], the technology of making holograms of front-lighted moving objects has rapidly advanced.

Single-mode pulsed laser output has increased from 10 millijoules to 20 joules. Holographic recording emulsions have improved from a sensitivity of 7000 ergs/cm^2 to 200 ergs/cm^2 with little degradation in resolution, diffraction efficiency, and clarity.

The improved emulsions are available on special order in sizes up to 3 feet by 4 feet.

This paper describes the techniques used to make large pulsed laser transmission and reflection "white light" holograms. Techniques and operating practices to protect live subjects from skin and eye damage are discussed.

HOLOGRAM STABILITY REQUIREMENTS

A hologram records the spatial pattern produced by the interference of a reference beam with light reflected from (or transmitted through) an object. This interference pattern on the photographic (hologram) plate should not move during the exposure time; otherwise, a poor (or no) hologram will result. This requires that all components of the holographic apparatus remain stationary to a fraction of a wavelength during the exposure time. Figure 1 illustrates this requirement. $\overline{AB}$ is the reference beam illuminating the photographic plate. $\overline{AC}$ is the beam illuminating the object. $\overline{CB}$ is the wave reflected from the object onto the photographic plate. During the exposure

$$\Delta r_1 - \Delta(r_2 + r_3) \leq \frac{\lambda}{8}$$

For a fixed exposure time, the lowest allowable object velocity occurs when the direction of the front-lighting illumination on the object is parallel to the line of sight of the observer and the object is moving directly toward or away from the observer ($\alpha = 0^o$). For typical values of $\lambda = 6943$ Å and exposure time (τ) = 30 nsec, the maximum object velocity is 2.9 m/sec. If the object point moves on an ellipse with foci A and B, the maximum velocity is not restricted except by resolution considerations. Another method[2], suggested by Don Neumann, is to vary r_1 during the exposure to compensate for the change in $r_2 + r_3$.

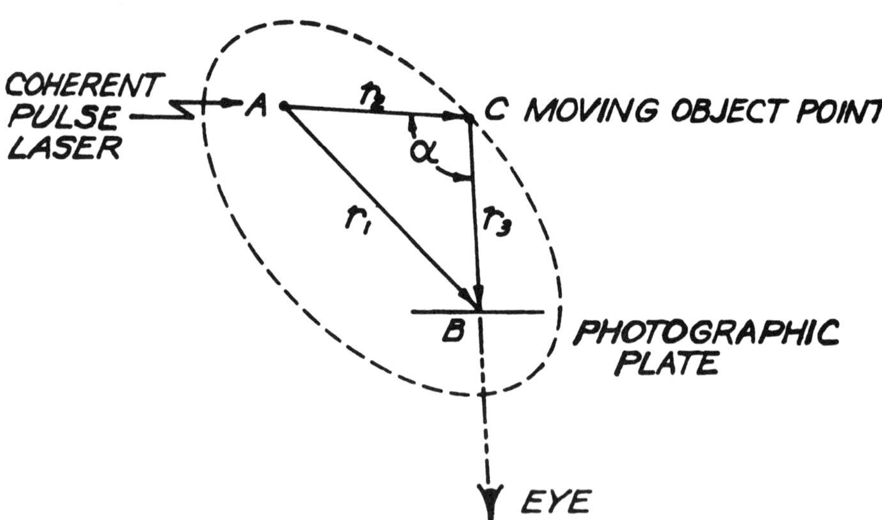

Fig. 1. Geometry for Calculation of Maximum Object Velocity

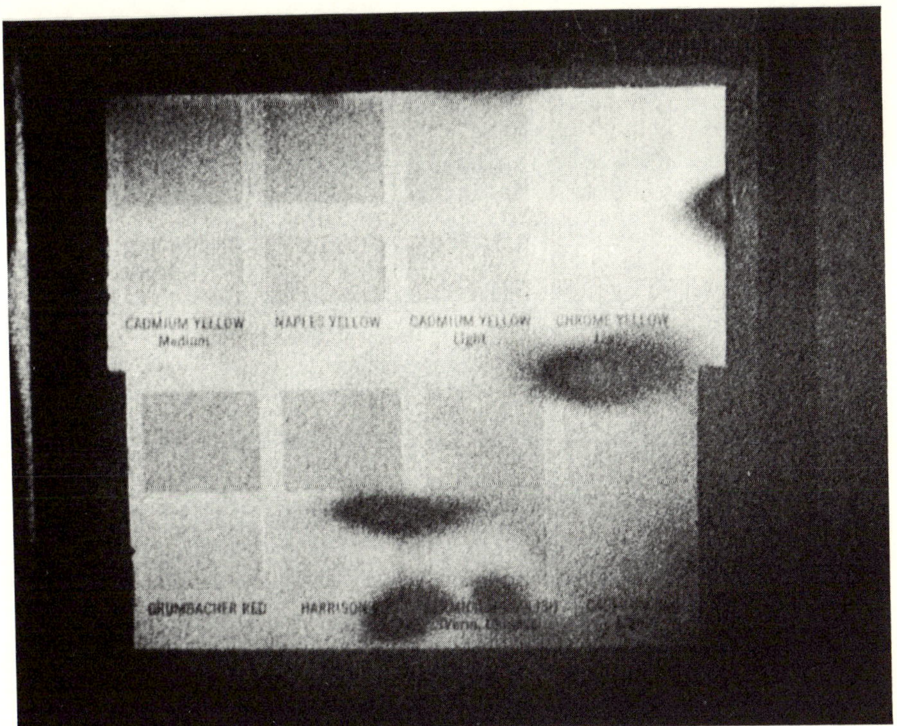

Fig. 2. Example of Object Motion During Holographic Exposure (Gas Laser)

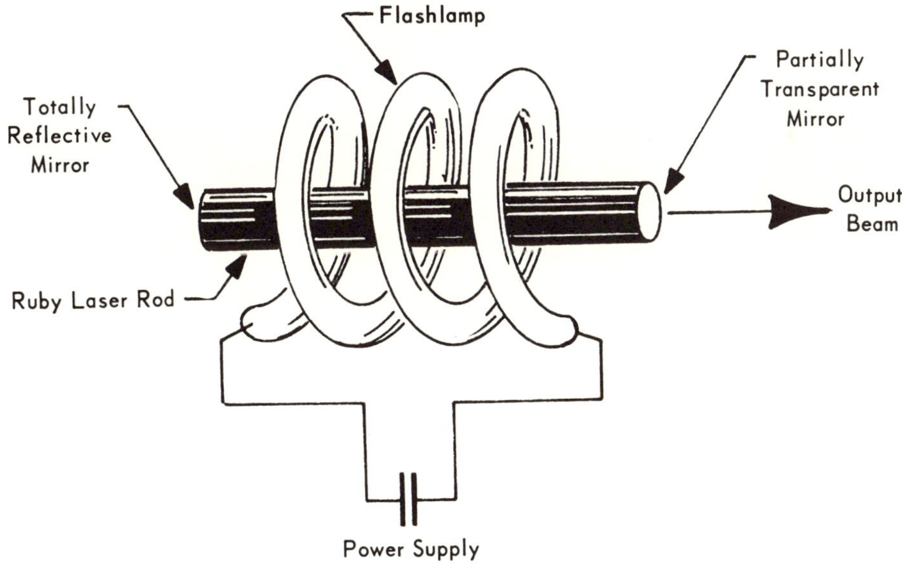

Fig. 3. Basic Components of Ruby Laser

Normal incidence back-lighting ($\alpha = 180^{o}$) is frequently used when the researcher does not need to view the object surface detail. The holographic stationarity requirement does not restrict the maximum allowable object velocity; rather, the researcher need only consider the conventional blurring due to object motion during the exposure. Table 1 summarizes the maximum allowable object velocities for normal incidence front-lighting ($\alpha = 0^{o}$), and a normal incidence back-lighting ($\alpha = 180^{o}$) for five different exposure times.

The sensitivity to motion can be illustrated by holograms made with a gas laser. Figure 2 is a photograph of the reconstruction of a color test chart in which motion occurred during the required 10 second period needed to record the hologram with a 0.6328 μ helium-neon gas laser of 50 milliwatts power. The color test chart was not firmly secured to the plate. Wherever there was movement, the holographic image was destroyed, resulting in black areas. Using the information in Table 1, it is apparent that movement in excess of 10^{-8} m/sec would degrade the holographic image. To keep the movement of all parts of the holographic apparatus less than this velocity requires elaborate precautions, such as granite slabs in temperature stabilized rooms to eliminate acoustical and seismic noise and convection air currents. This is the reason for using pulsed lasers to reduce the object stability requirements.

MULTI-MODE RUBY PULSED LASER

The basic components of a ruby pulsed laser are shown in Figure 3. These consist of a ruby rod (aluminum oxide with chromium ions), and a xenon flashlamp[3]. One end of the rod is coated with a dielectric mirror, the other end has a partially transparent dielectric mirror (reflectivity approximately 50%).

This type of laser is not truly coherent because it lases simultaneously in a number of transverse and axial modes. The natural line width of ruby at 20°C is approximately 7 Å (420GHz). The number of modes Δm which can lase is calculated from the resonance condition $m\lambda = 2L$ of a Fabry-Perot etalon

$$\Delta m = 2L \frac{\Delta\lambda}{\lambda^2}$$

where

L is the optical path length between reflectors

$\Delta\lambda$ is the resonance width of the laser transition.

TABLE 1

MAXIMUM ALLOWABLE OBJECT VELOCITY

Laser Pulse Time (T)	Normal Incidence Front Lighting ($\alpha = 0^o$)	Normal Incidence Back (Silhouette) Lighting ($\alpha = 180^o$)
1 second	8.7×10^{-8} m/sec	1×10^{-6} m/sec
1 millisecond	8.7×10^{-5}	1×10^{-3}
1 millisecond	8.7×10^{-2}	1
30 nanoseconds	2.9	33
1 nanosecond	87.0	1×10^{3}

NOTES: Ruby laser λ = 0.6943 micron.

Front-lighting velocity calculations based on hologram image brightness reduction of 19% ($\lambda/8$ movement).

Back-lighting velocity calculations based on hologram image blurring equal to 1 micron.

For a typical L = 17 cm

$$\Delta m \cong 500$$

each mode will lase if its gain exceeds its losses. The coherence length of a laser is inversely proportional to the frequency bandwidth of the axial modes.

$$\Delta C = \frac{\lambda^2}{\Delta\lambda}$$

for the above case.

$$\Delta C \cong 0.7 \text{ mm}$$

a very small coherence length indeed!

Holograms can be made with a multi-mode pulsed laser if care is taken to superimpose the same parts of the object and reference waves at the hologram plate[4,5]. Otherwise fringes appear on the holographic images and/or the holographic image is very low intensity and noisy.

BACK-LIGHTED HOLOGRAPHY

Figure 4 is a schematic of an experimental arrangement for recording back-lighted transmission holograms. The prism plate and the relay lens image the object wave onto the hologram plate with spatial and temporal characteristics nearly identical to the reference wave.* Although multi-mode, each point in the wave is coherent with itself. Since the object and reference waves are superimposed, they will therefore interfere and produce a hologram.

Holograms of transparent objects, such as gas flames, can be made using a technique called holographic differential interferometry. In this technique, two exposures are made on the same hologram plate: one exposure with the object, the other exposure without the object. Dark interference fringes are seen in the holographic image wherever the transparent object caused an optical path change of $\lambda/2$.

In Figure 5 is illustrated a holographic differential interferogram of airflow from a supersonic jet impinging on an inclined flat plate[6]. The entering air is at 50 psi with exhaust into a near vacuum. For comparison, a double-pass Schlieren photograph of the same flow field is shown. The low spatial frequency, roughly parallel fringes seen in the holographic interferogram indicate motion of the apparatus between exposures. This fringe pattern may be used to measure small phase differences. As compared to the Schlieren photographs, which show the shock structure near the jet exits in greater detail, the interferograms show more sensitivity in the low pressure areas away from the main flow. For example, in the interferogram of the supersonic jet, a density change (from the no-flow condition) is evidenced by the shift in fringe pattern under the inclined plate as compared to the background fringes above the same plate. The Schlieren photograph offers no density information in this area though it does show density changes near the jet that appear lost in the interferogram.

Three-dimensional phase information (rather than integrated two-dimensional phase information) may be obtained as shown in

---

* For TEM_{00} mode pulsed lasers, the prism plate, and relay lens are not required.

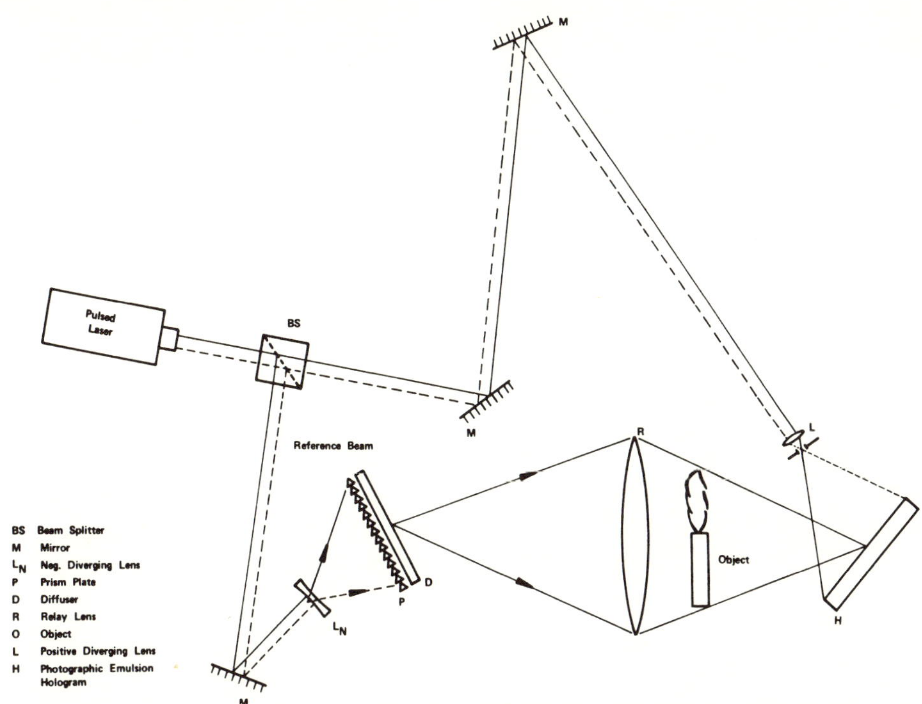

Fig. 4. Schematic of Experimental Arrangement for Recording Back-lighted Transmission Holograms

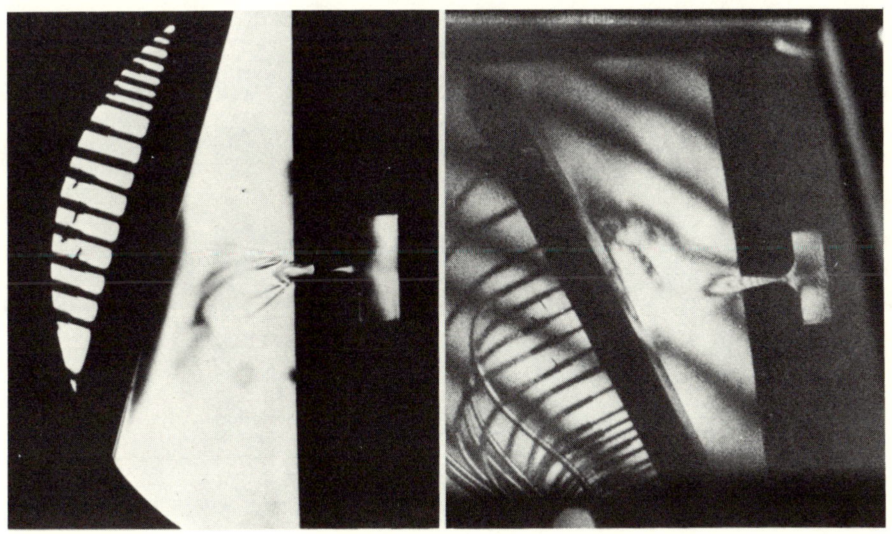

Fig. 5. Comparison of Schlieren Photography and Holographic Differential Interferometry

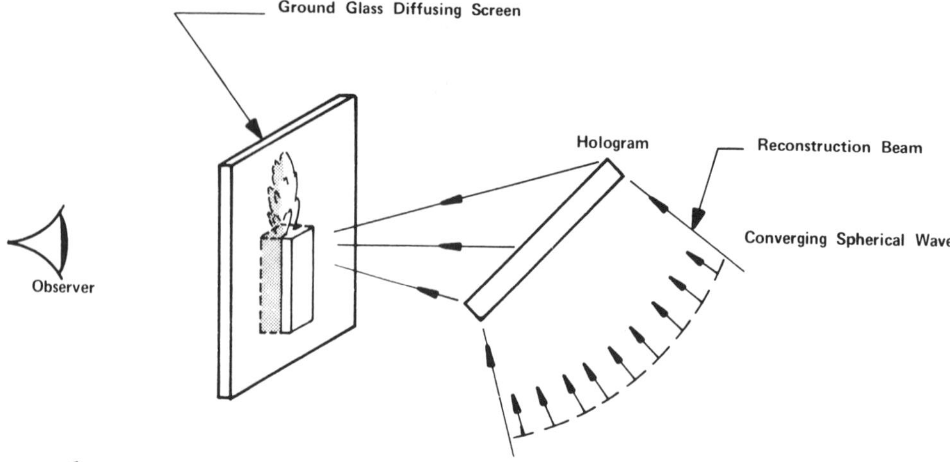

Fig. 6. Schematic of Experimental Arrangement for Obtaining Three-Dimensional Holographic Differential Interferometric Information

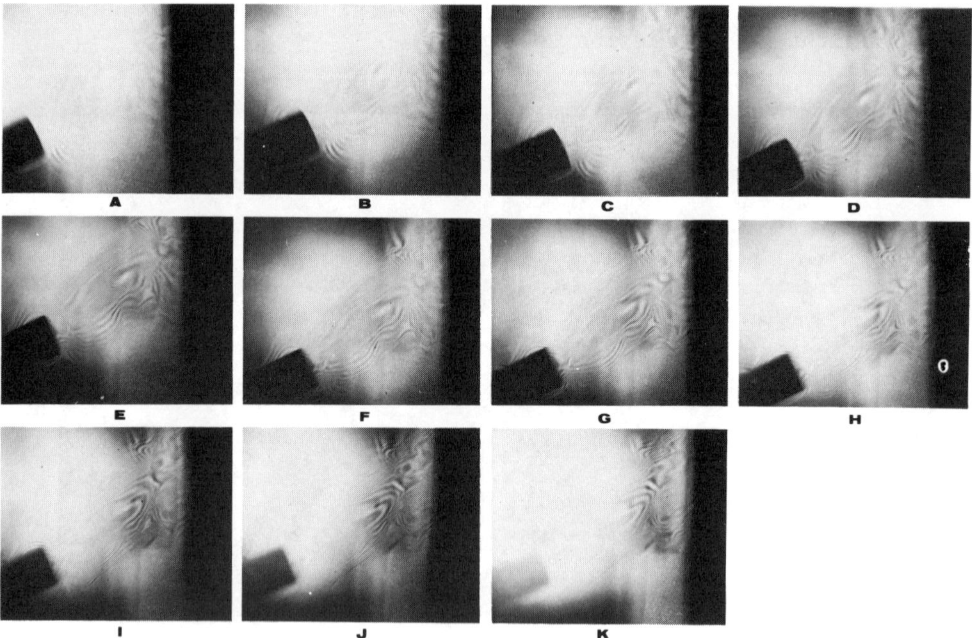

Fig. 7. Three-Dimensional Holographic Differential Interferometry – Propane Torch

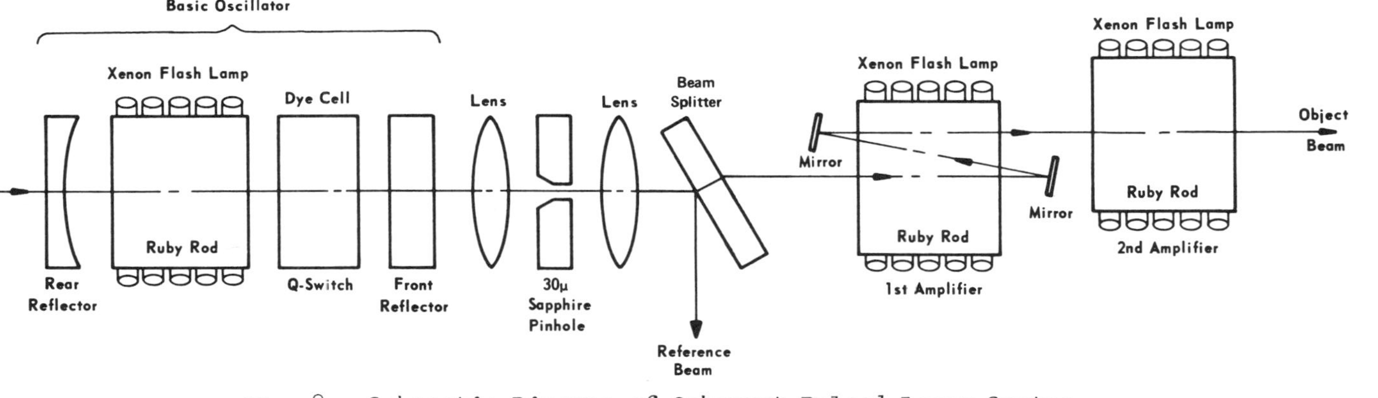

Fig. 8. Schematic Diagram of Coherent Pulsed Laser System

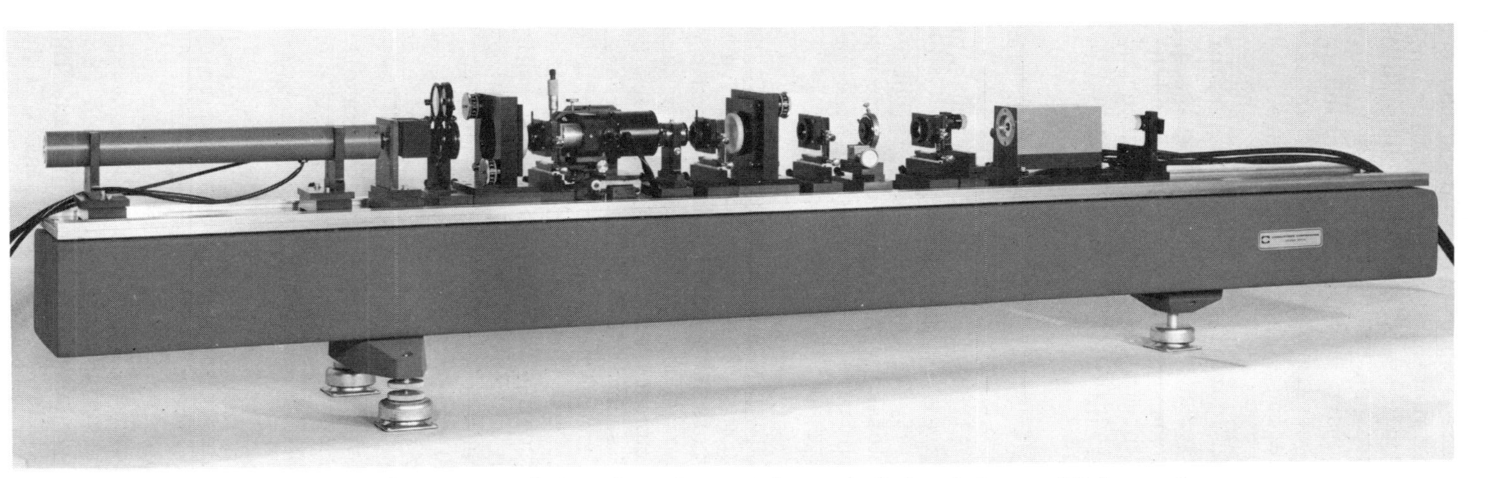

Fig. 9. Conductron Model C-10 Ultra-Coherent Pulsed Laser Without Cover

Figure 6. By illuminating the processed hologram plate with a reconstruction beam propagating in the reverse direction of the reference beam, a real image of the object is focused in space. The image can be viewed by use of a ground-glass viewing screen. Only the part of the image focusing in the plane of the viewing screen is seen in sharp focus. This technique may be used to find the location of the phase change.

Figure 7 shows the results of three-dimensional holographic differential interferometry of a propane torch. The viewing screen was placed at 11 positions in the volume, starting in front of the image and finishing behind the image. Notice the variation in the interference patterns.

SINGLE-MODE PULSED LASER

Although wavefront superpositioning devices have been demonstrated for making holograms of diffusely reflecting objects, the quality of the holograms is not comparable to those obtained with CW lasers. Thus, work was started in June 1966 to modify the basic ruby pulsed laser, shown in Figure 3, to obtain single transverse and axial mode output.

Figure 8 is a schematic diagram of a coherent pulsed laser system[1]. The totally-reflective and partially-reflective end mirrors have been removed from the ruby rod and spaced up to a meter apart. The mirror configuration is stable, i.e. the radius of curvature of one mirror lies between the radius of curvature of the second mirror and its surface. The laser is operated near threshold to reduce the number of modes with net gain. A bleachable dye (crypto-cyanine and methyl alcohol) is used both for Q-switching and transverse mode selection[7]. Longitudinal mode selection is also simultaneously obtained by selective wavelength (frequency) bleaching in the dye. In some instances, a partially reflective plano-parallel glass etalon is inserted in the cavity for additional axial mode selection. The coherence length of the laser is thus determined by the frequency bandwidth of its single axial mode. The lens and pinhole assembly is used to optically decouple the amplifier from the oscillator to eliminate regeneration.

Typical energy output of a single-mode laser is 10-20 millijoules. Amplifiers are used to obtain the 1-3 joules used to make front-lighted holograms of diffusely reflecting objects. A typical coherent pulsed laser is shown in Figure 9. Technical specifications for the same laser are shown in Table 2. Single transverse mode is determined from laser burn patterns in unexposed and developed polaroid paper. The single axial mode output of the laser is verified by using a Tektronix Model 519 oscilloscope and

TABLE 2

TECHNICAL SPECIFICATION FOR CONDUCTRON CORPORATION'S ULTRACOHERENT PULSE LASER SYSTEM C-10

Wavelength:	6943 Å
Mode of Operation:	Single Transverse and Axial Mode (Tem_{oon})
Energy Output:	1 Joule Per Pulse Nominal
Pulse Width:	Less Than 50 Nanoseconds
Coherence Length:	Greater Than 2 Meters
Repetition Rate:	45 Seconds
Power Requirement:	115 Volt 60 Cycle Power With 1% Voltage Regulation
Cooling:	Low Pressure 3CFM Air Supply

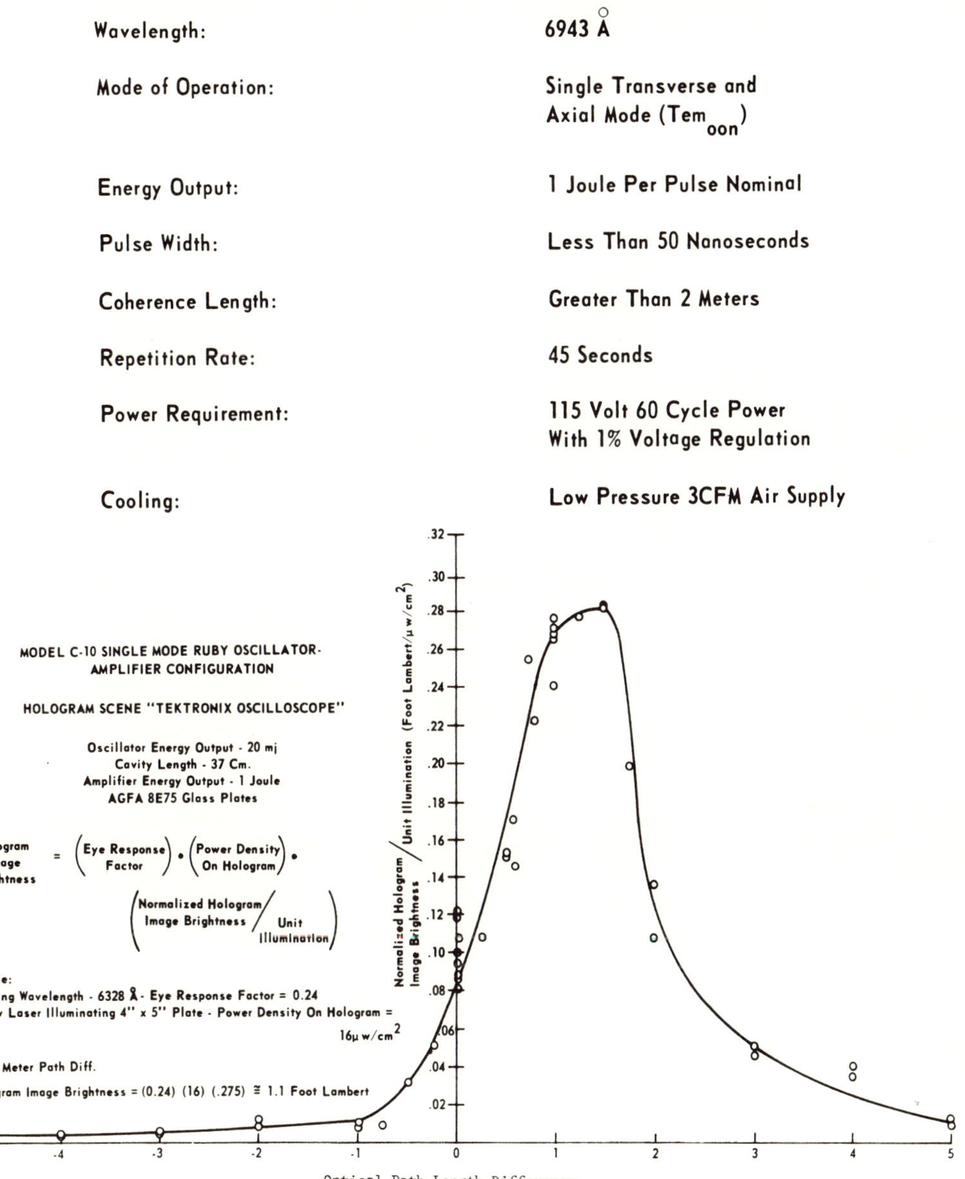

Fig. 10. Normalized Hologram Image Brightness/Unit Illumination Vs. Optical Path Length Difference (Object Beam - Reference Beam)

a Korad Model K-D1 photo detector having a rise time of 0.3 nsec. The scope/detector bandwidth is greater than the frequency difference of adjacent modes. The oscilloscope trace is a smooth Gaussian function with 3 db points at approximately 30 nsec for single mode operation. This technique has been collaborated by Fabry-Perot interferometry of the laser output spectrum.

MEASURING COHERENCE LENGTH

It is necessary to describe the technique used to measure coherence and define coherence length, since coherence decreases with increasing path length difference between object and reference beams. The coherence length of the Coherent Pulsed Laser was verified by making holograms at different path length differences (the experimental setup is kept the same). The path length is changed by adjusting the reference beam length before it is diverged by the negative lens) and measuring the brightness of diffusely reflecting objects in the holographic images. If the intent is to use the Coherent Pulsed Laser to make holograms, this is a far more accurate measure of coherence than the more common technique of using a Michelson or Tyman-Green interferometer. The reason that interferometry data is not valid is because the two waves, although time delayed, are still superimposed and will interfere with themselves even when not coherent with adjacent points in the waves, since the laser output has non-uniform spectral characteristics throughout the beam cross section. We have found from experience that pulsed laser outputs which produce good contrast interferograms do not necessarily produce high brightness holograms (holograms of diffusely reflecting objects require coherence over the entire wave because of the scattering effect of the objects).

Figure 10 shows that for a path length difference of ±2 meters (measured from the nominal position: object beam optical path length minus reference beam optical path length equals +1.5 meters)* the holographic image brightness is approximately 10% maximum brightness.

Hologram brightness at 5 meters is approximately 3% vs. 50% theoretical. However, useful data can still be extracted from the

---

*Peak brightness is at a path length difference of +1.5 meters. This is caused by doppler shifting in the amplifier (due to changing optical path length) since the reference beam does not pass through the amplifier. In later tests, using Czochralski coreless amplifier rods, both the reference and object illumination beams passed through the amplifier. In this case, the peak hologram brightness was at zero path length difference. The shapes of the curves were very similar.

hologram since the range in brightness of a typical holographic scene is generally from 1000 to 10,000:1.

Lower than expected image brightness is probably caused by frequency shifting (FM'ing) of the single-axial mode from a design line width of 15 MHz to over 50 MHz. Further research is being done in this area and will be the subject of a paper to be published soon by L. D. Siebert of our laboratory.

LASER SAFETY

Since the early days of laser research, there has been concern for the health hazards (particularly eye-damage) of lasers. One of the leaders in the area of laser safety is Martin-Marietta Corporation of Orlando, Florida who sponsored the first conference on laser safety[8] in 1966. Hamm and Geeraets, et al.[9], in 1965 proposed a biological damage threshold of 0.07 joule/cm^2 for a 30 nsec pulse duration*. This is the same value as recommended by the American Conference of Governmental Industrial Hygienists (1968), Committee on Industrial Hygiene Codes and Regulations (shown in Tables 3, 4, and 5)**.

Table 4 shows the maximum permissible exposure levels for laser radiation at the cornea for direct illumination or specular reflection at a wavelength to 6943 Å. We use the night condition with a 7 mm pupil to calculate the safe level for Q-switched pulsed laser holograms. For normal eye in a worst-case situation, the focusing power of the eye increases the energy/density of the ruby laser light striking the cornea by a factor of 600,000 (12 micron diameter spot for a 7 mm pupil). In this case, the safety factor is approximately 10, i.e., an energy density of 1.0×10^{-8} joule/cm^2 will produce an energy density of 0.007 joule/cm^2 on the retina.

Table 5 shows the maximum illuminance permitted from a diffuse surface reflection as measured at the reflecting surface. Figure 11 shows the technique used to keep the energy from a diffuse surface within the values recommended in Table 5. The reference beam

---

*Same value prescribed by the British Ministry of Aviation (in its Code of Practices for Laser Systems). U.S. Air Force (Regulation 161-24, January 1967) recommends 0.125 joule/cm^2.

**A copy of the original report, designated Supplement No. 7 to "A Guide for Uniform Industrial Hygiene Codes and Regulations," is available for $0.50 a copy from the Secretary/Treasurer ACGIH, 1014 Broadway, Cincinnati, Ohio 45202.

TABLE 3

THRESHOLDS FOR BIOLOGICAL DAMAGE BY LASER RADIATION AT THE RETINA OF THE EYE

Type Laser	Wavelength	Pulse Duration	Level
Q-switched	6943 Å	30 nsec	0.07 j/cm^2
Non-Q-switched	6943 Å	200 μsec	0.85 j/cm^2
CW	White light	---	6.00 w/cm^2

TABLE 4

MAXIMUM PERMISSIBLE EXPOSURE LEVELS FOR LASER RADIATION AT THE CORNEA FOR DIRECT ILLUMINATION OR SPECULAR REFLECTION AT A WAVELENGTH TO 6943 Å

Conditions	Q-Switched 1 nsec to 1 μsec Pulse PRF $\leq$ 10 (j/cm^2)	Non-Q-Switched 1 nsec to 0.1 sec Pulse PRF $\leq$ 10 (j/cm^2)	Continuous Wave Exposure $>$ 0.1 sec (w/cm^2)
Daylight - 3 mm pupil	5.0×10^{-8}	5.0×10^{-7}	5.0×10^{-5}
Laboratory - 5 mm pupil	2.0×10^{-8}	2.0×10^{-7}	2.0×10^{-5}
Night - 7 mm pupil	1.0×10^{-8}	1.0×10^{-7}	1.0×10^{-5}

TABLE 5

MAXIMUM ILLUMINANCE PERMITTED FROM A DIFFUSE SURFACE REFLECTION AS MEASURED AT THE REFLECTING SURFACE AT A WAVELENGTH TO 6943 Å

Type Laser	Daylight (3 mm pupil)	Laboratory Environment (5 mm pupil)	Night Environment (7.5 mm pupil)
Q-Switched pulse (joules/cm^2)	0.45	0.15	0.070
Non-Q-Switched pulse (joules/cm^2)	5.5	2.0	0.9
Continuous-wave laser (watts/cm^2)	13.5	6.0	2.5

is diverged by a negative lens, L_n. The area of illumination on the ground-glass screen is calculated by the formula

$$A_T = \frac{\text{laser energy output}}{0.07 \text{ j/cm}^2} = \frac{F - S}{F} \cdot A_L$$

for a 3 joule output, $A_T \geq 43 \text{ cm}^2$.

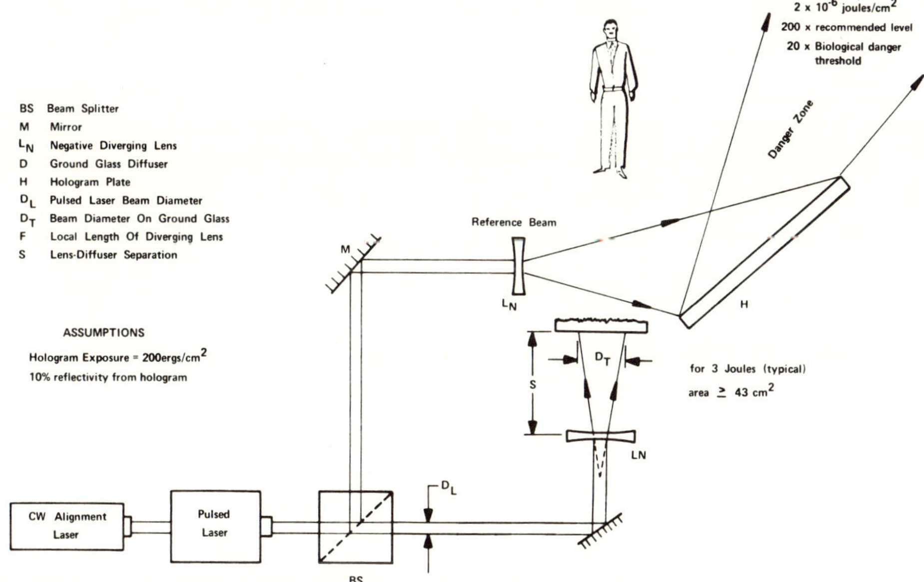

Fig. 11. Experimental Configuration for Making Transmission Holograms of Humans

Maximum permissible exposure levels for the skin should not exceed by a factor of more than 10^5 the daylight levels for the eye (Table 4)[10]. Therefore, the subject should be at least 10 cm from the diffuser (for a 3 joule output).

The reference beam, reflected from the hologram, can cause eye damage. For Agfa Scientia 8E75 emulsions, the illumination at the hologram plate should be 200 ergs/cm^2. The reflected reference light (assuming 10% reflection and reference light of much greater intensity than object light on the hologram) is 200 times the recommended level and 20 times greater than the biological damage threshold. By using a diverging reference beam, the energy density decreases with increasing distance from the hologram plate, though the size of the danger zone also increases.

The danger is even greater from another type of hologram - the reflection or "white-light" hologram - because the reference beam illuminates the hologram plate from the side opposite the subject as shown in Figure 12. Here the reference beam light transmitted through the hologram plate may be as much as nine times greater than the reference beam light reflected from the hologram in the previous case. Thus, in the danger zone, the energy density may be 180 times the biological damage threshold.

Precautions must obviously be performed to avoid possible injury to the subject. These include:

- Bringing the reference beam in at a steep angle so that the light will be reflected or transmitted away from the subject.

- Determining the illumination pattern of the reference beam and the object beam before exposure through use of a CW alignment laser propagating in the same direction as the pulse laser beam.

- Physical barriers and shields should be erected to prevent the subject from accidentally viewing the reference beam illumination.

Other hazards associated with making pulse laser holograms are:

- Ultra-violet radiation from the flashlamps.

- Glare from the high-intensity visible light of the flashlamps.

- Electrical Hazards. Pulsed lasers utilize capacitor banks for energy storage of many thousands of joules at several

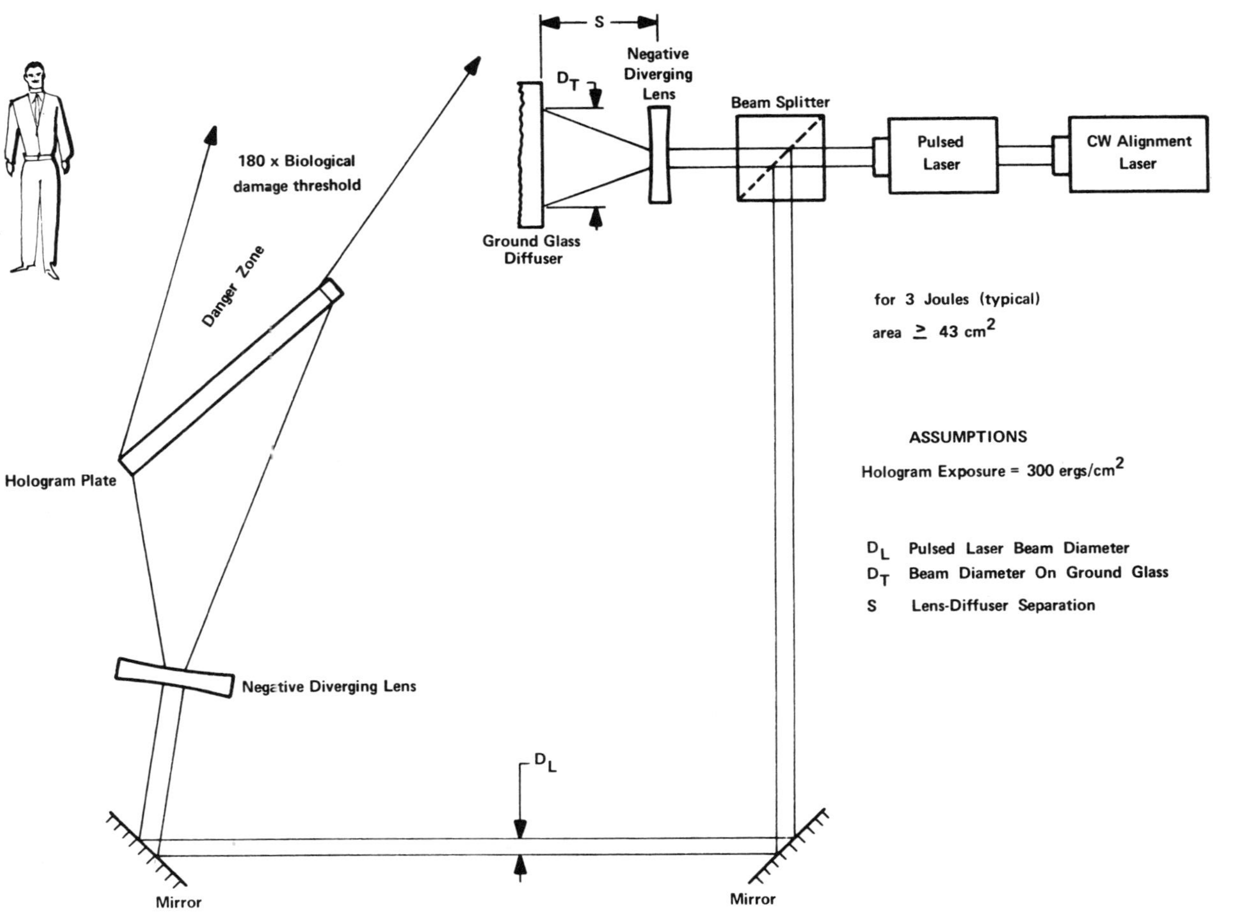

Fig. 12. Experimental Configuration for Making Reflection "White Light" Holograms of Humans

TABLE 6

COMPARATIVE PROPERTIES OF HOLOGRAPHIC PLATES

Film Designation	Mfg.	Form	Wavelength of Peak Sensitivity	Photo-recording Sensitivity	(From Mfg. Information) ($ergs/cm^2$)	ASA Speed Rating	Resolving Power (lines/mm)
14C70	Agfa*	Film	0.64 μ	---	3 (0.632 μ)	---	1500
10E70	Agfa*	Plate	0.64 μ	---	50 (0.632 μ)	---	2800
8E70	Agfa*	Plate	0.64 μ	---	200 (0.632 μ)	---	3000
10E75	Agfa*	Plate	0.69 μ	---	50 (0.694 μ)	---	2800
8E75	Agfa*	Plate	0.69 μ	---	200 (0.694 μ)	---	$\lesssim 3000$
649F	Eastman†	Plate Film	P	0.002	1000 (0.632 μ) 7000 (0.694 μ)	0.025	$\lesssim 4000$
SO-243	Eastman†	Film	P (to 0.7 μ)	0.1	1 (in red)	1.6	200

*Agfa-Gevaert, Antwerp, Belgium

†Eastman Kodak, Rochester, New York

kilovolts. To accidentally touch a charged capacitor bank would be fatal.

- Potential for explosions at the capacitor bank or optical pump system exists in operation of high energy pulse lasers.

If the pulse laser hazards are recognized and properly controlled, pulsed laser holograms of human subjects can be safely made.

HOLOGRAPHIC EMULSIONS

Initial pulsed laser holography experimentation was done using Eastman 649F plates. It is panchromatic and is extremely slow (by photographic standards, ASA 0.025). It starts losing its sensitivity in the red at about 0.67 microns*. At the ruby wavelength, its sensitivity is about an order of magnitude less than at helium neon wavelength.

In 1967 Agfa Gevaert released photographic emulsions specially sensitized for different laser wavelengths. The resolution was not as high as that of 649F, but was adequate for most holographic applications.

Properties of the different emulsions taken from both Eastman and Agfa Gevaert published information[11], is presented in Table 6. The 30x greater sensitivity between the 8E75 emulsion and the 649F emulsion constitutes a major breakthrough in holographic recording.

EXPERIMENTAL RESULTS

Figure 13 is the first pulsed laser hologram of a human (L.D. Siebert, made October 31, 1967). Approximately 250 millijoules of energy and Agfa 10E75 film was used[12].

Examples of recent holograms are shown in Figures 14 and 15. The photograph of the holographic reconstruction of Figure 14 was taken with a Nikon camera with a 35 mm focal length f/2 lens at f/5.6 setting. This was a compromise setting to obtain maximum

---

*See "Kodak Plates and Films for Science and Industry," Eastman Kodak Company, First Edition, Third Printing, p. 23d (1967).

Fig. 13. Photograph of the Reconstruction of the First Pulsed Laser Hologram of a Human (L. D. Siebert, October 31, 1967)

Fig. 14. Photograph of the Reconstruction of a Hologram of a Dog

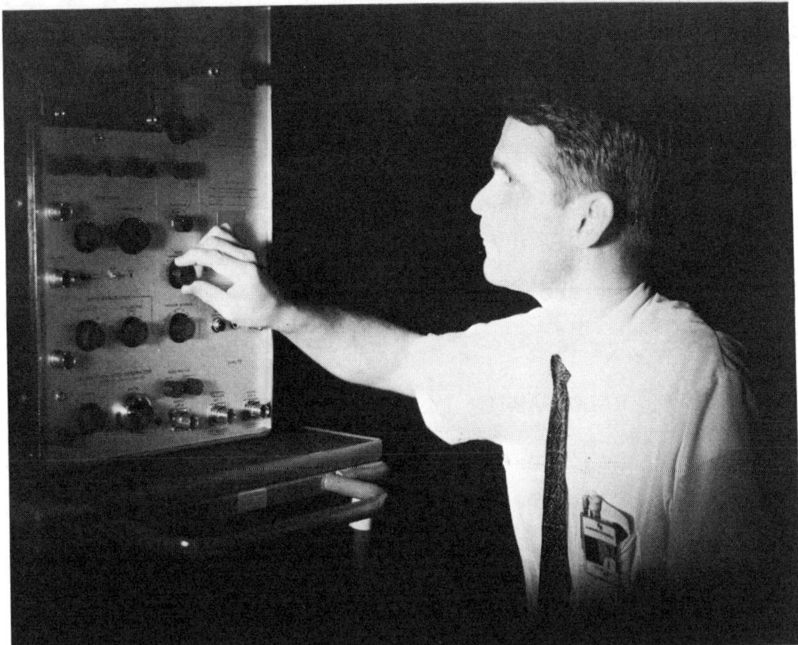

Fig. 15. Photograph of the Reconstruction of a Hologram of a Man (L. D. Siebert) and Oscilloscope

Fig.16. Photograph of the Reconstruction of a Reflection "White Light" Hologram of R. G. Zech

depth of focus with minimum speckle (increasing the f number increases both the depth of focus and the speckle pattern on the holographic image). Figure 15 was taken at a lower f number. Note the almost total absence of speckle.

Figure 16 is an example of a reflection hologram. Recording medium was Agfa 8E75, 15 microns thick.

Other researchers[11] have operated their pulsed laser with two axial modes and obtained contouring of the holographic image. This is useful for many metrological applications.

HOLOGRAPHIC IMAGE RESOLUTION

There are a number of causes of holographic image degradation. They include:

Coherent Noise and Speckle

Interference of light scattered from diffuse surfaces is a major problem in holograms. The noise extends through space and does not average out as in noncoherent imaging. For high resolution work, speckle can be reduced by using direct illumination instead of a ground-glass diffuser.

Phase of Adjacent Objects

Resolution of co-phase objects in coherent light is 26% worse than noncoherent imaging, since

$$\delta = \frac{1.54\ \lambda}{2\ \sin\ \frac{D/2}{R}}$$

where

D = hologram diameter

R = object-hologram separation.

Two objects π out of phase can theoretically be resolved even if infinitely small. Thus, the resolution is dependent on the phase of the adjacent object.

Motion During Holographic Exposure

This was previously discussed for back-lighted holograms; the image is blurred by an amount equal to the velocity times the exposure time. For front-lighted holograms, movement of the holographic apparatus or object reduces the brightness and contrast of the hologram image.

Extended Source Reference Beam

If the auto-correlation function of the reference beam is not a δ function, the holographic image will be blurred by the convolution of the reference beam with the object. Example: if the effective point source of the reference beam subtends an angle of 1 milliradian, the maximum angular resolution of the holographic image will be 1 milliradian.

Non-Uniform Reference Beam Illumination on Hologram

Only part of the hologram is properly exposed. This reduces the effective aperture of the hologram.

Extended Source Reconstruction Beam

The same applies as per the extended source reference beam discussion.

Wavelength of Reconstruction Beam Different Than Wavelength of Reference Beam

Using a different wavelength produces a shift in image location, a magnification change, and a degradation of resolution.

Polychromatic Reconstruction Beam

Each wavelength reconstructs the holographic image in a slightly different spatial position; thus, if a polychromatic reconstruction beam is used, a multi-color blurred image will result.

Photographic Emulsion Modulation Transfer Function

If the film cannot record the higher spatial frequencies at the edge of the diffracted light cone, resolution will be lost.

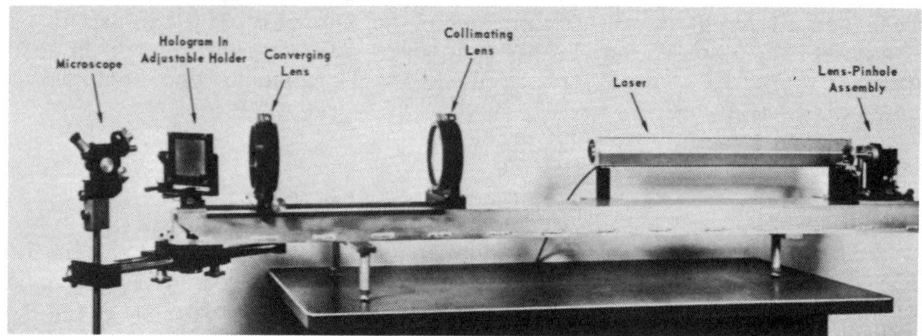

Fig.17. Hologram Viewer

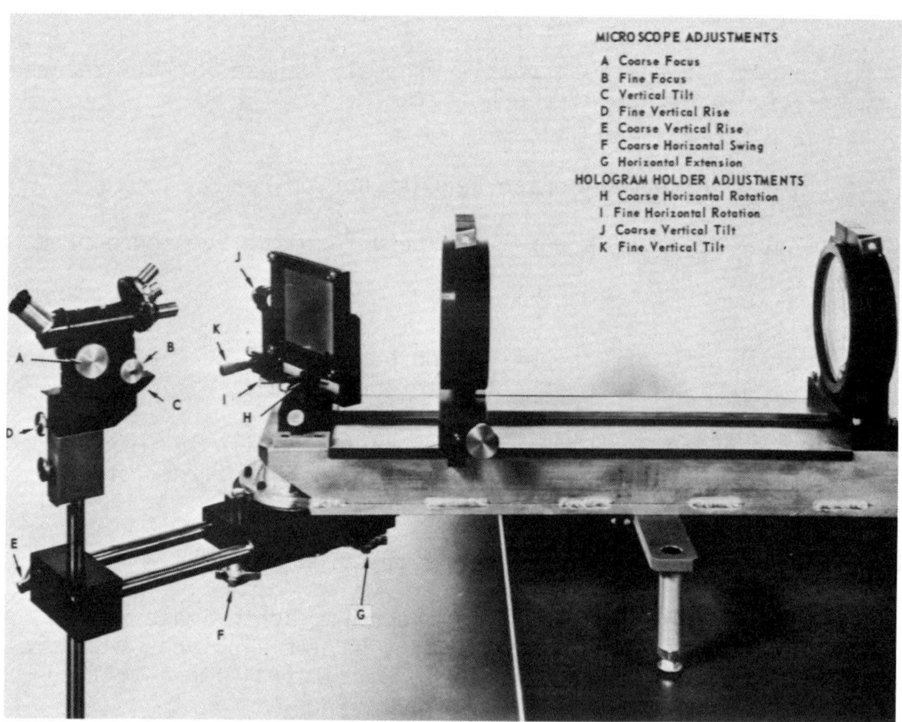

Fig.18. Hologram Viewer Closeup

Stress in Photographic Emulsion/Photo Processing Techniques

If the photographic emulsion is not optically flat before and after processing, aberrated deformation of the image wavefronts will occur. The recommended technique is to pre-soak before exposure to remove stress and then preharden the emulsion before fixing to prevent reticulation problems.

Tilted Hologram Plate

In many cases it is necessary to use an off-axis spherical reference beam. The spherical reconstruction beam illuminating the tilted flat glass hologram produces a holographic image with astigmatism and coma. This can be partially corrected by corrector plates.

Optically Imperfect Support Glass or Film

If not optically flat, the image wavefronts are deformed and focused to an aberrated image.

Reflected Light From Back Surface of Support Glass or Film (Woodgrain)

Woodgrain acts as a screen grating and produces multiple images, broadens and reduces the contrast of the primary image.

Misalignment of Hologram Relative to Reconstruction Beam

To obtain maximum resolution, the hologram must be aligned to several seconds of arc relative to the reconstruction beam.[13]

Nearly diffraction limited imagery is possible by proper attention to the causes of holographic image degradation.

The apparatus shown in Figures 17 and 18 is used to obtain micron-range resolution of the holographic image. Since most microscopes have a relatively short working distance, the reconstruction beam illuminates the hologram in the reverse direction of the reference beam. Thus, as before, a real image is reconstructed in space. The microscope can be moved within the entire image field to focus on particles or other points of interest.

The optical quality of standard hologram plates is not adequate for high resolution work. They degrade the resolution by a factor of 10 in the 10 micron range. We have obtained best results with Agfa 8E75 on special plate glass (4 fringes/inch). For CW holograms we have obtained 3 micron resolution at 6 in. from a 4 x 5 in. hologram plate. This is approximately two times the diffraction limit. Alignment of the hologram plate is extremely critical. Best resolution obtained using the TEM_{00} pulsed ruby laser was 10 microns. The poorer resolution was due to reconstruction of the holographic image using a CW laser of a different wavelength (6328 Å rather than 6943 Å)*.

SUMMARY

The pulsed laser is tremendously useful in making holograms of transparent, opaque, and diffusely reflecting man-sized objects. It reduces the engineering effort necessary. For example, a high quality "white-light" hologram which formerly required 12 minute exposure and 40 manhours setup time to isolate all vibration, can now be made after only a one hour setup time. Cosmetic quality and brightness is equal to or better than that of the CW hologram.

This paper has described the techniques used to make holograms of people. Major emphasis has been given to the safety aspect of pulsed laser holography.

If the hazards are recognized and properly controlled, pulse laser holograms of human subjects can be safely made, thereby applying holography to a number of new and useful applications.

Because pulsed lasers are semi-portable, they can be taken where the application is. This will expand the use of holography to a number of new applications previously unobtainable with CW lasers.

REFERENCES

1. L. D. Siebert, "Front-Lighted Pulse Laser Holography," Appl. Phys. Letters, 11, pp. 326-328 (15 November 1967).

2. D. B. Newmann, "Moving Object Holography with Unblurred Images of Normal Brightness," SPIE Holography Seminar Proceedings, San Francisco, California (1968).

---

*A CW ruby laser is available commercially from Siemans America, Incorporated, 230 Ferris Avenue, White Plains, New York.

3. T. H. Maiman, et al., "Stimulated Optical Emission in Fluorescent Solids, II. Spectroscopy and Stimulated Emission in Ruby," Phys. Rev. 123, pp. 1151-1157 (1961).

4. R. E. Brooks, L. O. Heflinger, R. B. Wuerker, and R. A. Briones, "Holographic Photography of High Speed Phenomena with Conventional and Q-Switched Lasers," Appl. Phys. Letters 7, pp. 92-94 (1965).

5. R. E. Brooks, et al., "Pulsed Laser Holograms," IEEE Journal of Quantum Electronics, QE-2, pp. 275-279 (1966).

6. L. D. Siebert, D. E. Geister, 1968, "Pulsed Holographic Interferometry vs. Schlieren Photography," AIAA Journal, 6 (11), pp. 2194-2195.

7. B. H. Soffer, "Giant Pulse Laser Operation by a Passive, Reversibly Bleachable Absorber," J. Appl. Phys., 35, pp. 2551, August 1964.

8. Martin-Marietta Corporation (Co-sponsored with U.S. Surgeon General's Office), "Proc. of the First Conf. on Laser Safety," Graham W. Flint, Ed., Orlando, Florida, May 19-20, 1966.

9. W. J. Geeraets, W. T. Ham, Jr., R. C. Williams, Harold A. Mueller, J. Burkhart, D. Guerry III, and J. J. Vos, "Laser Vs. Light Coagulator: A Funduscopic and Histologic Study of Chorioetinal Injury as a Function of Exposure Time," Federation Proceedings 24, pp. 5-60 (1965).

10. "A Guide for Uniform Industrial Hygiene Codes and Regulations for Laser Installation," American Conference of Governmental Industrial Hygienists, Cincinnati, Ohio (1968).

11. R. F. Wuerker, "Pulsed Laser Holography," Technical Report AFAL-TR-68-371, USAF Avionics Lab.

12. L. D. Siebert, "Large-Scene Front-Lighted Holograms of a Human Subject," Proc. of the IEEE, 56, 7, pp. 1242-1243 (1968).

13. E. G. Champagne, "A Qualitative and Quantitative Study of Holographic Imaging," Technical Report AFAL-TR-67-107 (1967).

APPLICATION OF NON-LINEAR HOLOGRAM TO INTERFEROMETRY

Kazuo Sayanagi and Kazuya Matsumoto

Canon Inc.※, 30-2, Shimomaruko 3 Chome, Ohta-ku

Tokyo, Japan

Introduction

Holography has been applied to many fields of interferometry by virtue of its ability in recording phase information of light waves. Some remarkable works about this holographic interferometry have been already reported; for example, holographic shearing interferometry[1),2)], holographic schlieren method[3),4)], diffusely reflecting surface interferometry[5)-8)], holographic multiple-beam interferometry[9)-11)], measurement of OTF with holography, etc.

In the above-mentioned examples only faithfully reconstructed wavefronts are used and other wavefronts that appear at reconstruction are generally excluded because they are considered as noise[12)-15)]. However, it was found that these undesirable wavefronts are usable in holographic interferometry. Such wavefronts are "the conjugate wavefront with respect to the recorded wavefront" and "higher order diffracted wavefronts" caused by non-linear characteristics of photographic recording.

In this paper we will discuss three examples of the application of the "undesirable wavefronts" such as;

I) Phase-Difference Amplification by Non-Linear Holograms[17)].
II) Holographic Multiple-Beam Interferometry[10)].
III) Holographic Test Plate.

※ Mailing address : P.O. Box #50, Tokyo Airport

Introducing the first part of this report, the interference of a wavefront with its conjugate wavefront generated by a hologram has been worked by K. Matsumoto and T. Ose[18], and O. Bryngdahl[19]. The feature of this interferometry is that the interferogram has doubled sensitivity to detect the phase deformation of the wavefront compared with a conventional interferogram using a plane wavefront as a reference. Further improvement of sensitivity was achieved by using the higher order diffracted wavefronts by Bryngdahl and Lohmann[20]. They obtained interferograms of higher order diffracted wavefrontsby using a plane reference wavefront. A use of a higher order conjugate conbination was already suggested in Bryngdahl's recent paper[19]. The authors have been working along the same line to amplify the phase-difference of the recorded wavefront. In this paper we will show the experimental result of this work.

We can also obtain the multiple-beam interferograms using the "undesirable wavefronts" and this is the second part of this paper. In this case the higher order diffracted wavefronts and its higher order diffracted conjugate wavefronts are all used to form a multiple-beam interferogram.

Since a "hologram is an interferogram", we can make an artificial hologram by drawing it by hand or with a plotter. Accordingly, we can get an arbitrary wavefront from the artificial hologram. This is the third part of the present paper and this technique will be very important for an aspheric lens testing, because it is generally difficult to have a reference wavefront for testing an aspheric surface. It can be thought that the binary artificial hologram is a strongly non-linear hologram. Therefore from this hologram many higher order diffracted wavefronts are obtained. When we use such a higher order diffracted wavefront as a reference wavefront, we can reduce the number of fringes in the artificial hologram.

I. PHASE-DIFFERENCE AMPLIFICATION BY NON-LINEAR HOLOGRAMS

Principle

Let us suppose that there is a wave whose wavefront was deformed by an optical system or an optical element, and that the disturbance at the point (x,y) on the plane of a photographic plate is given by

$$U_t = a_t \exp ik\{\varphi(x.y) + x\sin\theta_t + \delta_t\}, \quad k = 2\pi/\lambda, \qquad (1.1)$$

where a_t is a constant absolute amplitude, $\varphi(x,y)$ is the optical path length given by the deformation, θ_t is the angle between the wavefront normal and the normal of the plate, δ_t is the retardation of the wavefront in this optical-path, and λ is the wavelength.

We make a hologram of this wavefront using a plane reference wave;

$$U_r = a_r \exp ik(x \sin\theta_r + \delta_r), \tag{1.2}$$

where a_r is a constant absolute amplitude, θ_r is the angle between the wavefront normal and the normal of the plate, and δ_r is the retardation in this reference path. Then the illuminance on the photographic plate becomes

$$I = |U_t + U_r|^2$$
$$= a_t^2 + a_r^2 + 2a_t a_r \cos k\{\varphi(x,y) + x\sin\theta_c + (\delta_t - \delta_r)\}, \tag{1.3}$$

where

$$\sin\theta_c = \sin\theta_t - \sin\theta_r, \quad \text{and} \quad \theta_c \neq 0 .$$

After processing the exposed photographic plate with a gamma that is greater than 2.0, the amplitude transmittance of the plate is described by

$$T = \sum_{m=-N}^{N} b_m \exp imk\{\varphi(x,y) + x\sin\theta_c + (\delta_t - \delta_r)\}, \tag{1.4}$$

where b_m is an absolute amplitude transmittance of mth-order diffracted wave from the hologram and is defined by a_t and a_r and also by the processing condition. In the above expression, the order of diffraction is taken into account up to N^{th} order. Then the hologram is illuminated by two plane waves that are incident on the hologram from opposite directions,

$$B = C_1 \exp ik\{x\nu\sin\theta_c + \Delta_1\} + C_2 \exp ik\{-x\mu\sin\theta_c + \Delta_2\}, \tag{1.5}$$

$$\nu, \mu = 0, 1, \cdots\cdots, N$$

where C_1 and C_2 are constant absolute amplitudes. With the above illuminating waves, we have diffracted waves,

$$D = BT = \sum_{m=-N}^{N} b_m (P_m + Q_m), \tag{1.6}$$

where

$$P_m = C_1 \exp ik\{m\varphi(x,y) + x(m+\nu)\sin\theta_c + m(\delta_t - \delta_r) + \Delta_1\} \tag{1.7}$$

and

$$Q_m = C_2 \exp ik\{m\varphi(x,y) + x(m-\mu)\sin\theta_c + m(\delta_t - \delta_r) + \Delta_2\}. \tag{1.8}$$

The angles of diffraction from the normal of the plate, $\theta_{i,j}$ are given by

$$\sin\theta_{m,\nu} = (m+\nu)\sin\theta_c \tag{1.9}$$

and

$$\sin\theta_{m,\mu} = (m-\mu)\sin\theta_c . \tag{1.10}$$

Now we select just two diffracted waves, which are diffracted by a hologram in the same direction, and make a two-beam interferogram. To select those waves from other diffracted waves a spatial filtering technique can be used with a small pinhole on a back focal plane of a lens which is introduced behind a hologram (Fig.1.1 (b)). For example, a hologram is illuminated by two illuminating waves with $\nu = \mu = N$ and we select by a filtering pinhole the waves which are diffracted to the perpendicular direction to a hologram plate, that is, $\theta_{m,\nu} = \theta_{m,\mu} = 0$. Then two selected waves are given by

$$V_1 = C_1 b_{-N} \exp ik\{-N\varphi(x,y) - N(\delta_t - \delta_r) + \Delta_1\} \tag{1.11}$$

and

$$V_2 = C_2 b_N \exp ik\{N\varphi(x,y) + N(\delta_t - \delta_r) + \Delta_2\} . \tag{1.12}$$

This means that these two waves have phases which are N times larger than that recorded in the hologram and also that they are longitudinally reversed from each other. Therefore, when these two waves are brought into interference and when $C_1 b_{-N} = C_2 b_N = b$, we then have

$$I_f = |V_1 + V_2|^2$$

$$= 2b^2[1 + \cos k\{2N\varphi(x,y) + 2N(\delta_t - \delta_r) - (\Delta_1 - \Delta_2)\}] . \tag{1.13}$$

When the path-differences $(\delta_t - \delta_r) = (\Delta_1 - \Delta_2) = 0$, it follows from Eq.(1.13) that

$$I_f = 2b^2[1 + \cos k\{2N\varphi(x,y)\}] . \tag{1.14}$$

Comparing the above with the illuminance of the interferogram of two-beam interference using waves of equal absolute amplitude b,

$$I = 2b^2[1 + \cos k\varphi(x,y)] , \tag{1.15}$$

we find that the optical-path difference $\varphi(x,y)$ or phase-difference $k\varphi(x,y)$, given by deformation of wavefront that is caused by an optical system or an optical element is amplified.

When there is a slight relative tilt between two illuminating waves, then the phase-difference caused by this tilt is given by

$$\Delta_1 - \Delta_2 = x \sin\alpha + y \sin\beta , \tag{1.16}$$

where α and β are angles between illuminating waves around y- and x-axis respectively. Then it follows from Eq.(1.13) with $\delta_t - \delta_r = 0$ that

$$I_f = 2b^2[1 + \cos k\{2N\varphi(x,y) - (x\sin\alpha + y\sin\beta)\}] . \tag{1.17}$$

Then it is possible to change the fringe orientation and to map contours of amplified optical-path differences.

Experiment

An experiment has been made with a Mach-Zehnder interferometer, arranged as shown in Fig.1.1 (a). A parallel plate S with striae was introduced in one of the optical paths of the interferometer as a test specimen. To avoid interference that might be caused by internal reflections in the beam splitters BS_1 and BS_2, Brewster's angle settings were used in the interferometer and the plane of polarization of the incident beam was adjusted with fixed Q_1 and adjustable Q_2 quarter-wave plates.

Recording. A photographic plate H was placed at the position shown in Fig.1.1 (a) and an interferogram, to be called the hologram hereafter, was exposed. To avoid the influence of lateral shrinkage of the processed photographic emulsion when reconstructing the wave, the carrier spatial frequency was selected as small as possible, that is about 5 lines/mm. Also, to avoid the effects of thickness shrinkage the two interfering beams were symmetrical with respect to the normal to the photographic plate.

In placing the photographic plate, care was taken so that an emulsion layer of the plate was faced toward an object. With this precausion the two diffracted waves, which were brought into interference later at reconstruction, went through the same portion of the base plate of the hologram, therefore the effect of striae in the base plate was cancelled out. Fuji Process Panchromatic plate was used to make the hologram and was processed with D-19 developer for 8 minutes.

Reconstruction. After processing, the plate was again placed in the apparatus (Fig.1.1 (b)) and the specimen S was removed. Then, on the back focal plane of lens L two sets of diffraction patterns were found. By adjustment of the angles of reflection at the mirror M_1 and beam splitter BS_2 two diffracted waves, which correspond to the parameters ν and μ in Eq.(1.5), were selected by a filtering pinhole P. In this way an interferogram that was made with ($\nu+\mu$) times amplified phase difference, was exposed in camera C.

Interferograms. The interferograms shown in Fig.1.2 were made in the manner described. To show the result of phase-difference amplification clearly, the interferograms were photographed with almost same fringe spacings adjusting tilt angles α and β. The interferogram in Fig. 1.2 (a) was made with 0th- and 1st-order diffracted waves, (b) with -1st and +1st, (c) with -3rd and +3rd, (d) with -5th and +5th, and (e) with -7th and +7th. The maximum magnification of the phase differences is 14 in Fig.1.2 (e).

The structure of striae in the test specimen, which is not remarkable in (a), is clear in (e). The maximum optical path difference caused by the striae amounts to 0.23 λ with an estimated error of $\pm$ 0.04 λ. These values were obtained by reading the fringe deviation in (e), with a reading accuracy of 1/4 of a fringe spacing.

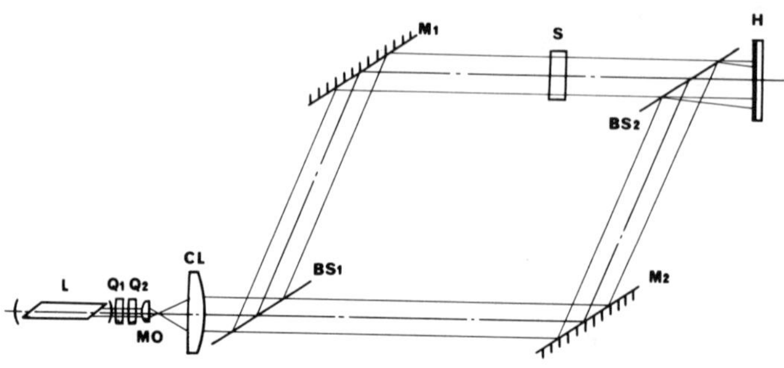

Fig.1.1 (a)

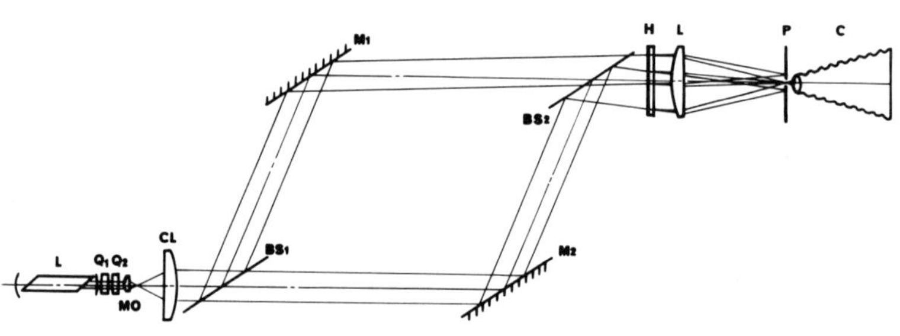

Fig.1.2 (b)

Fig.1.1. Experimental arrangements.
(a) for making a hologram and (b) for phase-difference amplification. L, He-Ne laser (0.6328μ); Q_1 and Q_2, fixed and adjustable quater-wave plates; MO, micro objective; CL, collimator lens; M_1 and M_2, mirrors; BS_1 and BS_2, beam splitters; S, specimen under study; H, Holographic plate (its emulsion layer faces towards the interferometer); L, lens; P, filtering pinhole; C, camera.

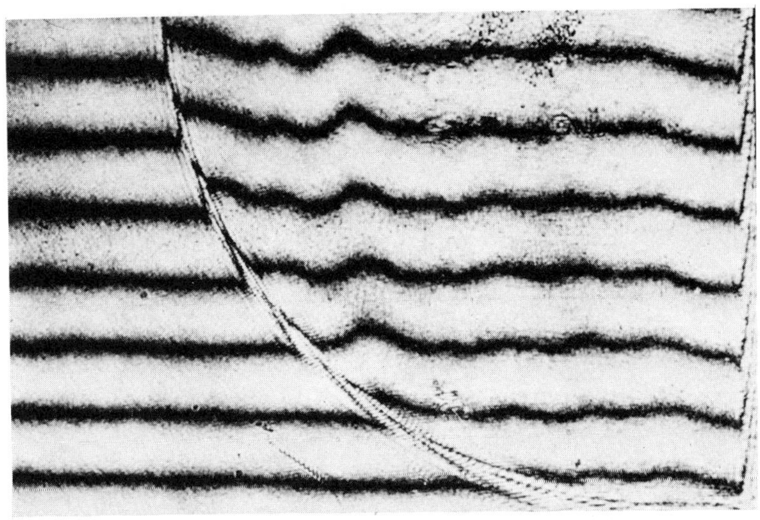

Fig.1.2 (a)

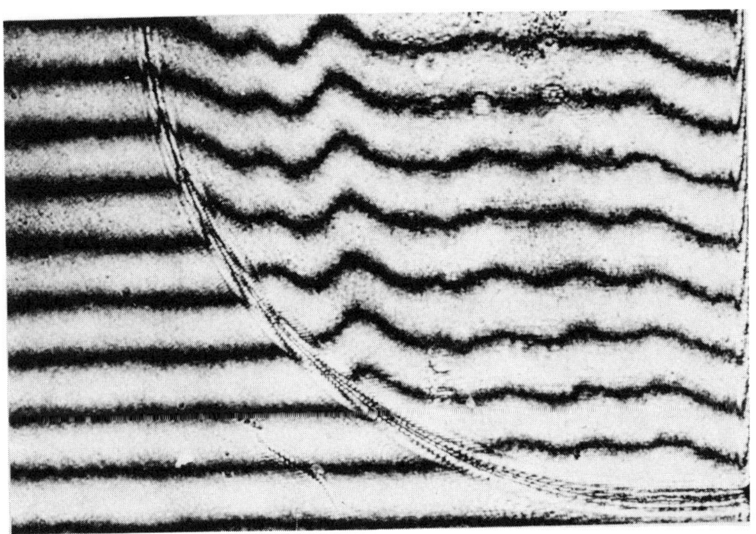

Fig.1.2 (b)

Fig.1.2. Interferograms exposed with diffracted waves from a hologram.
(a) with 0th and +1st order diffracted waves, (b) with -1st and +1st.

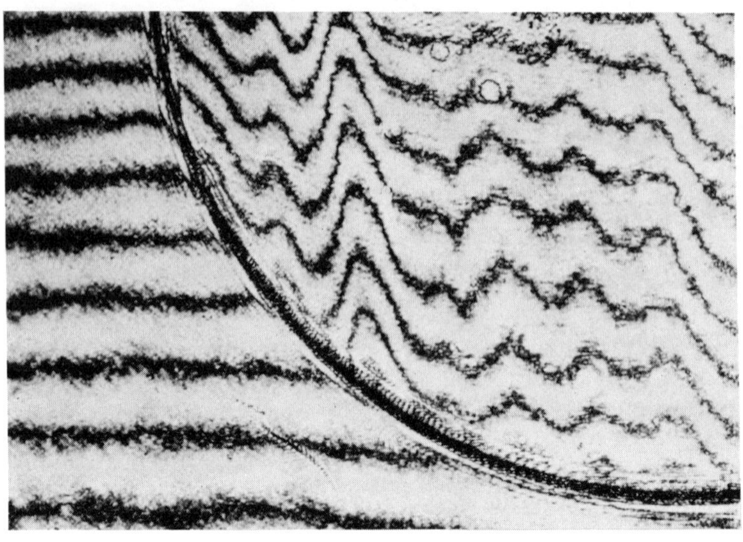

Fig.1.2 (c)

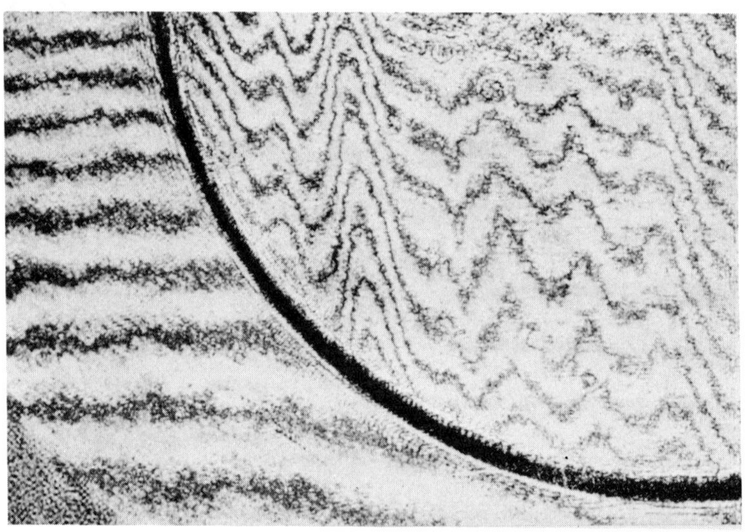

Fig.1.2 (d)

Fig.1.2. Interferograms exposed with diffracted waves from a hologram.
(c) with -3rd and +3rd, (d) with -5th and +5th.

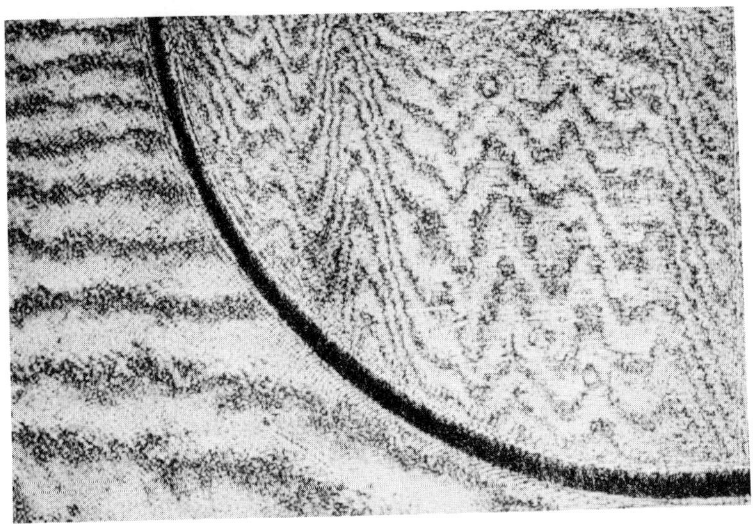

Fig.1.2 (e). Interferograms exposed with diffracted waves from a hologram with -7th and +7th.

Influence of Defects in Recording and Reconstruction System

As it is described in the forgoing section, the influence of striae in a hologram plate cancels out or is negligible when the emulsion layer of the plate is faced toward an object. However, any intrinsic optical path difference in the Mach-Zehnder interferometer, ($\delta_t - \delta_r$) influences the amplified results. It is amplified by the same magnification of optical-path difference $\varphi(x,y)$ as is shown in Eq.(1.13). Another source of error is the intrinsic optical-path difference $(\Delta_1 - \Delta_2)$ in the reconstruction system. This also depends on the tilt angles of the mirror M1 and a beam splitter BS2.

Adding these two errors for the case of the maximum order of diffraction N, we find that the total system error is

$$\varepsilon_s = 2N(\delta_t - \delta_r) - (\Delta_1 - \Delta_2) . \tag{1.18}$$

When the carrier frequency of the hologram is low, then $(\delta_t - \delta_r) \simeq (\Delta_1 - \Delta_2)$ and $\varepsilon_s = (2N-1)(\delta_t - \delta_r)$. If we can read a fringe position in an interferogram of 2N times amplified phase difference with the accuracy up to 1/e of the fringe spacing, then the reading error is $\varepsilon_r = \frac{\lambda}{2Ne}$ and the total error is

$$\varepsilon = \varepsilon_r + \varepsilon_s \simeq (2N-1)(\delta_t - \delta_r) + \lambda/2Ne . \tag{1.19}$$

To make the system error less than the reading error, it is necessary to fulfill the condition

$$\delta_t - \delta_r < \lambda/2N(2N-1)e . \tag{1.20}$$

For the striae shown in Fig.1.2 (e), the maximum deviation of a fringe was read up to a quarter of a fringe spacing (e=4) and $\varepsilon_r = \pm\lambda/56$. The system error was roughly estimated from the local deviation of background fringes and $\varepsilon_s = \pm\lambda/56$, which corresponds to ±1/4 fringe spacing. Therefore the total error amounts to $\varepsilon = \pm\lambda/28$.

Features of This Method

With a photographic plate exposed to fine interferometric fringes and processed to high gamma, it is possible to amplify the original phase-difference recorded in the plate. Hence we can read with better accuracy interferograms obtained even with transmission objects. In such a case the feasibility of multiple-beam interferometry is limited by the walk-off effect caused by large spacing of reflection surfaces. This technique can be applied to testing wave front aberration of an optical system, and to testing surface figure and surface irregularity of optical elements. Especially the Twyman-Green interferometer has advantages in that it enables us to have higher magnification by virtue of the doubled paths in the interferometer, and also to make use of the surface or optical system under investigation themselves for the amplification of phase-difference.

However, there is a disadvantage in this technique; the intrinsic phase-difference of the interferometer is also amplified with the same magnification as that of the wavefront deformation under study. Therefore it is necessary to have good-quality mirrors and beam splitters in the interferometer in which the original interferogram is made. Otherwise, it is not possible to make the total experimental error small, no matter how much the reading error is reduced.

II. HOLOGRAPHIC MULTIPLE-BEAM INTERFEROMETRY

Introduction

Two-beam interferometry, which enables us to measure the phase variations of an object, is an old and well-established branch of interferometry. However for precise measurement, it is desirable to have sharper and narrower fringes than sinusoidal two-beam fringes, because sharper fringes give more exact fringe location. Such fringes can be obtained with multiple-beam interferometry.

Burch et al[9)] attempted to obtain sharper and narrower fringes with holographic interferometry. Their method was restricted to objects that can be bent. It has been found that a multiple-beam interference effect can be introduced by the non-linear characteristics of photographic materials and that this effect can be applied to measure the phase variations of any object.

Principle

Suppose that the object is a pure phase object, with the phase $k\varphi(x,y)$, that the reference wave is a plane wave which is represented by $\exp[ikx\sin\theta]$, and also that the recording is non-linear, then the following wavefronts are generated from the hologram[12),13),14)],

$$\sum_{m=-\infty}^{\infty} a_m \exp(i\delta_m)\cdot \exp imk[x\sin\theta + \varphi(x,y)] . \qquad (2.1)$$

In Eq.(2.1) a_m denotes amplitude ratio of the mth-order diffracted wavefront and δ_m represents the phase retardation of the mth-order diffracted wavefront at diffraction. If the transmittance of the hologram is real, the retardation δ_m takes only value 0 or π . However, when the transmittance of a hologram is complex, it is found in our experiment that the retardation is not restricted to 0 or π , but takes the values between 0 and 2π . Equation (2.1) shows that the mth-order diffracted wavefront has m times larger phase than that originally recorded, and propagates in the direction $m\theta$ when θ is small. If these wavefronts could be superposed in one direction, then we can obtain multiple-beam interference fringe. The total irradiance of these superposed wavefronts can be written as,

$$I(x,y) = \left|\sum_{-\infty}^{\infty} a_m \exp[mik\varphi(x,y) + \delta_m + \varepsilon_m]\right|^2, \qquad (2.2)$$

where ε_m is the phase variation which appears in each diffracted wave when all of the propagation directions of higher order diffracted wavefronts are brought into the same direction. The term $mkx\sin\theta$ which is related to the propagation direction does not remain in the above expression of irradiance, since each wavefront is brought in to propagate in the same direction. Generally the fringe irradiance distribution given by the above expression does not become sharp. However, we can make it sharper than sinusoidal two-beam fringe by taking ε_m properly with respect to the values δ_m and a_m , which are inherent to the hologram.

Method of Superposition

We used a diffraction grating to superpose all diffracted wavefronts in the same direction.
A grating having the same spacing as that of the hologram diffracts each wavefront that is generated from the hologram and it brings wavefronts of every order to travel in the same direction, as shown in Fig.2.1 (only the wavefronts in the +1-order direction are described). In this case, the phase variation ε_m is related to the spacing between the hologram and the grating, and it is expressed

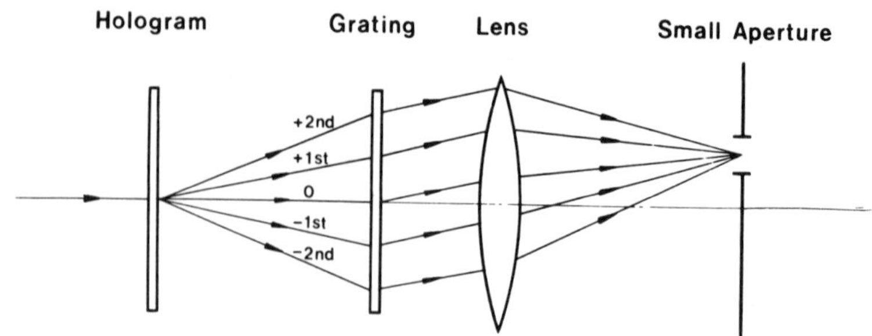

Fig.2.1. Optical system for holographic multiple-beam interferometry.

by the following relation,

$$\varepsilon_m = \frac{1}{2} k d m^2 \sin^2\theta \quad . \tag{2.3}$$

The deduction of this expression is given in Appendix. Here we assume that the complex transmittance of a grating is given by

$$T_G = \sum_{n=-\infty}^{\infty} G_n \exp(iknsin\theta), \tag{2.4}$$

where

$$G_n = g_n \exp(i\delta_n') \quad .$$

δ_n' is the phase reterdation at diffraction. Then the total irradiance observed from the jth direction of diffraction grating can be written as

$$I_j(x,y) = \left| \sum_{m=-\infty}^{\infty} a_m g_{j-m} \exp[mik\varphi(x,y) + \delta_m + \delta_{j-m}' + \varepsilon_m] \right|^2 . \tag{2.6}$$

By this grating method many higher order waves are diffracted from the grating, therefore we apply a collimator lens and a small aperture as a spatial filter shown in Fig.2.1. Through the small aperture we can observe the holographic multiple-beam fringe. It is possible to vary a fringe pattern, as in ordinary interferometry, by giving a slight tilt to a testing wavefront with respect to a reference wavefront. The fringe without tilt can be obtained when the orientations of the fringes in the hologram and of the grating are adjusted to be parallel. Tilting effect in x and y are obtained by varying the grating constant and by rotation of the grating around the normal to the hologram, respectively.

Experimental Results

We used a glass wedge as a test object. Figure 2.2 shows the two-beam interference fringes of the object observed with Mach-Zehnder interferometer. This photograph was processed with the same development condition applied to make non-linear holograms. Therefore this fringe has not sinusoidal but rather rectangular distribution. The fringes in non-linear hologram have the same density distribution as in Fig.2.2.

The hologram was made using the same optical arrangement shown in Fig.1.1. A hologram, which was photographed by the same optical set-up without an object, was used as a grating. Figure 2.3 shows multiple-beam fringe observed from +1-order direction when d=18mm. One can recognize the difference of fringe sharpness between the holographic multiple-beam fringe and an ordinary two-beam interference fringe. Figure 2.4 shows the tilted fringe obtained with slight rotation of the grating used for Fig.2.3.

We can vary the irradiance distribution of the multiple-beam fringe by changing the spacing d between the hologram and the grating. Some examples are shown in Fig.2.5. Figure 2.5 (a) shows the fringes observed from 0-order direction when d=2.0mm and they are similar to those that can be obtained by "Äquidensitometrie[21]". Figure 2.5 (b) shows the doublet fringes observed from 0-order direction when d=4.0mm. Figure 2.5 (c) is the triplet fringes observed from the +1-order direction when d=5.5mm. Figure 2.5 (d) shows half period fringes observed from 0-order direction when d=18mm.

In these experiments the carrier frequency of a hologram is taken about 5 lines/mm, and Fuji Process Panchromatic Plate **was** used as photographic plate and processed with D-19 developer for 8 minutes. Up to ±8th order diffracted waves were observed from the above processed hologram with an unaided eye.

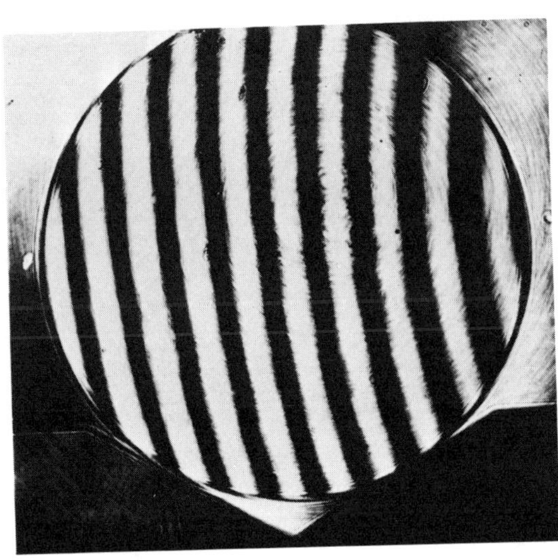

Fig.2.2. Two-beam interference fringes of the object observed with Mach-Zehnder interferometer.

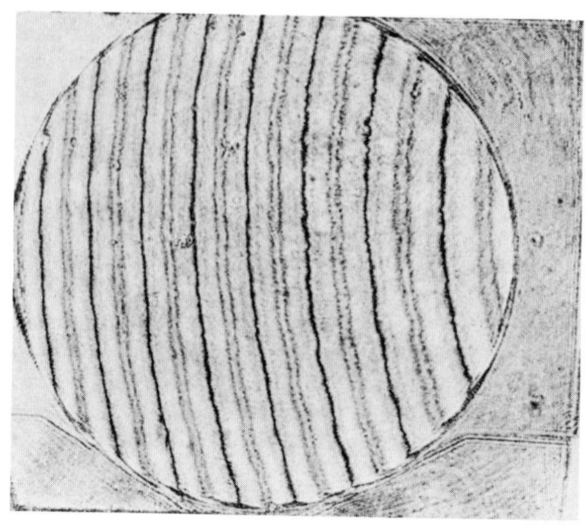

Fig.2.3

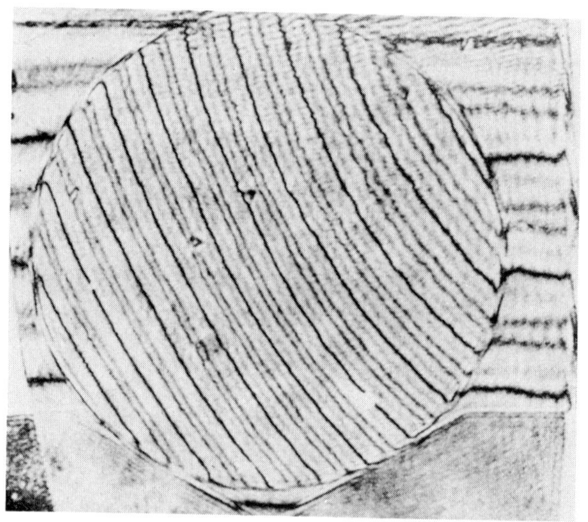

Fig.2.4

Fig.2.3. Holographic multiple-beam fringe observed from +1-order direction when d=18mm. Fig.2.4. Tilted fringe obtained with slight rotation of the grating used for Fig.2.3.

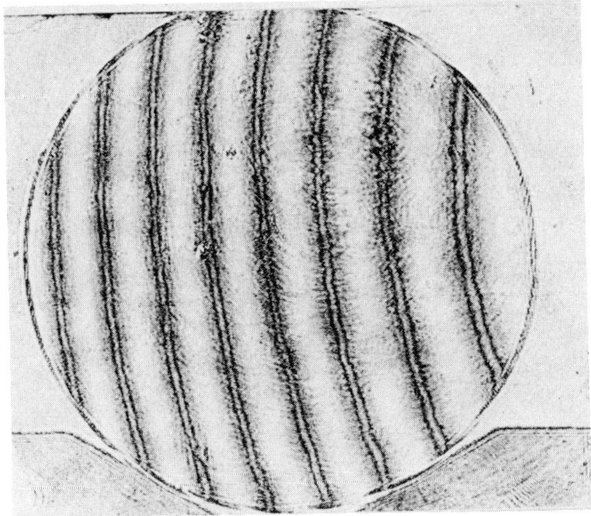

Fig.2.5 (a)

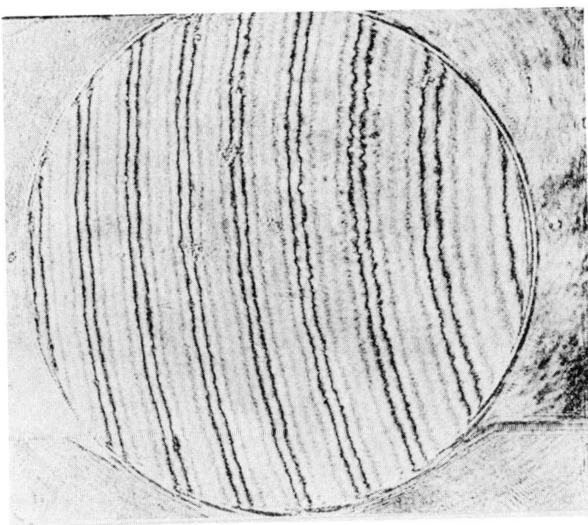

Fig.2.5 (b)

Fig.2.5. Holographic multiple-beam fringe obtained by varying the separation between a hologram and a grating. (a) Fringes observed from 0-order direction when d=2.0mm and they are similar to those that can be obtained by "Äquidensitometrie". (b) Doublet fringes observed from 0-order direction when d=4.0mm.

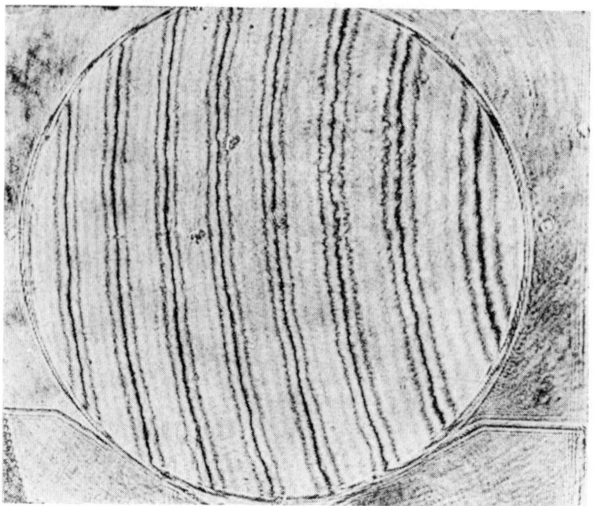

Fig.2.5 (c)

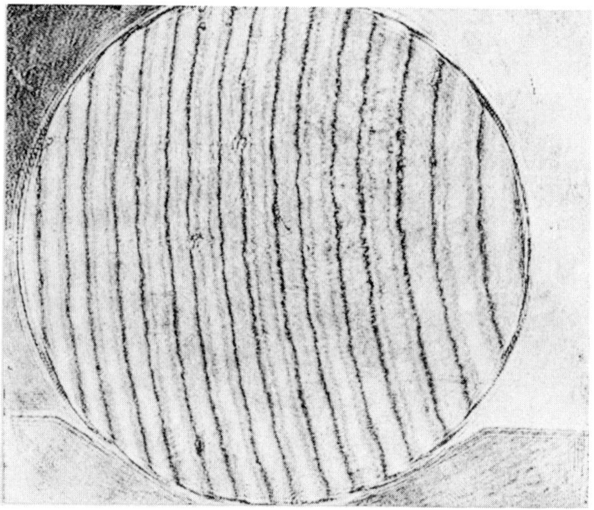

Fig.2.5 (d)

Fig.2.5. Holographic multiple-beam fringe obtained by varying the separation between a hologram and a grating. (c) Triplet fringes observed from the +1-order direction when d=5.5mm. (d) Half period fringes observed from 0-order direction when d=18mm.

III. HOLOGRAPHIC TEST PLATE

Introduction

A use of optical test plate and an observation of Newton fringes are common practices to test polished lens at optical shops. With this technique, we can check the deviation of a finished lens surface from designed surface figure. It is also possible to measure this deviation with an artificial hologram which functions like a test plate. For this reason the hologram is named "holographic test plate" in this paper. There are two applications of holographic test plates and they can be used as

(I) wavefront generator, and
(II) aberration corrector of a lens under inspection.

In the former case we compare the wavefront from the lens under study with a generated wavefront from the holographic test plate. In the latter case the aberration of a lens under inspection is compensated with the holographic test plate and the aberration corrected wavefront is compared with a plane wave to detect residual wavefront deviation. In both cases the deviation of a lens figure from that designed can be obtained with interference fringes as it is done with Newton fringe.

To make an artificial hologram, it is desirable to reduce the number of fringes in the hologram. If it is binary, many higher order diffracted waves are obtained from this hologram since it can be seen as a strongly non-linear hologram. When we use the higher order diffracted wave, it is possible to reduce the number of fringes in the artificial hologram. In our study the artificial hologram of a reduced zone number was used. The principle of the reduction of zone number in the hologram and the interferograms obtained with holographic test plates will be shown.

Principle

Suppose that the wave from the lens without polishing error is described by $\exp[ik\varphi(x,y)]$, and that the optical path difference caused by polishing error is defined by $\Delta\varphi(x,y)$, where (x,y) is the coordinate of the point on a holographic test plate, then the wave from the lens under inspection can be written as

$$\exp ik[\varphi(x,y) + \Delta\varphi(x,y)] . \qquad (3,1)$$

Now, we consider the test plate whose transmittance is given by

$$T = 1 + 2\cos[ik\varphi(x,y)] . \qquad (3,2)$$

When this test plate is illuminated by the wave from the lens under inspection, the diffracted waves from this test plate are given by

$$T \exp ik[\varphi(x,y) + \Delta\varphi(x,y)]$$
$$= \exp ik[\varphi(x,y) + \Delta\varphi(x,y)] + \exp ik[2\varphi(x,y) + \Delta\varphi(x,y)]$$
$$+ \exp ik\Delta\varphi(x,y) . \qquad (3,3)$$

In the above expression there is a term which contains only the polishing error $\Delta\varphi(x,y)$. Therefore if we could separate a wave defined by the term from others and superpose this wave with a plane reference wave, then we can see the polishing error as fringe pattern. The method of separating the interference wave will be described later.

A binary holographic test plate for testing the same lens, which is defined by Eq.(3.1), can be written as follows using Eq. (1.4), and setting $\theta_c = (\delta_t - \delta_r) = 0$;

$$T_B = \sum_{m=-\infty}^{\infty} b_m \exp ikm\varphi(x,y) . \qquad (3.4)$$

If we sample the zones in the test plate and reduce the number of zones to $1/\ell$ times that of original binary hologram, then the transmittance of the sampled test plate becomes

$$T_B' = \sum_{m=-\infty}^{\infty} b_m' \exp[ik\, {}^{m}\!/_{\ell}\, \varphi(x,y)] . \qquad (3.5)$$

When the sampled test plate is illuminated by the wave from the lens under inspection, we have diffracted waves which are given by

$$T_B' \exp ik[\varphi(x,y) + \Delta\varphi(x,y)]$$
$$= \sum_{m=-\infty}^{\infty} b_m' \exp ik[({}^{m}\!/_{\ell} + 1)\varphi(x,y) + \Delta\varphi(x,y)] . \qquad (3.6)$$

In the above series we find the term with m = -1 and it is

$$b_{-\ell}' \exp ik\Delta\varphi(x,y) . \qquad (3.7)$$

This wave contains only the polishing error, therefore we can evaluate the error with an interferogram when the wave given by Eq.(3.7) is separated from other waves and it is brought in to interfere with a plane wave.

Fresnel Zone Type Hologram

In the previous section an important problem has not be mentioned, ... that is, how to separate the wave which is brought into interference from the others. There are two types of holograms, such as

(I) on-axis hologram originated by Gabor[22], and
(II) off-axis hologram originated by Leith and Upatnieks[23].

In the former case all the diffracted waves go to the same direction as shown in Eq.(3.3), and the separation of one wave from the others is difficult. In the latter case the separation is easy as shown in chapter I and II. However, a lens has generally rotationally symmetrical shape, and therefore the zones in the test plate are also rotationally symmetrical. This symmetry makes preparation easier of a test plate with just one-dimensional data of zone spacings than for an off-axis hologram. For this reason on-axis type holographic test plates have been selected.

These test plates can be made supposing the following analogous cases of preparation of hologram.

(I) A plane wave, which is deformed by the aberration of a lens, interfereswith a spherical reference wave, and

(II) A spherical wave, which is deformed by the lens interferes with a plane reference wave.

In both cases the test plates resemble ordinary Fresnel zone plates in shape, but differ from their spacings which are specified by the wavefront aberration of a lens under inspection.

Since the second type of holographic test plate has been applied for testing lens surface figures, the separation of single diffracted waves generated by this type of test plate will be described hereafter.

Suppose that the wave from a lens under inspection can be written as

$$\exp ik[(x^2+y^2)/2R + \psi(x,y) + \Delta\varphi(x,y)], \qquad (3.8)$$

where R is the radius of the approximate spherical wave, $\psi(x,y)$ is the wavefront aberration, which is calculated from lens design data and $\Delta\varphi(x,y)$ is the polishing error. If the holographic test plate for this lens is made binary and the 1st order diffracted wave from this plate is used for testing the lens, then using Eq.(3.5) the transmittance of the plate is written as

$$T_B' = \sum_{m=-\infty}^{\infty} b_m' \exp ik\, m/\ell\, [(x^2+y^2)/2R + \psi(x,y)]. \qquad (3.9)$$

When the plate is illuminated by the wave from the lens under inspection, we have the diffracted waves given by

$$\sum_{m=-\infty}^{\infty} b_m' \exp ik\{(m/\ell+1)[(x^2+y^2)/2R + \psi(x,y)] + \Delta\varphi(x,y)\}. \qquad (3.10)$$

In the above series the term with m = -1 is the wave which is used for testing the lens. The above expression shows that each diffracted wave has its own curvature in its wavefront. Therefore each wave images on a different plane as shown in Fig.(3.1). At the plane on which the -1st order diffracted wave focuses the other waves are defocused and make blurred point images. If the size of the defocused images are much larger than the width of the point

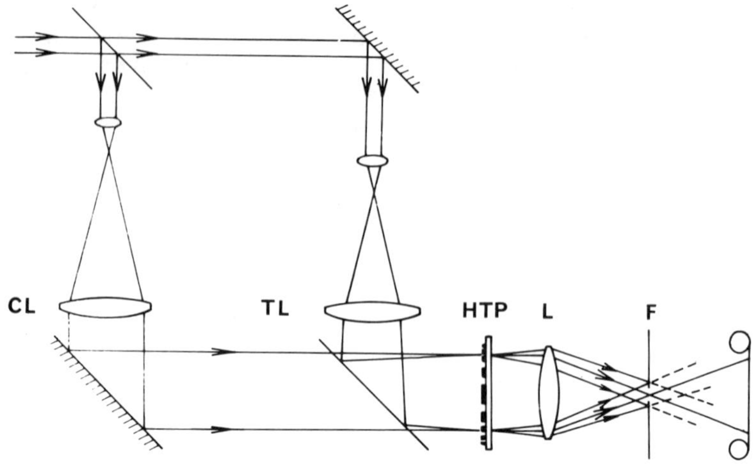

Fig.3.1. Experimental arrangements, HTP, holographic test plate; TL, lens under inspection; CL, collimator lens for a reference wave; F, small aperture for filtering.

spread function made with the test wave which is influenced by polishing error, then we can separate the test wave from others using a small aperture as a spatial filter. To apply this type of separation of test wave we must determine the radius R beforehand. When the filtered -1st order diffracted wave is brought into interfere with a plane wave which passes directly through the holographic test plate and also the small aperture, we can observe the interference fringe through the aperture. The polishing error $\varphi(x,y)$ of the lens makes the fringe irradiance distribution

$$2 + 2\cos[K\Delta\varphi(x,y) + x\sin\alpha + y\sin\beta] , \qquad (3.11)$$

where b'm and the amplitude of a plane reference wave in Eq.(3.10) were set to be unity, and α and β are wedge angles between the filtered wave and the reference wave. The reason why the plane wave is let to pass through the hologram is that both waves suffer from the effect of the strie in the test plate and that the effect cancels out at interference.

Experiment

Holographic Test Plate. The original chart which is 50mm wide and 600mm long fan shape was drawn by hand and it was photographed on a plate. During exposure the original chart was rotated. The reduction ratio of original chart is 1/10 and the radius of the prepared holographic test plate is 32.0mm. The amount of the radius of wavefront curvature, which R is in Eq.(3.8), is taken about 10m. The number of zones in the test plate is reduced to 1/3 of complete zone plate and it is 42. This prepared holographic test plate is shown in Fig.3.2.

Optical Arrangement. In Fig.3.1 the optical arrangement used for testing lens surface is shown, and HTP and TL denote the holographic test plate and the lens under inspection respectively. The test plate is illuminated by two waves, that is, a wave from the test lens and a plane wave from a collimator lens CL. A small aperture F for filtering is placed on the back focal plane of a lens L, and through the aperture the interference fringes which show polishing error is observed.

Result. Figure 3.3 shows the interference fringes obtained with the above mentioned optical arrangement. Figure 3.3 (a) shows tilted fringes in x direction and (b) corresponds to tilted fringes in y direction.

Fig.3.2. Holographic test plate used in our experiment.

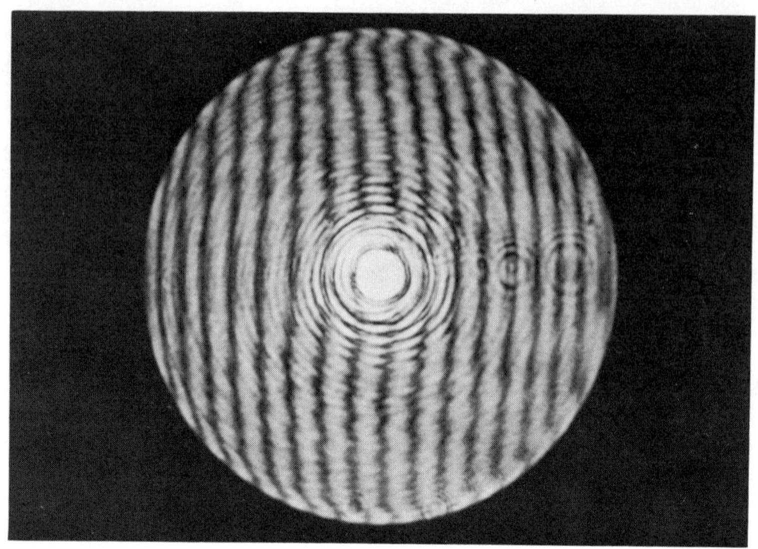

Fig.3.3 (a)

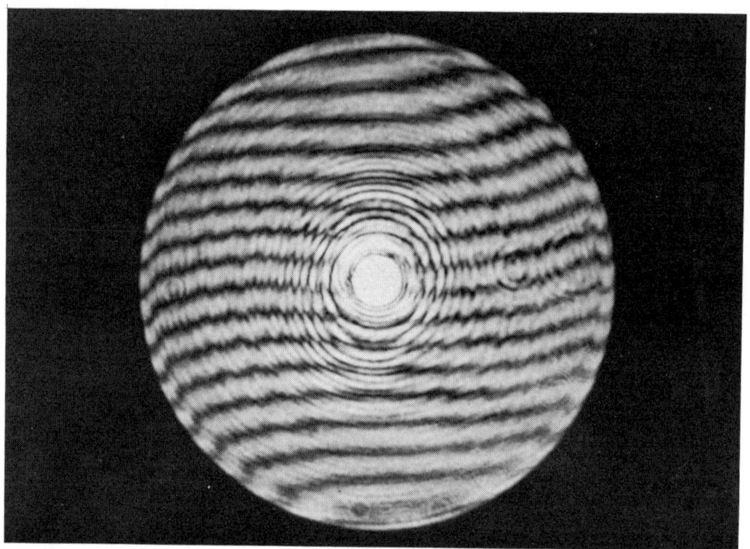

Fig.3.3 (b)

Fig.3.3. Interference fringes obtained with the optical arrangement shown in Fig.3.1.
(a) Tilted fringe in x direction. (b) Tilted fringe in y direction.

Conclusion

In this paper we discussed our subjects in rather restricted ways because of our demands to establish the methods of higher accuracy in optical work shop for interferometric measurements. We found that applications of non-linear hologram promise us future possibility to achieve practical work shop procedure, and especially that our method to amplify phase-difference could be used widely.

From the above practical point of view, we have to work continuously for instrumentation, e.g. to build convenient interferometers which realize our discussion in this paper easily under many different circumstances.

Studies on holographic materials and processing will be needed as the next step. To obtain relatively stronger higher order diffractions, we found that the combination of Fuji Process Panchromatic plate and D-19 developer was the best among several different combinations of materials and processings. Our experiences were limited and there will be still many other combinations which will enable us better results. For instance, a use of the adjaciency effect in development seems one of the interesting subjects to be studied.

Furthermore, applications of higher order diffracted wavefront from the non-linear holograms for different purposes are also very interesting. Lohmann and Paris[25] discussed the use of higher orders in binary holograms. A future possible use of this concept is not yet well understood, but it seems to be an interesting subject.

Acknowledgments

Parts of these works in this paper were done by our colleagues, Dr. Matsuo Takashima and Mr. Takashi Suzuki. We are grateful to their support and helpful discussions during the work, and also during the preparation of this manuscript.

APENDIX

Here the expression for the phase retardation ε_m, which is shown in Eq.(2.3) is derived. This retardation occurs because the diffracted wave from a grating travels a finite distance before the second diffraction takes place by another grating. Suppose that the transmission of the first grating is written as

$$G(x) = \sum_{m=0}^{\infty} a_m \cos(k m u_0 x),$$

where u_0 is the fundamental frequency of the grating. When this grating is illuminated by a plane wave, many plane waves are diffracted in different directions. On the plane which is distance d apart from the grating those waves are superposed on each other after suffering from the phase retardation ε_m. The resultant disturbance on the plane is derived with Fresnel diffraction formula and

$$U(x_1) = \frac{1}{\sqrt{\lambda d}} \exp(ikd) \exp(ikx_1^2/2d) \cdot \int_{-\infty}^{\infty} G(x) \exp(ikx^2/2d) \exp(-ikxx_1/d)\, dx \qquad (A.2)$$

$$= \frac{1}{\sqrt{\lambda d}} \exp(ikd) \exp(ikx_1^2/2d) \cdot \int_{-\infty}^{\infty} \frac{1}{2} \sum_{m=0}^{\infty} [a_m \exp(ikmu_0x) + a_m \exp(-ikmu_0x)] \cdot \exp(ikx^2/2d) \exp(-ikxx_1/d)\, dx \qquad (A.3)$$

In the above Eq.(A.3) the integral can be calculated as

$$\int_{-\infty}^{\infty} \exp(\pm ikmu_0x) \exp(ikx^2/2d) \exp(-ikxx_1/d)\, dx$$

$$= \int_{-\infty}^{\infty} \exp(ikx^2/2d) \exp[-ikx(x_1/d \mp mu_0)]\, dx$$

$$= \sqrt{\lambda d} \exp(i\pi/4) \exp(-ikx_1^2/2d) \exp[i\tfrac{k}{2}(2ux_1 - dm^2u_0^2)]. \qquad (A.4)$$

Substituting Eq.(A.4) into Eq.(A.3), we have

$$U(x_1) = \exp i(kd + \tfrac{\pi}{4}) \sum_{m=0}^{\infty} a_m \cos(kmu_0x_1) \exp(-i\tfrac{k}{2}dm^2u_0^2). \qquad (A.5)$$

From the above Eq.(A.5) it follows that each diffracted wave which is generated by the grating has the phase retardation

$$\Delta\varphi_m = kd + \pi/4 - k/2\, dm^2u_0^2. \qquad (A.6)$$

Then the phase difference in each higher order diffracted wave with respect to the 0th order wave is given by

$$\varepsilon_m = \frac{k}{2} d m^2 u_0^2. \qquad (A.7)$$

In chapter II, $u_0 = \sin\theta_c$, then

$$\varepsilon_m = \frac{k}{2} d m^2 \sin^2\theta_c. \qquad (A.8)$$

REFERENCES

1). O. Bryngdahl, J. Opt. Soc. Am. 58, 865 (1968).
2). S. Mallick and M. L. Roblin, Appl. Phys. Lett. 14, 61 (1969)
3). J. B. Story, G. S. Ballard and R. H. Gibbons, J. Appl. Phys. 37, 2183 (1966).
4). T. Tsuruta and Y. Itoh, Japanese J. Appl. Phys. 8, 96 (1969).

5). B. P. Hildebrand and K. A. Haines, Appl. Opt. 5, 172 (1966).
6). K. A. Haines and B. P. Hildebrand, Appl. Opt. 5, 595 (1966).
7). B. P. Hildebrand and K. A. Haines, J. Opt. Soc. Am. 57, 155 (1967).
8). N. Shiotake, T. Tsuruta and Y. Itoh, Japanese J. Appl. Phys. 7, 904 (1968).
9). J. M. Burch, A. E. Ennos and R. J. Wilton, Nature 209, 1015 (1966).
10). K. Matsumoto, J. Opt. Soc. Am. 59, 777 (1969).
11). O. Bryngdahl, J. Opt. Soc. Am. 59, 1171 (1969).
12). A. A. Friesem and J. S. Zelenka, Appl. Opt. 6, 1755 (1967).
13). J. W. Goodman and G. R. Knight, J. Opt. Soc. Am. 58, 1276 (1968).
14). O. Bryngdahl and A.W. Lohmann, J. Opt. Soc. Am. 58, 1325 (1968).
15). A. Kozma, Optica Acta 15, 527 (1968).
16). K. Matsumoto and T. Ose, Japanese J. Appl. Phys. 7, 621 (1968).
17). K. Matsumoto and M. Takashima, J. Opt. Soc. Am., to be published.
18). K. Matsumoto and T. Ose, Seisan-Kenkyu (Monthly Journal of Institute of Industrial Science, Univ. of Tokyo) 19, 18 (1966).
19). O. Bryngdahl, J. Opt. Soc. Am. 59, 142 (1969).
20). O. Bryngdahl and A. W. Lohmann, J. Opt. Soc. Am. 58, 141 (1968).
21). E. Lau and W. Krug, "Die Äquidensitometrie (Akademic-Verlag, Berlin, 1957).
22). D. Gabor, Proc. Roy. Soc. A197, 454 (1949).
23). E. Leith and J. Upatnieks, J. Opt. Soc. Am. 53, 1377 (1963).
24). J. W. Goodman, "Introduction to Fourier Optics" (MacGrawhill 1968).
25). A. W. Lohmann and D. P. Paris, Appl. Opt. 6, 1739 (1967).

DIGITAL PICTURE PROCESSING AND HOLOGRAPHY

Yoshiki Ichioka, Masaharu Izumi and Tatsuro Suzuki

Department of Applied Physics, Faculty of Engineering
Osaka University, Suita, Osaka, Japan

1. Introduction

Digital picture processing by an electronic computer has heretofore occupied the important field in the image science technology. Backgrounds in this field seem to exist in the development of large computers and the requirement from the space science[1]. Digital picture processing has, recently, been required in many other fields such as medical research, electron microscopy, coherent optical information processing, the real time picture processing and so on.

As it is well known, optical systems act as two-dimensional analog computers. Coherent or incoherent optical information processing seems to be mainly explained with the help of this concept. Spatial filtering, image restoration and holography are typical examples. Merits in picture processing by an ordinary optical technique are as follows. First, the detection and memory of optical signals can be achieved at the same time. Second, the data accumulation is very large. Third, optical signals are treated two-dimensionally, so that the processing speed is very fast. However, the actual technique has unavoidable faults such as precise control of the brightness is very difficult. Difficulty of fabrication of complicated complex filters which control the amplitude and phase components simultaneously and the difficulty of restoration from degraded image are another examples. These demerits seem to be taken off with the help of electronics and computer operation in some extent.

Advantages obtained by using a digital computer for picture processing are large flexibility and high degree of precision in it. It seems that there are many problems in picture processing which may be solved by combining ordinary optical techniques,electronics and computer operations. If so, it is expected to compensate their faults mutually. Also there will be another possibility to develop new technique that is very useful, but has not been done yet, though it should be expected in future. If the computer action of an optical system is substituted by that of a digital computer, it will also be possible to simulate optical information processing using no real optical system. In this paper, we deal with such a field in which a digital computer is used together with an ordinary optical system.

Pictures inherently consist of two-dimensional analog signals concerned with position, brightness and color. However, data obtained from a computer are sequential digital values, so that they must be transduced to two-dimensional analog signals - that is, continuous-tone pictures. For this purpose, we have developed the CRT display called "Halftone Plotter" which operates as off-line with a computer. Digital data are displayed as two-dimensional patterns on CRT by this plotter. This plotter can display any discrete continuous-tone patterns by an irregular scanning as well as by a regular scanning such as the TV system. Capability of displaying pictures with wide brightness levels has enabled us to promote study on digital picture processing. This flexibility of display apparatus is useful for our study to handle the information from the digital computer. A digital computer and the "Halftone Plotter" are main experimental devices in our study.

On the other hand, the reduction of computer time of the Fourier transform by the new algorithm, so-called "the Fast Fourier Transform[2]" (FFT) has reduced difficulties in digital picture processing. The new algorithm has made it possible to compute in a short time the Fourier spectrum of objects, convolution integrals, correlations and so on by using even a middle class computer. In this paper, we report our technique on digital picture processing and some applications. First, the outline of the "Halftone Plotter" developed by us is explained. Second, we describe computer generated holograms. Here we represent the principle of digitized continuous-tone hologram with the help of the notion of the binary hologram[3]. Digitized holograms are displayed as the sequence of two dimensional continuous-tone patterns by the "Halftone Plotter". Some experimental results are shown at the same time. A complex spatial filter is also obtained with the help of technique constructing digitized hologram. Third, the basic concept of digital processing of Fourier optics is described. Then the computer-simulation of spatial filtering and the image synthesis are treated as examples. Finally some experimental results are shown. In our study, the middle class computer NEAC 2206 was used.

2. Halftone Plotter

The "Halftone Plotter" developed by us is inherently a three parameters (X, Y, Z) CRT display system. This device is the improved one that we had developed previously for plotting spot diagrams[4]. Digital signals X and Y representing coordinate numbers and Z representing the brightness of individual points are converted into corresponding analog values. Two converted analog signals concerning X and Y are fed to the horizontal and vertical axes of a cathode ray tube. The beam deflection can be controlled to 2^{12} discrete horizontal and vertical positions. The brightness of individual points can be displayed by 2^8 discrete levels. Flexibility of the spot brightness seems to be enough to display a continuous pattern for digital picture processing. This device operates as off-line with a computer. The data processing speed is about 20 points per second, which is limited by a photo-tape-reader (PTR).

Figure 1 shows the block diagram of the "Halftone Plotter". This device consists of a photo-tape-reader, two blocks of discriminators and gate circuits, three shift registers and D-A converters, and analog gate and amplifier. Digital signals punched on the paper tape sequentially are read by PTR according to the order of X, Y and Z. These digital data are discriminated into X, Y and Z signals by two blocks of discriminators after shaping the pulse forms. These digital signals are converted to corresponding analog values (voltages) through individual shift registers and D-A converters. Converted ana-

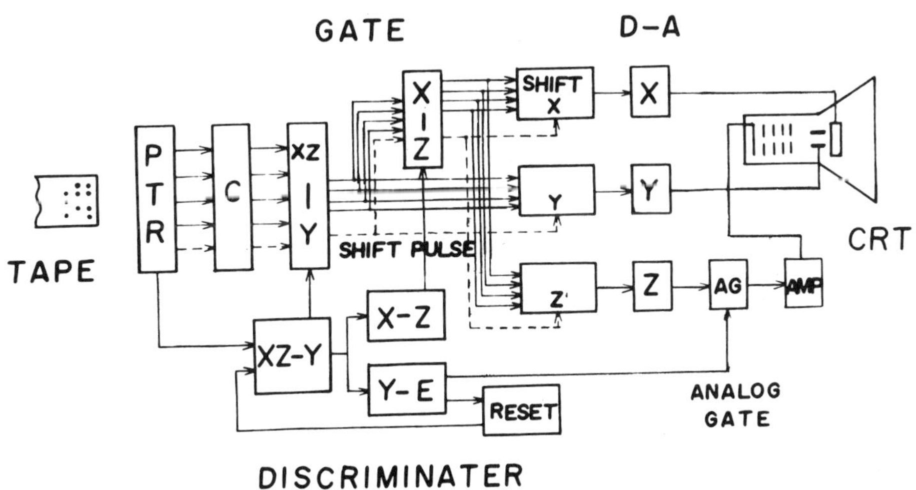

Fig. 1 Block Diagram of the "Halftone Plotter".

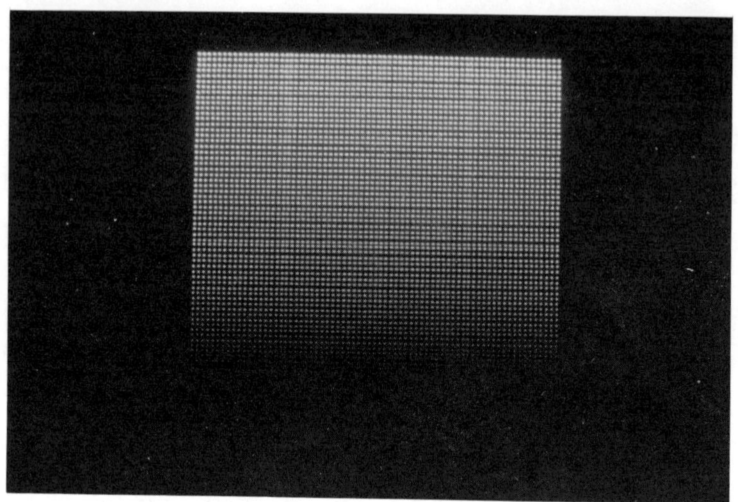

Fig. 2 Test pattern with 64 x 64 points and 64 brightness levels plotted by the "Halftone Plotter".

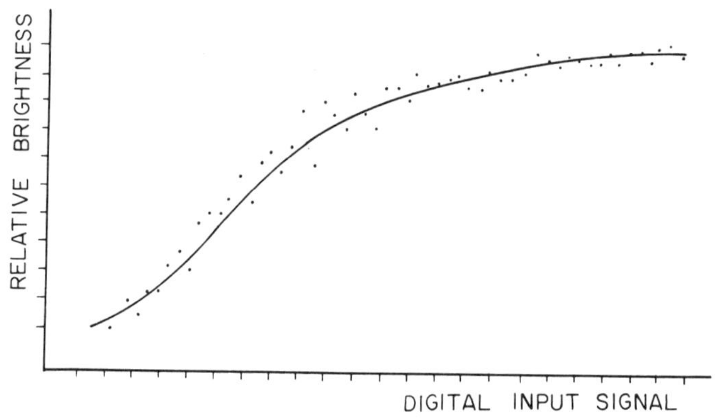

Fig. 3 Exposure curve.

log signals X and Y are directly fed to the horizontal and vertical axes of the cathode ray tube. Analog signal Z is fed to the analog gate with the control pulse, and amplified. Resulting analog signal Z presented by pulse form is fed to the cathode of the cathode ray tube. This analog signal Z fills the role of unblanking as well as that of brightness modulation of individual points. This operation is repeated spot by spot. Accordingly this plotter can plot various patternswith irregular scanning as well as with regular scanning like the TV system.

A precision of this device is 0.5% of full scale for X and Y coordinate numbers. Reproducibility and stability are also satisfactory for practical use. Figure 2 shows test patterns plotted by this device in order to examine the precision concerning coordinate numbers and to evaluate an efficiency of brightness modulation at each sampling point. This pattern was displayed by 64 x 64 lattice points with 64 brightness levels. The total efficiency on intensity modulation obtained by combining circuits of brightness modulation and photographing processing was measured by a microphotometer. Figure 3 shows this result. In this figure, the vertical axis means the relative brightness and the abscissa shows the digital number corresponding to discrete brightness signal Z. Nonlinearity depends on that of the photographic film and the adjustment of a controlled circuit. This nonlinearity is not so harmful for displaying continuous-tone pattern because of the human eye. Figure 4 shows one example of continuous-tone perspective drawing. Display in Figs. 3 and 4 suggest that the "Halftone Plotter" can be used for data obtained by computer

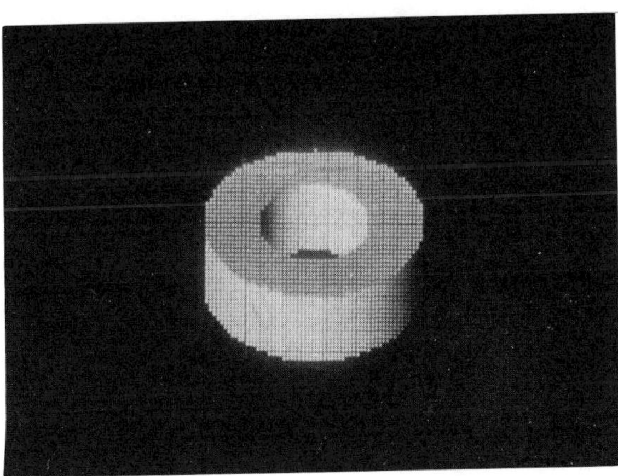

Fig. 4 Example of halftone perspective drawing by the "Halftone Plotter".

processing to draw continuous-tone patterns. The "Halftone Plotter" can, of cause, plot patterns with a constant brightness level.

3. Computer Generated Holograms

Computer generated holograms form a most important branch of digital optical information processing. Objects, which do not exist actually but can be expressed mathematically, can be reconstructed from computer generated holograms. Holograms of large objects which coherent light cannot illuminate over the full size are also synthesized. Computer generated holograms seems to be also utilized for processing the information carried by non-optical media such as sound waves, microwaves, and electron beams.

Many studies 3, 5-11 have been made for synthesizing a digitized hologram. So-called Fourier hologramsseem to be very easily synthesized by calculating the Fourier transform of any digitized object with the help of FFT. Here the synthesis of Fourier holograms was attempted. Its fundamental construction was accomplished by modifying the unique notion for the binary hologram. The difference between the display apparatuses produced that of the constructing process between the binary hologram and ours. The "Halftone Plotter" which can display the computer generated hologram with 2^8 gray scale levels is used for our study. This affects the difference between the constructing process of our hologram and that of the binary one.

The hologram is constructed by the interference of the reference beam and the diffracted one from an object. Information on the object can be expressed by the amplitude and phase modulation of carrier waves in the hologram. Suppose that the carrier waves are substituted by square waves, from the extension of this concept to two-dimensional one, holograms can be shown as an assembly of many spots. Intervals between individual sampling points are related to the carrier frequency.

The starting point of the binary hologram is to divide a hologram plane with a finite extension into N^2 square cells. Each cell has a rectangular aperture. The amplitude and phase signals concerning each cell can be given by area and the shift of the center of this rectangular aperture respectively. Then the transmittance of the aperture is assigned by black and white signals. The amplitude coding method in the binary hologram is similar to pulse width modulation[11]. On the other hand, the amplitude component of our computer generated hologram is given by modulating the transmittance of an aperture, where, in our hologram, the hologram plane is divided into N^2 cells with circular aperture. Difference of transmittance in each aperture is attained by spot brightness on CRT by the "Halftone Plotter". The means of phase coding is the same as that of the binary hologram. The amplitude coding method of our hologram is analogous to the pulse amplitude modulation.

Note that the desired complex amplitude of the reconstructed image can be described by the two-dimensional Fourier transform of a hologram. Figure 5 shows the optical arrangement illustrating the reconstruction of the image from the computer generated hologram. Figure 6 shows the schema illustrating the transmittance and shift of the center of the aperture in the cell (n, m).

Let the complex amplitude of the diffracted beam from the cell (n, m) be

$$u(n\delta\nu,\ m\delta\nu) = A_{nm} \exp[i\phi_{nm}], \tag{1}$$

where n and m mean cell numbers, $\delta\nu$ the cell size in the hologram plane H, A_{nm} is the amplitude component of the complex amplitude and ϕ_{nm} is the phase component of that in the cell (n, m). Suppose that the desired reconstructed image U(x, y) is expressed as the following Fourier series,

$$U(x, y) = \sum_n \sum_m u(n\delta\nu,\ m\delta\nu) \exp[2\pi i(nx + my)] \quad \text{for } |x| < \frac{\Delta x}{2},\ |y| < \frac{\Delta x}{2} \tag{2}$$
$$= 0 \qquad \text{otherwise,}$$

where x and y show coordinate numbers in image plane, Δx is

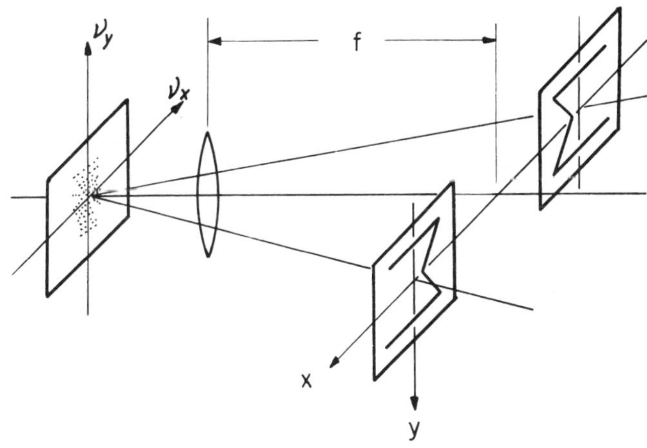

Fig. 5 Optical arrangement illustrating reconstruction from the computer generated holograms.

the finite extension in the image plane (now $\Delta x = \Delta y$). If the transmittance of the circular aperture located in each cell can be described by Gaussian, the complex amplitude of the hologram $h(\nu_x, \nu_y)$ is shown as,

$$h(\nu_x, \nu_y) = \sum_n \sum_m I_{nm}\, g\{\nu_x - (n + P_{nm})\delta\nu,\ \nu_y - m\delta\nu,\ a\delta\nu\}, \quad (3)$$

where $g(\xi, \eta, \sigma) = \{1/2\pi\sigma^2\} \exp\{-(\xi^2 + \eta^2)/2\sigma^2\}$,

ν_x and ν_y are spatial frequency variables, I_{nm} is spot brightness, P_{nm} is the relative shift off center and $a\delta\nu$ is the point spread of each bright spot. If coherent plane waves illuminate this hologram, the resultant complex amplitude displayed on the image plane is,

$$\iint h(\nu_x, \nu_y) \exp[2\pi i\{\nu_x(x + x_0) + \nu_y y\}]\, d\nu_x\, d\nu_y$$
$$= \exp\{-(2\pi r a\delta\nu)^2/2\} \sum_n \sum_m I_{nm} \exp\{2\pi i\delta\nu[(x + x_0)(n + P_{nm}) + my]\}, \quad (4)$$

where x_0 means the distance from the origin in the image plane to the optical axis and $r = \sqrt{(x+x_0)^2 + y^2}$. Comparing eqs.(1) and (2) with eq.(4), following two relations,

$$I_{nm} = A_{nm}, \qquad P_{nm} = \phi_{nm}/2 \quad (5)$$

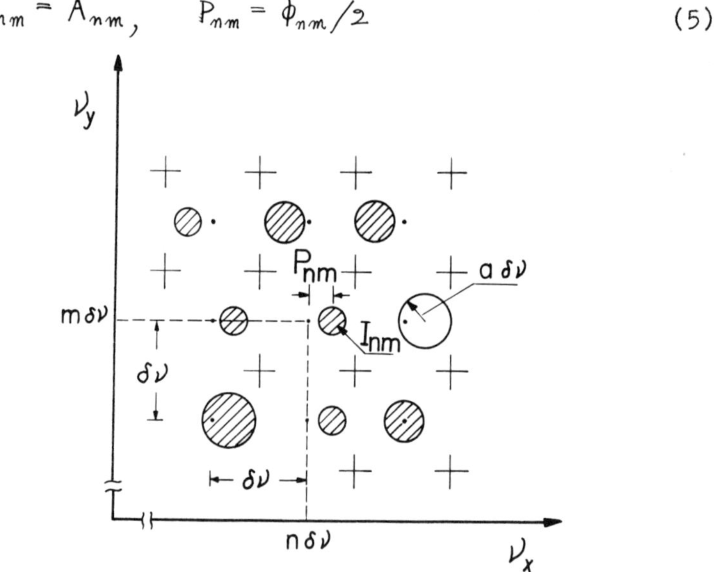

Fig. 6 Illustration on the cell (n, m) in hologram plane. Transmittance of circular aperture is expressed by Gaussian in each cell. $\delta\nu$ means cell size, I_{nm} is spot brightness, $a\delta\nu$ shows the point spread of the spot on CRT and P_{nm} is relative shift off center of cell (n, m).

must be satisfied in order that eq.(4) shows the desired image, where the following two approximations have to be justified at this time,

$$\exp\{-(2\pi ra\delta\nu)^2/2\} \approx const$$
$$\exp(2\pi i x P_{nm}\delta\nu) \approx 1 \qquad (6)$$

The first approximation affects the brightness in the image plane. If this approximation is not satisfied, the brightness on the image plane gradually alternates with Gaussian according with the distance from optical axis and does not distribute uniformly over full size. But this term hardly affects details in reconstructed image and image quality. The effect of the second approximation has been illustrated in reference 3 in details. These approximations can be avoided by the more rigorous solution for I_{nm} and P_{nm}[3].

Digitized objects are used as input data of the Fourier transform in our experiment. The Fourier spectrum of input data is computed by FFT. The computer time was about 6 minutes for objects with 64×64 sampling points by NEAC 2206. The computed spectrum is arranged by the format with the relation in eq.(5) and is displayed as continuous-tone digitized hologram on CRT with the "Halftone Plotter". This hologram is photographed and photoreduced. Reconstructing process is made with an ordinary optical technique. Objects with 64×64 sampling points are only used as examples of our experiment.

Figure 7 shows a computer generated continuous-tone hologram of letters "M. I.", letter "OPTICS" and corresponding reconstructed images. Figure 8 shows the hologram of the object (face) with five brightness levels and reconstructed image. This hologram is calculated with random phase. Figure 9 shows the hologram corresponding to phase object "H" and reconstructed image. The nonlinearity in photographing processing acts as Schlieren effect, so that the contour of "H" becomes visual and the edge of the letter is enhanced.

4. Spatial Filter Construction by a Computer

Fabrication of spatial filters is important in optical information processing, but is one of the most difficult problems in the optical technique. Especially it is very difficult to make spatial filters which can control amplitude and phase components simultaneously. Large flexibility of a digital computer and the "Halftone Plotter" makes it possible to remove the above difficulties, and to calculate rather complicated spatial filter functions and display them. Spatial filters composed of amplitude and phase components can be obtained by the same technique applied in the construction

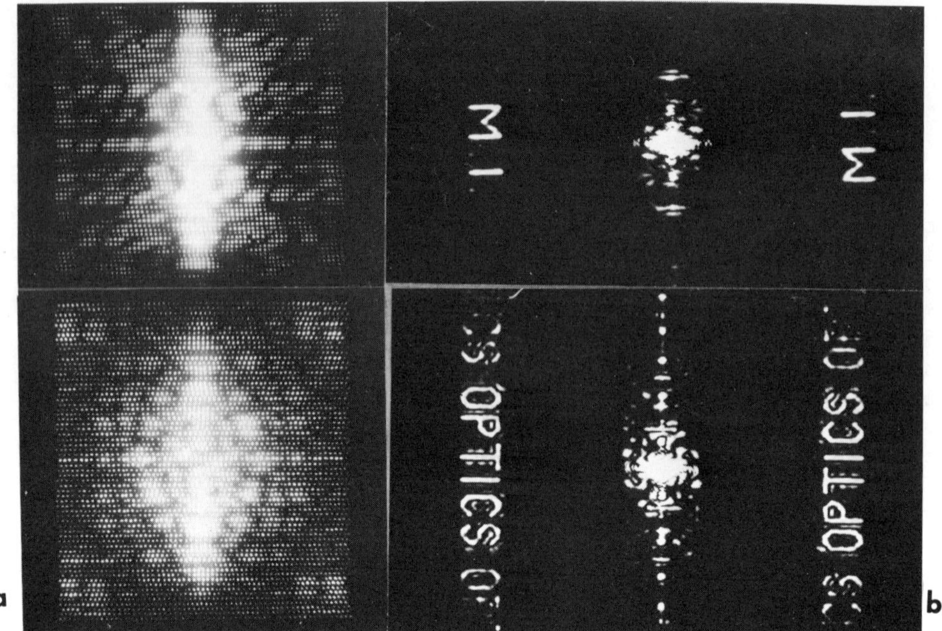

Fig. 7 Display of computer generated holograms and reconstructed images.
Column (a) shows halftone drawn holograms of letter "M. I." and word "OPTICS".
Column (b) shows corresponding reconstructed images.

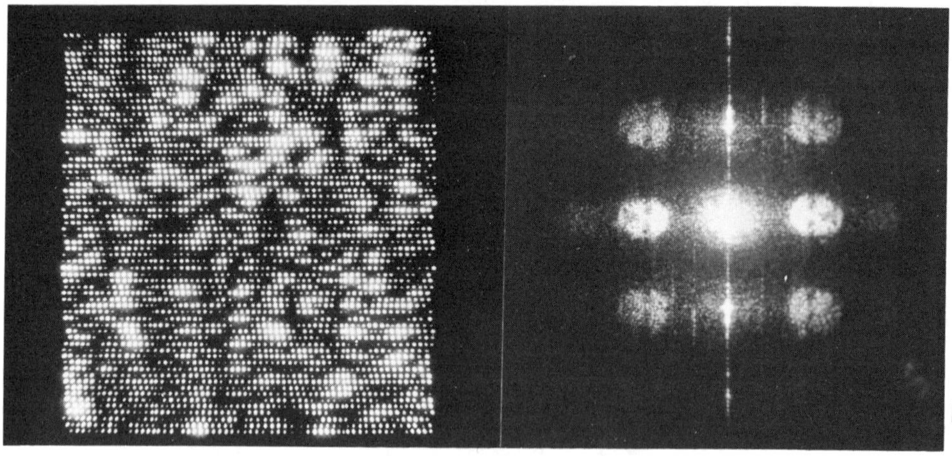

a b

Fig. 8 Same as Fig. 7 but for halftone object (face) with five transmittance levels.

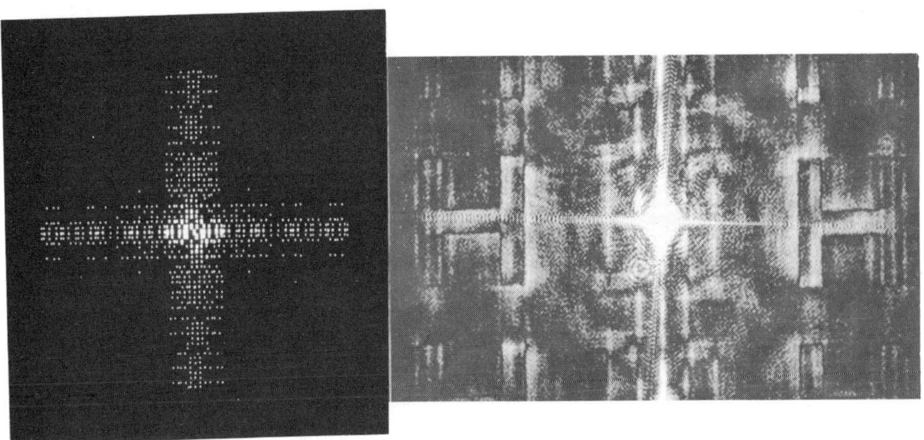

a b

Fig. 9 Same as Fig. 8 but for phase object (letter H). Reconstructed image is affected by Schieren effect.

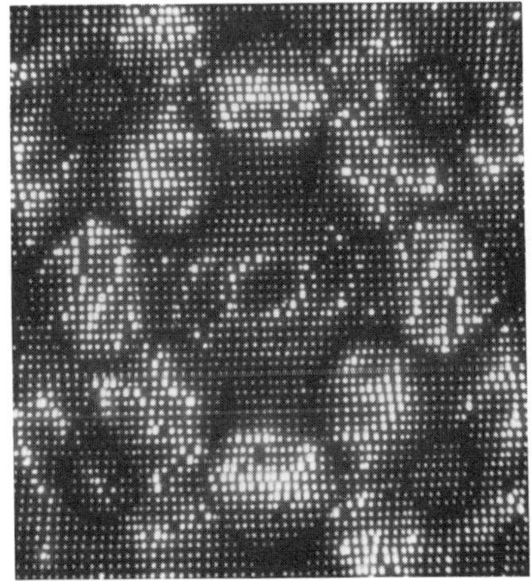

Fig. 10 Example of cord transform filter f_+/f_5, where f_+ means spectra of object + and f_5 spectra of letter 5.

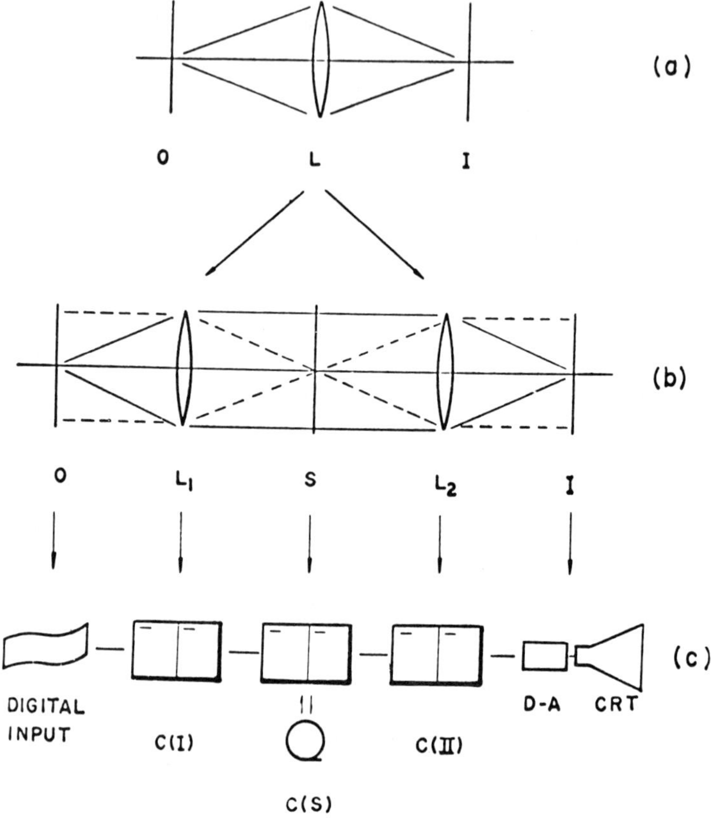

Fig. 11 Schematic diagram of image forming system, coherent optical system and corresponding computer processing system.

of computer generated holograms. Differential filters, code transform filters and inverse filters can be easily synthesized by computer processing[13]. Line plotting operation by "Halftone Plotter" also permits us to construct such a filter as $i\omega^2$, $\exp\{i(\omega^2+\omega)\}$ and so on. If these spatial filters can be made with high degree of precision, they will become greatly useful for optical information processing. Figure 10 shows one example of computer systhesized spatial filters. This is a code-transform filter f_+/f_5 and was displayed by continuous-tone patterns, where f_+ means Fourier spectrum of the object + and f_5 is that of letter 5.

5. Basic Concept of Digital Picture Processing

An image forming system shown in Fig.11(a) can be reduced to a corresponding optical system in Fig.11(b), in which a two-step image forming can be described with the help of computer operation in an optical system. The first step is that, if an object with complex amplitude O(x, y) is set at the front focal plane of aberration free lens L_1, and is illuminated by a coherent monochromatic plane wave collimated by a lens, object spectra are displayed in the back focal plane S of the lens L_1. These spectra are described on the two-dimensional Fourier transform of O(x, y)[14], that is,

$$o(\nu_x, \nu_y) = \iint O(x, y)\exp[-2\pi i(\nu_x x + \nu_y y)]\,dx\,dy, \tag{7}$$

where ν_x, and ν_y mean spatial frequencies respectively. The second image forming step is as follows. If the back focal plane S of the lens L_1 displaying object spectra is set in the front one of the lens L_2, resultant light distribution I(x',y') in the back focal plane of the lens L_2 is also described as the Fourier transform of spectrum $o(\nu_x, \nu_y)$, that is,

$$I(x', y') = \iint o(\nu_x, \nu_y)\exp[-2\pi i(x'\nu_x + y'\nu_y)]\,d\nu_x\,d\nu_y. \tag{8}$$

From eqs.(7) and (8), the image I(x', y') can be given as the result of sequential Fourier transforms of O(x, y). The role of each optical system is to perform two-dimensional Fourier transform of input signals, and this action can also be achieved by substituting by computer operation. In this case the role imposed by a digital computer is merely to compute a two-dimensional discrete Fourier transform of O(x, y) or $o(\nu_x, \nu_y)$ corresponding with eq.(7) or (8). It is shown as[15],

$$o(\nu_x, \nu_y) = \frac{1}{N^2}\sum_n\sum_m O(n, m)\exp\left\{-\frac{2\pi i}{N}(\nu_x n + \nu_y m)\right\} \tag{9}$$

or

$$I(x', y') = \frac{1}{N^2}\sum_n\sum_m o(n, m)\exp\left\{-\frac{2\pi i}{N}(x' n + y' m)\right\}. \tag{10}$$

The schematic diagram of digital picture processing system is shown in Fig.11(c). This corresponds to the coherent image forming system in Fig.11(b).

Computation of discrete Fourier transform for eqs.(9) and (10) is easily attained by using FFT subroutine. All digitized picture data obtained by computer processing can be displayed as discrete continuous-tone patterns on CRT with the "Halftone Plotter". Hence it is regarded that these displayed patterns are equivalent to the optical information given by the practical optical system in finite region.

6. Simulation of Spatial Filtering

The coherent optical system makes it possible to modify optical frequencies in so-called spectral plane S in Fig.11(b). The amplitude and phase components of Fourier spectrum $o(\nu_x, \nu_y)$ can be altered by spatial filtering at each frequency. If the spatial filter having complex transmittance $f(\nu_x, \nu_y)$ is set in this plane, a complex disturbance appearing from this plane is expressed as the product $o(\nu_x, \nu_y) \times f(\nu_x, \nu_y)$. Hence the resultant image formed in the image plane I can be described by the Fourier transform of the product $o \times f$,

$$I(x', y') = \iint o(\nu_x, \nu_y) f(\nu_x, \nu_y) \exp[-2\pi i(\nu_x x' + \nu_y y')] d\nu_x d\nu_y. \quad (11)$$

This relation is rewritten by the expression in the spatial frequency domain,

$$i(\nu_x, \nu_y) = o(\nu_x, \nu_y) f(\nu_x, \nu_y). \quad (12)$$

Spatial filtering by using a digital computer can be easily performed with the help of a discrete expression of eqs. (11) and (12). One of the merits of spatial filtering by computer operation is to remove technical difficulties contained in the real optical technique, such as setting error and photo reduction. The other merit is that the large flexibility of a digital computer and the high degree of precision make it possible to perform precise filtering by even such spatial filters expressed in complicated analytical forms. Especially, amplitude and phase components can be stored in magnetic tapes separately, so that the phase component can be controlled precisely. This is the unrealizable merit in the ordinary optical technique.

The procedure of a computer-simulation of spatial filtering is as follows. First, digital input data $O(x, y)$ sampled by 64×64 points are fed to a computer. This data scanning was performed by vidicon camera or human eyes in our experiment.

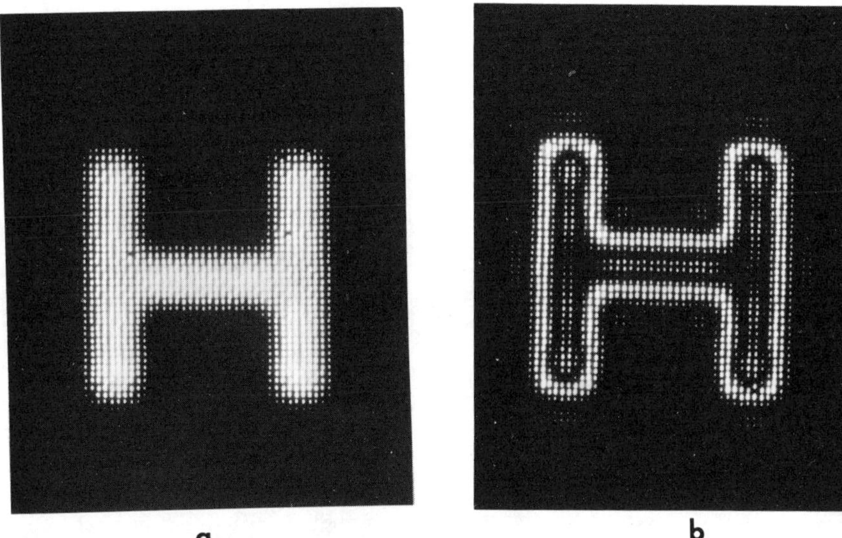

Fig. 12 Examples of spatial filtering by a digital computer.
(a) Example of low pass filtering by square mask with 16 x 16 points
(b) Example of high pass filtering by square mask with 16 x 16 points

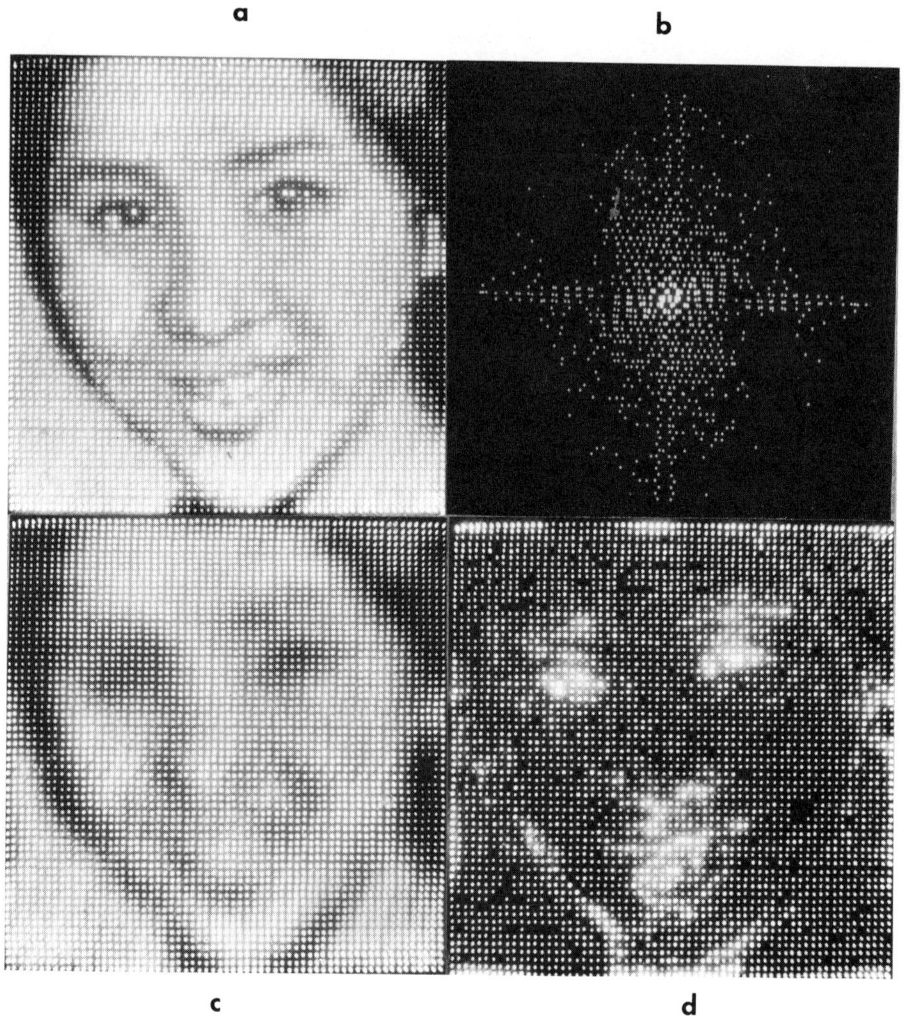

a b

c d

Fig. 13 Examples of spatial filtering for a halftone object;
(a) original digitized picture,
(b) corresponding spectrum of the object,
(c) example of low pass filtering by using the mask in Fig. 12(a), and
(d) example of high pass filtering by using the mask in Fig. 12(b).

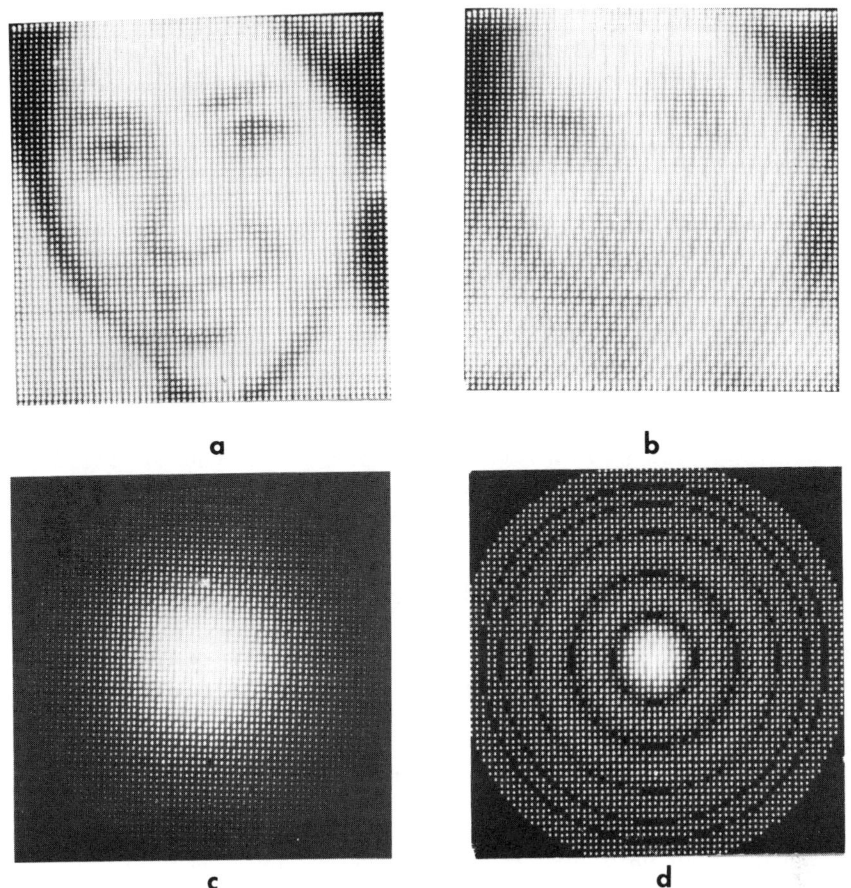

Fig. 14 Examples of image synthesis;
(a) simulated image by defocussing,
(b) same as (a) but for other situation,
(c) display of optical transfer function corresponding to the state in (a), and
(d) same as (c) but for (b).

Then the Fourier spectrum $o(\nu_x, \nu_y)$ corresponding to input data are computed and stored in the magnetic tape, where the real and imaginary parts of these spectra are stored separately. After regaining these spectra from the magnetic tape, the processing filter function f is multiplied by these spectrum o at the corresponding spatial frequency. This complex computation of product $o \times f$ can be achieved with a high degree of precision in a digital computer. No error term can be introduced at this time, so that we need not take account of inconvenience in ordinary optical experiments. This product in a spatial frequency region is fed to the FFT subroutine again. The resultant data are complex values I(x', y') in the space domain. This I(x', y') is the result produced by the computer-simulation of the spatial filtering. This is displayed on CRT with the "Halftone Plotter" and photographed. Figure 12 shows one example of computer-simulation of spatial filterings. Figure 12(a) shows the result of low pass filtering, where high frequency components of spectra of the object H with size of 64×64 is obstacled by the square mask with size of 16×16. Figure 12(b) shows that of high pass filtering by the square mask with the same size in Figure 12(a). Low frequency components are sheltered by the square mask of 16×16. Image quality of two reconstructed images is greatly affected by such defects of spectral components contained in original object. These two pictures tell us that the effect of a simple square low pass filter contributes to broadening of the contour, and the high pass filter enhances the edge of the picture. Figure 13 shows the result obtained by a low pass and high pass filter, for continuous-tone object. Figure 13(a) is the display of the original data. Figure 13(b) is the corresponding spectrum of it. Figures 13(c) and (d) are filtered images which are obtained by a square low pass mask with a size of 16 x 16 and by a square high pass mask with the same size respectively. From the pattern processed by the high pass mask, edge enhancement can be recognized and the feeling of that picture is similar to the negative because of the obstruction of zero frequency component.

7. Simulation of Image Synthesis

From the extension of the concept of computerized filtering, we can simulate the image quality of the optical system whose constructing parameters are known but have not been manufactured. These are useful for studies of image evaluation or lens designing. Image forming under incoherent illumination can also be described with the help of Fourier optics. The procedure of image synthesis can be shown with the same technique as that of spatial filtering under coherent optics described by Sec. VI. The Fourier transform of a point spread function for an image forming system is well known as the optical transfer function. This optical transfer function corresponds to filter function f in previous treatment. Spatial frequency contents of an ideal image correspond here to object spectrum o in the coherent system. Thus, product $o \times f$, in this case, shows a Fourier spectrum of an image affected by characteristics of an image-forming system. Figure 14 shows examples of the simulated images affected by defocussing. These patterns (a) and (b) are simulated images suffered from defocussing of outfocus and infocus. The optical transfer function corresponding to these situations are given by such a formula as $f=Ae^{i\phi}$.[6] Figure 14(a) shows an example of outfocusing, and Figure 14(c) shows a display of the corresponding optical transfer function, where $\phi = 0$, and A is described by a positive component. Figure 14(b) shows an example of infocusing and Figure 14(d), a display corresponding to optical transfer function, where ϕ becomes 0 or π in accordance with positive and negative components of amplitude A. This treatment is valid in a so-called isoplanatic region, so that the above results might be described only for the beam coming from a specific field angle. Hence, it is a very time-consuming task to evaluate all image characteristics described over full field angles.

Summary

1. The "Halftone Plotter" developed by us may be greatly useful for digital picture processing. Any continuous-tone drawing of two-dimensional data can become possible. Capability of plotting the bright spot on CRT with 2^8 brightness levels greatly helps the treatment of digitized computer output. Ability of repetition of the "Halftone Plotter" by the off-line operation saves computer time spent on data displaying.

2. The continuous-tone digitized Fourier holograms could be synthesized and were reconstructed optically. Display of these continuous-tone digitized holograms becomes possible by using the "Halftone Plotter". Merits of our hologram are as follows. (1) We need not pay attention to overlap fault, because overlapped points can be plotted on the same coordinate on CRT, and if the nonlinearity of the emulsion is ignored, the hologram transmittance is proportional to the addition of the brightness of these overlapped spots. (2) The amplitude component of holograms can be described not by the size of the circular aperture, but by the brightness of the spot, so that it is possible that the hologram can be recorded in a lateral position as well as in a vertical one. (3) Plotting time of hologram is very short. However, we have to also con-

sider the nonlinearity of the emulsion in the ordinary hologram because of the treatment of the gray-scale signals. And also we have to take care of the change of the size of spot on CRT.

3. Computer-simulation of spatial filtering is useful in understanding the characteristics of the Fourier spectrum of objects. Experiments of image synthesis are meaningful for lens designing and image assessment. Most important thing in this computerized filtering is to control precisely the Fourier spectrum of objects. This cannot be achieved in the ordinary optical experiment. The purpose of digital picture processing is not to substitute the optical technique developed already by computer processing but to promote larger applications by combining the computer technique and the optical technique. To do so, the digital picture processing becomes useful in so-called optical information processing.

References

1. F.C.Billingsley, SPIE Proceedings, 1967 Computized Imaging Technique Seminar, II-1 (1967).
2. J.W.Cooey and J.W.Tukey, Math. Compt., 19, 297 (1965).
3. A.W.Lohmann and D.P.Paris, Appl. Opt., 6, 1739 (1967).
4. T.Suzuki, Y.Ichioka and M.Koyama, J. Appl. Phys. Jap. 4, Supple. I, 128 (1965).
5. L.B.Lesen et al., Proceedings, 1967 Fall Joint Computer Conference, p.41 (1967).
6. J.P.Waters, Appl. Phys. Letters, 9, 409 (1966).
7. J.P.Waters, J. Opt. Soc. Am., 58, 1284 (1968).
8. T.S.Huang and B.Prasada, Quartely Progress Report No.81, Research Laboratory of Electronics, M.I.T., p.199 (1966).
9. W.H.Lee, Quartery Progress Report No.88, Research Laboratory of Electronics, M.I.T., p.310 (1968).
10. W.H.Lee, Quartery Progress Report No.93, Research Laboratory of Electronics, M.I.T., p.231 (1969).
11. B.R.Brown and A.W.Lohmann, IBM J. Res. Develop., 13, 160 (1969).
12. J.L.Harris, Sr., J. Opt. Soc. Am., 56, 569 (1966).
13. A.W.Lohmann and D.P.Paris, Appl. Opt., 7, 651 (1968).
14. A.Vander Lught, Opt. Acta, 15, 1 (1968).
15. H.C.Andrews and W.K.Pratt, Computer Group News, Nov. 12 (1968).
16. Y.Ichioka and T.Suzuki, Appl. Opt., 7, 927 (1968).

COMPUTER SYNTHESIS OF HOLOGRAMS AND SPATIAL FILTERS

B. R. Brown

IBM Research Laboratory

San Jose, California

The computer generated hologram is a popular outgrowth of efforts to develop spatial filters for coherent optical data processing systems. Needless to say, the intense coherent light produced by lasers made possible the use of spatial filtering systems for processing pictorial information; however, versatility of such systems initially was restricted because only a few filter functions could be produced by conventional interference holography. Now there are several techniques employing digital computers to synthesize filters with an arbitrarily general complex transmission function.

A typical spatial filtering system, as shown in Fig. 1, has two intrinsic properties that facilitate processing pictorial information. First, the whole input image is processed simultaneously in a parallel manner. In contrast, the numerical processing of an image requires the transmittance value of each resolvable point to be digitized and processed sequentially, generally at the expense of lengthy computation to analyze the content of a relatively simple image.

The second attractive feature of the coherent system is that the Fourier transform of the wavefront in the front focal plane of any positive lens is generated automatically in the rear focal plane. Thus in Fig. 1, in which the input plane is illuminated by a uniform plane wave, the Fourier transform of the input transmittance is displayed in the filter plane. Since the Fourier transform is the basis of the correlation and convolution theorems of communication theory dealing with signal detection, the ability of a coherent system to perform this transform lends it naturally to optical processing tasks of pattern recognition and image analysis.

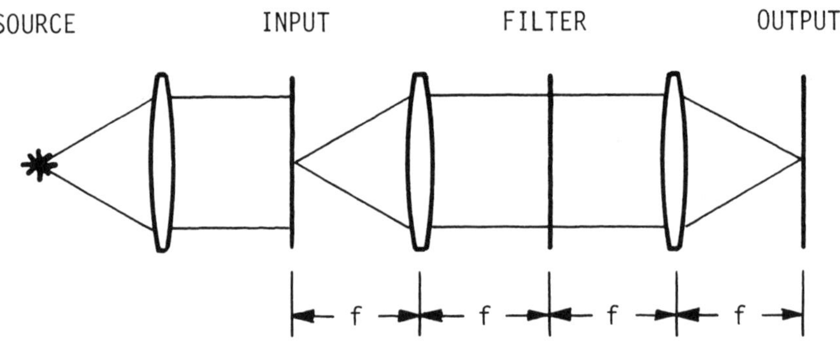

Fig. 1 Coherent spatial filtering system

To achieve a given result in the output plane, a filter must be inserted with an appropriate transmittance which is generally complex, that is, it must not only modify the intensity of the light as does a photographic transparency, but also must modify the phase angle. Before techniques were developed to synthesize spatial filters with computers, the only complex filters, except for relatively simple masks, phase plates and gratings, were holograms produced by the two beam interference techniques. Holographic techniques, however, are limited in practice almost entirely to production of so-called matched filters. These are actually holographic recordings of the Fourier transform of an object. Nevertheless, the matched filter is quite powerful for pattern detection because it forms auto-correlation and autoconvolution peaks in the output whenever the matched object is placed in the input of the system.

Production of matched filters was also the entry point of computer synthesis into the field of holography even though the synthesized versions offer no great advantage over those produced holographically. Unlike holography though, the synthesis techniques were extendable to the construction of filter transmittances involving quotients, differences or other relations between complex functions. Now filters can be made to perform any linear filtering operation on an image, including such dramatic feats as code translation.

To the author's knowledge, the first reported application of computers to the synthesis of complex spatial filters was the "hard clipped matched filter" of Kozma and Kelly.[1] Their filter was produced by calculating the continuous Fourier transform of an one-dimensional object followed by a clipping procedure which produced a two level or binary transmittance function. This function was plotted as a series of parallel black and white lines of varying width. A photo reduction was made to obtain a filter transparency at an appropriate scale.

Next, Lohmann applied the Cauchy sampling theorem used in communication theory to the synthesis of spatial filters and proposed several techniques for structuring binary transmittance masks that could reconstruct an arbitrarily general complex two-dimensional wavefront.[2] Two of the techniques assigned a single rectangular aperture to each computed sampling of the wavefront. Aperture position relative to the sampling point was determined by the phase of the wavefront such that light diffracting in the reconstruction direction would obtain the proper phase delay by virtue of the distance traveled from the offset aperture. When comparing with the uniform first order phase delay of 2π between adjacent slits of a regular diffraction grating, Lohmann appropriately terms this variable phase delay the "detour phase" effect. Encoding the wavefront amplitude modulus was slightly different in these two methods. One method varied the height of the aperture, the other the width. He also proposed a third scheme in which all apertures are identical, but with two per sampling point. In this case the average position of the pair is proportional to the phase, but the separation is varied to control the amplitude from the pair by mutual interference of the light from each aperture. This author was fortunate in having assisted in the construction of the first filters using these techniques and having performed some matched filtering experiments with them. Figure 2 shows the plotted artwork of a latter hologram synthesized by the variable aperture height method.

Fig. 2 Binary hologram artwork

Even though the computational economy of the sampling method allowed filters to be synthesized for very simple two-dimensional objects such as alphabetic characters, it is possible that little further interest would have developed if the Cooley Tukey Fast Fourier Transform Algorithm[3] had not become available. The Fourier Transform, which occurs in the mathematical description of light propagation, is the essential computational procedure occurring for all synthetic holograms except for the rather trivial case of an image hologram where the image itself is the function encoded.

To evaluate the Fourier Transform of an object of N elements at N sampling points according to the usual notation requires N^2 complex multiplications and additions. However, by judiciously dividing and sequencing the computation, Cooley and Tukey reduced the number of arithmetic operations to $2N\log_2 N$. For a modest image of 128 x 128 resolvable points, the time savings amounts to a factor of 600 or the difference between a minute and a day of computation time for a medium speed computer

Paris applied the Cooley Tukey Algorithm to the computation of binary holograms and, with Lohmann, carried out several spatial filtering experiments.[4, 5] They demonstrated the advantage of synthesizing filters which involve inverse, quotient or other mathematical functions which can be performed easily by computer but not by optical and photographic techniques.

The speed at which the Cooley Tukey Algorithm handled light propagation calculations made attempts to synthesize holograms of 3-D objects possible. Apparently with this goal, several groups entered the field and introduced additional techniques for synthesizing hologram transmittance functions. Waters[6] introduced a fixed position, fixed size binary aperture format in which the spatial density of apertures was varied to represent the transmittance function. Then Lee[7], going directly from a real and imaginary representation of the complex amplitude, devised a scheme which represented each computed wavefront sample with as many as two out of four possible fixed position variable sized apertures per sample. In his scheme, the four aperture positions correspond to the positive or negative real and imaginary parts of the computed complex amplitude thereby eliminating the need to convert the amplitude to a modulus and phase representation.

With variable intensity cathode ray plotters which expose film to levels of grey, direct simulation of the optical interference holographic recording process is possible and has been investigated by Lesem, Hirsch and Jordan,[8] and Meyer and Hickling[9] and possibly others. Subsequent study of the hologram synthesis problem[10] shows, however, that the grey level simulations offer less computational economy than the binary, and have inherently more sensitivity to photographic non linearities and grain noise. In this regard, the binary hologram compares to the grey hologram much like pulse code modulation compares to analog signal transmission in that the required resolution or channel bandwidth is higher for a binary code, but accuracy can be preserved despite noise and non linearity in the channel.

The common feature of all the synthetic holograms discussed so far is that the phase modulation of the reconstruction beam is accomplished by the "detour" effects of diffraction as with conventional holograms. The unfortunate consequence of diffraction is that only a small portion of the illuminating light can be directed to form the desired image. The rest is undiffracted or forms conjugate images which are usually of no value. Lesem, Hirsch and Jordan[11] in producing what they call a kinoform, avoided using diffraction detour phase effects by creating a mask of varying dielectric thickness. Their kinoform is, in fact, a phase plate synthesized with heretofore unprecedented complexity. Furthermore, their fabrication technique has no optical analog. It uses a computer controlled CRT plotter to expose a photographic emulsion in anticipation of a bleaching process that will render the gelatin thickness proportional to the phase delay required to reconstruct the desired wavefront. Despite non linearities and spatial frequency dependence of photographic relief images as well as the relatively high accuracy requirements of this system, they have produced kinoforms with high efficiency and low noise.

By and large though, the goal to produce a practical three-dimensional display for computer generated data using synthetic holograms has not been achieved, even though most people working in the field have shown some limited three-dimensional reconstructions. The major obstacle is the number of computations involved to determine the wavefront propagating from a three-dimensional object as well as the difficulty of synthesizing a hologram with an aperture large enough to satisfy the parallax and binocular requirements of human vision.

The usual method for computing the light amplitude propagating from a 3-D object requires a Fourier transform for each resolution element in depth, and, if hidden line effects are to be included which certainly they must for a useful display, then two transforms are required per resolvable depth unit. Combine this with need for a practical minimum of about 256 x 16,000 sampling points in the hologram to create a viewing aperture of 2 by 120 mm and we are stuck with an unreasonable amount of computation for even a simple object.

Recently, however, King, Noll and Berry[12] have viewed the 3-D computer display problem from a different point of view. They combined computer and optical processing in a more optimum manner and used some simplifications permitted by human 3-D vision to produce large aperture holograms of relatively complicated synthetic objects. By using the computer to write transparencies showing projections of the object as viewed from an incrementally changing angle, and then using these transparencies sequentially as the objects in a normal two beam hologram recording set-up to expose adjacent strips of the hologram plate, they have made large composite holograms which show good parallax effects. With relatively little computation required, these holograms produce a pleasing display.

The omission of vertical parallax to further reduce fabrication difficulty, and the lack of differential depth of focus effects are not considered to be serious faults for the purpose of 3-D display. Furthermore, they are able to capitalize on the two dimensionality of the actual reconstructed image to form an image hologram which reconstructs with high efficiency in white light. The latter characteristics, of course, increases the utility of this method for display purposes many fold by eliminating the need for special monochromatic light sources for reconstruction.

To summarize the capabilities and applications of computer generated holograms, we can probably say that those synthesized entirely by computer are of little value for image display purposes because of the enormous number of calculations now required to compute the light propagation from any reasonable three-dimensional object. Nevertheless, the reconstruction of an image

Fig. 3 Two dimensional reconstruction from binary hologram in Fig. 2

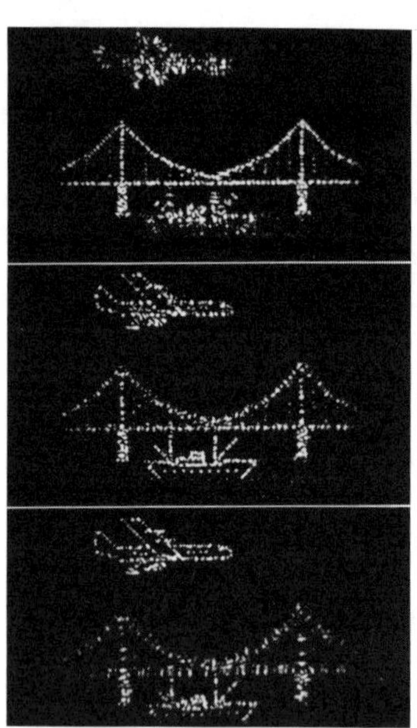

Fig. 4 Successive planes of 3-D reconstruction from a binary hologram.

provides a means for judging the wavefront quality obtained with a given synthesis method.

Figures 3 and 4 are examples of the reconstruction of binary holograms such as that in Figure 2. The reconstructed line widths are approximately that due to the hologram aperture diffraction. The reconstruction of Figure 4 occurs at different planes in depth which were photographed individually for the purpose of illustrating a limited three-dimensional capability. No parallax could be demonstrated due to the small hologram aperture of 2 x 2 mm.

At present, within a limited aperture at least, the wavefronts reconstructed from computer generated holograms can have quite good fidelity with the computed wavefront. In fact a study of their use as interferometric prototypes for testing optical elements has been undertaken.[13]

Practical application of computer generated spatial filters to such problems as image enhancement, or pattern recognition, although possible, is somewhat deterred by the photographic processes now necessary to produce the filters. The attendant photographic processing delay is generally incompatible with computer interaction speed and the alternative of mechanically sequencing a repertoire of previously constructed filters would also limit the application. Consequently most effective uses of computers in image processing have continued to follow the path of digitizing the image, then performing all filtering operations numerically within the computer and plotting the result. Should it become possible to write filters at electronic speeds, the computer generation concepts would likely play a larger role in image processing. Nevertheless, the synthetically produced filters are capable of performing any linear filtering function and it is safe to say that their full capabilities have not yet been exploited.

REFERENCES

1. A. Kozma and D. L. Kelly, "Spatial Filtering for Detection of Signals Submerged in Noise", Appl. Opt. 4, 387 (1965).

2. B. R. Brown and A. W. Lohmann, "Complex Spatial Filtering with Binary Masks", Appl. Opt. 5, 967 (1966).

3. J. W. Cooley and J. W. Tukey, "An Algorithm for the Machine Calculation of Complex Fourier Series", Math. Comp. 19, 297 (1965).

4. A. W. Lohmann and D. P. Paris, "Binary Fraunhofer Holograms Generated by Computer", Appl. Opt. 5, 1739 (1967).

5. A. W. Lohmann, D. P. Paris, and H. W. Werlich, "A Computer Generated Spatial Filter Applied to Code Translation", Appl. Opt. 6, 1139 (1967).

6. J. P. Waters, "Holographic Image Synthesis Utilizing Theoretical Methods", Appl. Phys. Letters 9, 405 (1966).

7. W. H. Lee, "Sampled Fraunhofer Hologram Generated by Computer", J. Opt. Soc. Am. 58, 729A (1968) and Applied Optics 9, 639 (1970).

8. L. B. Lesem, P. M. Hirsch, and J. A. Jordan, Jr., "Computer Generation and Reconstruction of Holograms", Proceedings of the Symposium on Modern Optics, J. Fox, Ed., Polytechnic Institute of Brooklyn, distribured by Interscience Publishers, New York, p. 681 (1967).

9. A. J. Meyer and R. Hickling, "Holograms Synthesized on a Computer Operated Cathode Ray Tube", J. Opt. Soc. Am. 57 1388 (1967).

10. B. R. Borwn and A. W. Lohmann, "Computer-Generated Binary Holograms", IBM J. Research and Development, 13, 160 (1969).

11. L. B. Lesem, P. M. Hirsch, and J. A. Jordan, Jr., "The Kinoform: A New Wavefront Reconstruction Device", IBM Publication No. 320.2324 (1969).

12. M. C. King, A. M. Noll, and D. H. Berry, "A New Approach to Computer-Generated Holography", Applied Optics, 9, 471 (1970).

13. Alan J. MacGovern and James C. Wyant, "Computer-Generated Holograms for Testing Optical Elements", JOSA 59, 1511 (1969).

PATTERN CLASSIFICATION USING CORRELATION WITH RANDOM MASKS — AN OPTICAL PAPA DEVICE

Shun-ichi Tanaka, Hisatoyo Kato,* Satoshi Ishihara, and Toyohiko Yatagai

Department of Applied Physics, Faculty of Engineering

University of Tokyo, Bunkyo-ku, Tokyo, Japan

1. INTRODUCTION

Recently, information processing by optical methods has been used widely, the reason of which may be mainly due to the invention of the laser and that of elaborate holographic techniques. In such processing pattern recognition is one of the most important subjects.

In this paper, transformations of information using holographic techniques are summarized firstly[1] and then an optical pattern classifier using correlation technique is reported.

2. CLASSIFICATION OF HOLOGRAPHIC TRANSFORMATIONS OF INFORMATION

Fig. 1 shows the well-known arrangement of one-dimensional Fourier transform holography, the upper half of which is a hologram making setup and the lower half, a reconstruction setup. In the case of conventional holography, a point source, that is, a δ function is used as the amplitude distribution of either u(x) and v(x) as well as that of w(x). In this figure, a filter with the amplitude transmittance of $G(\omega)$ is also placed in the spectrum plane of the reconstruction process. Then, the amplitude transmittance $T(\omega)$ of the hologram and the amplitude distribution r(x) in the

* Present address: Fuji Photo Film, Co., Ltd., Ashigara, Kanagawa.

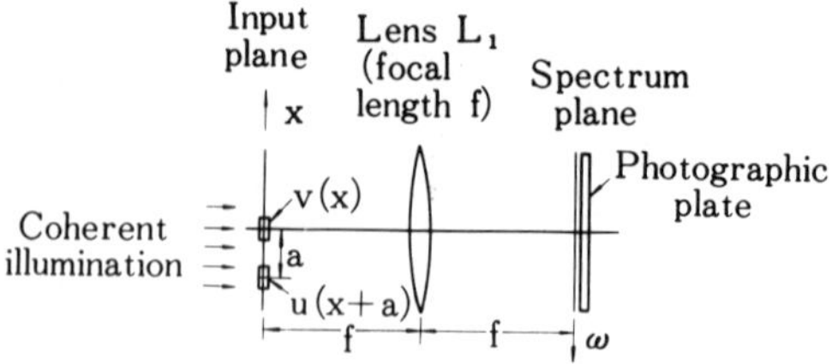

(a) Hologram making process.

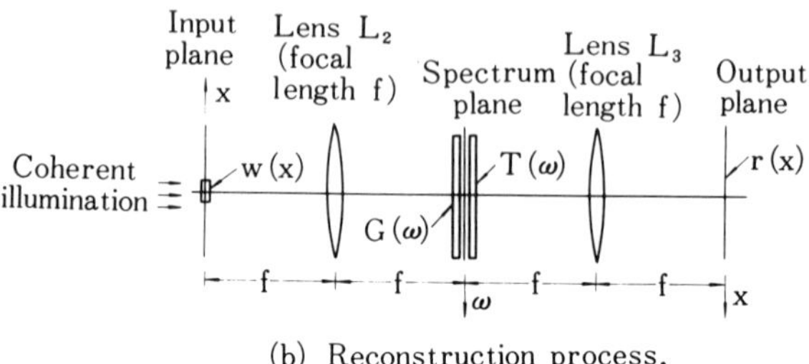

(b) Reconstruction process.

Fig. 1. Configuration of Fourier transform holography.

output plane of the reconstruction process are written as follows:

$$T(\omega) \propto |U(\omega)e^{ia\omega} + V(\omega)|^2$$

$$= |U(\omega)|^2 + |V(\omega)|^2 + U^*(\omega)V(\omega)e^{-ia\omega} + U(\omega)V^*(\omega)e^{ia\omega}, \quad (1)$$

$$r(x) \propto \mathscr{F}^{-1}[W(\omega)G(\omega)T(\omega)]$$

$$\propto \mathscr{F}^{-1}[W(\omega)G(\omega)\{|U(\omega)|^2 + |V(\omega)|^2\}]$$

$$+ \mathscr{F}^{-1}[W(\omega)G(\omega)U^*(\omega)V(\omega)e^{-ia\omega}]$$

$$+ \mathscr{F}^{-1}[W(\omega)G(\omega)U(\omega)V^*(\omega)e^{ia\omega}]. \quad (2)$$

Here, the spatial angular frequency $\omega = -2\pi\{x/(\lambda f)\}$, where λ is the wavelength of light. The capital letters U, V, W mean the Fourier transforms of the functions u, v, w, and the symbol $\mathscr{F}^{-1}$ denotes the inverse Fourier transform, that is,

$$\mathscr{F}[u(x)] = \int_{-\infty}^{\infty} u(x)e^{-i\omega x}dx = U(\omega),$$

$$\mathscr{F}^{-1}[U(\omega)] = \int_{-\infty}^{\infty} U(\omega)e^{i\omega x}d\omega = u(x). \quad (3)$$

The first, second and third terms of the right hand side of Eq. (2) mean the diffracted light of the zero and plus-minus first orders respectively and each of them is distributed around the point where x = 0, a and -a. If the value of a is properly selected, the three terms do not overlap. The setup of Fig. 1(a) can be replaced by that of the so-called lensless Fourier transform and this may also be replaced with a computer generated hologram.

Table 1 shows examples of several types of transformations of information when some different functions or patterns are used as u, v, w and G. This table is arranged in decreasing order of δ functions. By using a class a method, several authors[2-5] thors[2-5] made diffraction gratings of up to 3000 lines/mm.

As a modification of class b, Lu[6] treated the case of $u = \sum \delta(x + a_i)$. The outputs of the plus and minus first orders are then given by $\sum w(x \mp a_i)$ and the multiple images of an object are obtained.

Class c is the conventional Fourier transform holography originally proposed by Leith, Upatnieks[7] and Stroke.[8] Both the true image $v(x - a)$ and the conjugate image $v^*(-x - a)$ can be separately reconstructed.

In the case of class d, the outputs of the plus and minus first orders are equal to the convolution $\int w(x')v(x - a - x')dx'$ and the correlation $\int w(x')v^*(x' - a - x)dx'$ respectively, where these operations are simply denoted by the symbols $*$ and $\star$ in Table 1. When $w = v$, the output of the minus first order becomes the rotationally symmetrical bright pattern of the autocorrelation function distributed around the point $x = -a$ and this can be used for pattern identification.

The peak width α of the autocorrelation of any function $b(x)$ and the width β of its power spectrum $|B(\omega)|^2$ have the following relation:[9]

$$\alpha\beta \geqq \text{a constant of the order of unity} \tag{4}$$

so that the autocorrelation pattern has a finite width. When the slight difference between b and b' is to be compared, the cross-correlation pattern obtained by $B^*\varepsilon$ will sometimes be masked by the autocorrelation pattern obtained by B^*B, where B and $B + \varepsilon$ mean the Fourier transforms of b and b'. Huhn et al[10] showed that the output of the minus first order corresponding to B^*B can be transformed into a complete δ function if a filter with an amplitude transmittance proportional to $1/|B|^2$ is used as the filter G shown in Table 1, and that slight difference in patterns can be detected.

The filter represented by $1/|B|^2$ can be made as follows:[11] When a pattern with the amplitude transmittance of $b(x)$ is placed in the front focal plane of the lens L_1 of Fig. 1(a), the exposure E of the photographic plate placed in the back focal plane is proportional to $|B|^2$. From the characteristic curve of a photographic

Table 1. Transformations of information.

Class	u	v	w	G	Transformations	r		
						0 order	+1st order	-1st order
a	δ	δ	δ	——	(Diffraction grating)	δ	δ	δ
b	δ	δ	w	——	(Diffraction grating)	w	w	w
c	δ	v	δ	——	(Fourier transform holography)		v	v^*
d	δ	v	w	——	Convolution, correlation (Pattern identification)		$w \ast v$	$w \star v^*$
e	δ	s	s+n	$1/\|N\|^2$	Correlation (Matched filtering)			When $\|N\|^2=C$, $(s+n) \star s^*$
f	δ	b	$o \ast b$	$1/\|B\|^2$	Deconvolution (De-blurring)			o
g	δ	$o \ast b$	b	$1/\|B\|^2$	Deconvolution (De-blurring)			o
h	u	v	δ	——	Correlation (Pattern identification)		$u^* \star v$	$u \star v^*$
i	b	$o \ast b$	δ	$(1/\|B\|^2)$	Deconvolution (De-blurring)		o	o^*
j	u	v	u	$(1/\|U\|^2)$	(Generalized Fourier transform holography, code transformation, pattern recognition)		v	

emulsion, the following relation exists between the amplitude transmittance G of the processed plate and the exposure E.

$$G \propto E^{-\gamma/2}, \tag{5}$$

where γ denotes a photographic gamma. Accordingly, the desirable filter with G proportional to $1/|B|^2$ is obtained when the plate is processed with $\gamma = 2$.

The function of the filter represented by $1/|B|^2$ can also be included in the process of the first step.[12] Namely, by using the above method and processing with $\gamma = 1$, a filter with the amplitude transmittance proportional to $1/|B|$ is first made. Then this filter is inserted in front of the photographic plate placed in the back focal plane of the lens L_1 of Fig. 1(a) and the exposure of the hologram making process is made.

So far, a general case where $B(\omega)$ is a complex function and also where both the amplitude and phase are functions of ω is considered. If the amplitude of $B(\omega)$ is constant independent of the value of ω, $1/|B|^2$ becomes a constant and the use of the filter is not necessary.

Class e is matched filtering which detects the presence of a deterministic signal s submerged in additive, stationary noise n. This subject was first treated by Vander Lugt[11] and many papers followed.[13-23]

Using a class f method, Stroke, Zech[24,25] and Lohmann[12] apply the deconvolution operation to the de-blurring of the optical image. As a modification of class f, Stroke[26] also treated a class g method.

A class h method first proposed by Weaver and Goodman[27] as the counterpart of class d or e to reduce setting error in a hologram; it is useful for pattern identification.[28,29]

By using modified forms of class i, Goodman et al[30] and Gaskill[31] treated the resolution retrieval of images through random media and Noguchi, Ose[32] treated the correction of lens aberrations.

Class j is a generalized Fourier transform holography first treated by Stroke,[33] Pennington and Collier,[34] where neither u nor w is a δ function and u = w. In the case where both $|U|^2$ and $|V|^2$ are independent of ω, u or v is reconstructed when a hologram made with u and v is illuminated with v or u. This method is used for pattern recognition or code translation by Gabor,[35] Lohmann and so forth.[12,36,37]

So far, in order to make expressions simple, treatment was restricted to one-dimensional cases. The treatment, however, is easily extended to two-dimensional cases by increasing the number of variables. In addition, only the Fourier transform holography is considered. The Fresnel transform holography, however, is also used for many transformations of information.

In our case shown presently, a modified form of class d is used.

3. PRINCIPLE OF PAPA

The device reported here is an optical PAPA machine using a coherent optical system and correlation technique. PAPA, the **Italian** abbreviation for Automatic Programmer and Analyzer of Probabilities is a learning machine invented by Gamba.[38–45]

At the beginning, the principle of PAPA is briefly shown. Let us consider the case where the device separates the patterns into n classes, $C_1, C_2, \cdots, C_n$. In the training period, the jth training pattern s_{ij} which is known to belong to class C_i is placed in front of a random mask m_k ($k = 1, 2, \cdots, N$), as shown in Fig. 2. We assume that both the patterns and random masks consist of transparent and opaque zones.

Now, the following value

$$P_{ik} = \frac{\sum_j q_{ijk}}{\sum_j Q_{ij}} \tag{6}$$

is considered, where q_{ijk} denotes the light intensity falling on the detector and Q_{ij}, the light intensity when the random mask is removed. The summation is done for all patterns belonging to class C_i. Then, Eq. (6) means the degree of transparency of patterns belonging to class C_i through the random mask m_k and is thought to be the conditional probability of m_k when C_i is given. In the same

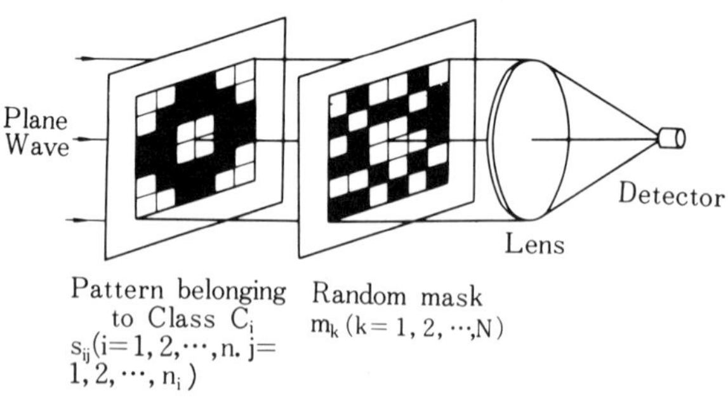

Fig. 2. Training process of PAPA

way, P_{ik} is measured and stored for the patterns of every class and for every random mask.

In the case of classification, an unknown pattern is placed and the light intensity x is measured for every m_k. Here, the observed vector $\underline{x}$ is expressed as follows:

$$\underline{x} = (x_1, x_2, \cdots, x_N). \tag{7}$$

As for the decision mechanism, Bayes' mechanism, an ideal observer is adopted.[46] If the values x_k are statistically independent of each other and if a priori probability $Pr(C_i)$ is equal on any class, a decision is done as follows: The equation

$$z_i = \sum_{k=1}^{N}\{x_k \log P_{ik} + (1 - x_k)\log(1 - P_{ik})\} \qquad (i = 1, 2, \cdots, n) \tag{8}$$

is calculated and when z_l is the maximum value of z_i, the decision that x belongs to class C_l is made.

From now, the case where the patterns are separated into two classes α and β is considered. From Eq. (8),

$$\begin{aligned} z_\alpha - z_\beta &= \sum_{k=1}^{N}[x_k(\log P_{\alpha k} - \log P_{\beta k}) \\ &\quad + (1 - x_k)\{\log(1 - P_{\alpha k}) - \log(1 - P_{\beta k})\}] \\ &= \sum_{k=1}^{N} x_k \log\{P_{\alpha k}(1 - P_{\beta k})\} - \sum_{k=1}^{N} x_k \log\{P_{\beta k}(1 - P_{\alpha k})\} \\ &\quad + \log\frac{\prod_{k=1}^{N}(1 - P_{\alpha k})}{\prod_{k=1}^{N}(1 - P_{\beta k})}. \end{aligned} \tag{9}$$

If the number of the random mask is sufficiently large and if the randomness of both patterns belonging to two classes α and β is same, the last term on the right tends to zero. So, if the following expressions

$$\begin{aligned} y_\alpha &= \sum_{k=1}^{N} x_k \log\{P_{\alpha k}(1 - P_{\beta k})\} \\ y_\beta &= \sum_{k=1}^{N} x_k \log\{P_{\beta k}(1 - P_{\alpha k})\} \end{aligned} \tag{10}$$

are used, the unknown pattern is determined to belong to class α

when y_α is larger than y_β.

4. AN OPTICAL PAPA DEVICE

In order to make $P_{\alpha k}(1 - P_{\beta k})$ or $P_{\beta k}(1 - P_{\alpha k})$ optically, several methods are proposed.[45,47] Here, a method using correlation technique is shown.

Fig. 3 shows the optical arrangement of this apparatus. Firstly, a Fourier transform hologram of a random mask is made with a setup of Fig. 3(a). When the random mask having a rotational symmetry around its center is used, we have

$$m(x, y) = m(-x, -y). \tag{11}$$

Fig. 3(b) illustrates the training period. When a pattern s(x, y) is placed in plane P_1, the output in plane P_3 is expressed

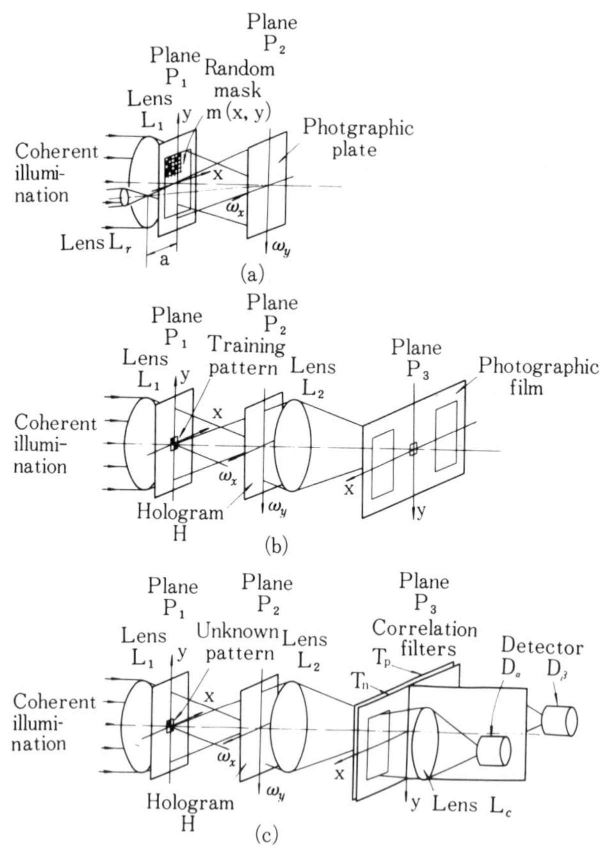

Fig. 3. Principles of optical PAPA device.

as follows:

$$r(x, y) \propto R^2 \iint_{-\infty}^{\infty} S(\omega_x, \omega_y) e^{i(x\omega_x + y\omega_y)} d\omega_x d\omega_y$$

$$+ \iint_{-\infty}^{\infty} S(\omega_x, \omega_y) |M(\omega_x, \omega_y)|^2 e^{i(x\omega_x + y\omega_y)} d\omega_x d\omega_y$$

$$+ R \iint_{-\infty}^{\infty} S(\omega_x, \omega_y) M(\omega_x, \omega_y) e^{i\{(x - a)\omega_x + y\omega_y\}} d\omega_x d\omega_y$$

$$+ R \iint_{-\infty}^{\infty} S(\omega_x, \omega_y) M^*(\omega_x, \omega_y) e^{i\{(x + a)\omega_x + y\omega_y\}} d\omega_x d\omega_y$$

$$\sim R^2 s(x, y) + R \iint_{-\infty}^{\infty} s(x', y') m(x - x' - a, y - y') dx'dy'$$

$$+ R \iint_{-\infty}^{\infty} s(x', y') m^*(x' - x - a, y' - y) dx'dy'. \qquad (12)$$

In this equation, capital letters S and M mean the Fourier transforms of small letters s and m and the amplitude of the reference $|R|$ is much larger than that of the random mask $|m|$. Furthermore, since the random pattern m is a real amplitude object and has rotational symmetry, Eq. (12) is reduced to the following equation:

$$r(x, y) \propto R^2 s(x, y) + R \iint_{-\infty}^{\infty} s(x', y') m(x' - x + a, y' - y) dx'dy'$$

$$+ R \iint_{-\infty}^{\infty} s(x', y') m(x' - x - a, y' - y) dx'dy'. \qquad (13)$$

The second and third terms on the right mean the plus and minus first orders of the diffraction patterns and represent the cross-correlations of s and m, the difference of which is their positions only.

If the dimension of the random mask is sufficiently larger than that of the training pattern, the plus or minus first order term is thought to be equivalent to $q_{\alpha jk}$ given in Eq. (6), that is,

$$\iint_{-\infty}^{\infty} s_{\alpha j}(x', y') m(x' - x + a, y' - y) dx'dy' \equiv q_{\alpha jk}, \qquad (14)$$

where suffixes α and j mean the jth pattern of class α. Here, the coordinates (x,y) and the suffix k, the number of the random mask used in Fig. 2 have the same function or meaning.

Successive exposures with equal exposure time are made for every training pattern of class α with a shutter placed before the half of the photographic film at plane P_3 and the same procedure is repeated for the patterns of class β with the shutter placed before the opposite half of the photographic film. The obtained transparency T_n

contains two cross-correlation patterns with the amplitude transmittance approximately proportional to $(1 - P_{\alpha k})^{1/2}$ and $(1 - P_{\beta k})^{1/2}$. When a contact print of the above transparency is made, the obtained transparency T_p also contains two cross-correlation patterns with the amplitude transmittance approximately proportional to $P_{\alpha k}{}^{1/2}$ and $P_{\beta k}{}^{1/2}$. Hereafter, these transparencies are named the correlation filters.

In the classification of unknown patterns, the setup shown in Fig. 3(c) is used. Here, the correlation filters T_n and T_p are placed in their original position at plane P_3 after T_p is rotated by 180 degrees of arc around the perpendicular axis positioned at its center. An unknown pattern to be classified is placed in plane P_1 and the light intensity falling on the two photomultipliers D_α and D_β is compared. If the photocurrent of D_α is larger than that of D_β, the unknown pattern is determined to belong to class α. In this case, however, certain approximation is made, that is Eqs. (10) are used without logarithms.

5. EXPERIMENTAL RESULTS

As shown in Fig. 4, the optical configuration after Lowenthal and Belvaux[18] was used for the experiment. Fig. 5 shows the random mask consisting of 60 × 40 matrix of transparent and opaque zones. For comparison, the dimension of the patterns for training and classification is also shown. The area of the patterns is a hundredth of that of the mask. As was expected, the amplitude transmittance of the random mask prepared as a transparency is certainly rotationally symmetrical but its phase is not because of the irregularity of the emulsion surface. To eliminate the phase effect, a liquid gate is used, that is, the transparency is put between the two optically parallel glass plates with a small amount of dibutyl phthalate.

Fig. 6 shows the output of matched filtering. The central pattern corresponds to the zero order term or the image of the random mask, the bright spot on the right, the minus first order term

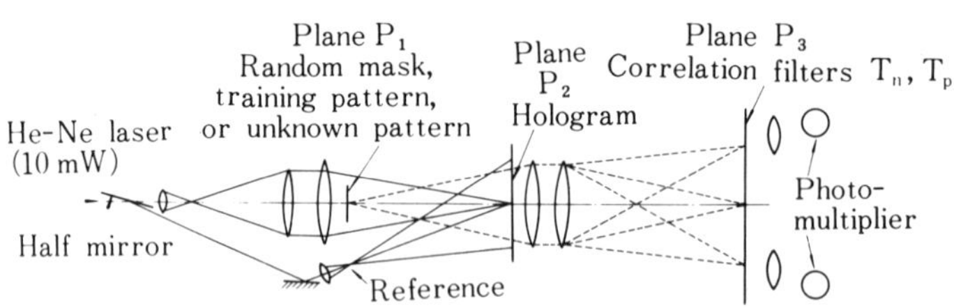

Fig. 4 Configuration of optical PAPA device.

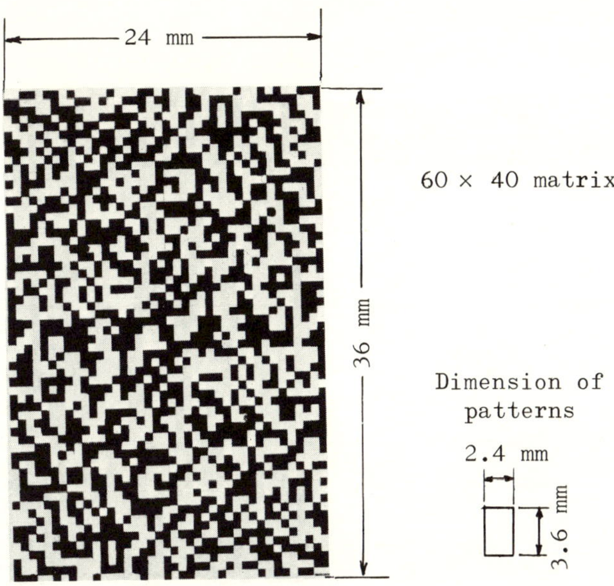

Fig. 5. Random mask.

or the autocorrelation of the random mask and the bright spot on the left, the plus first order term or the self-convolution of the random mask. Because of the incompleteness of phase compensation, the intensity of the plus and minus first order terms is not equal so far, which necessitates the use of unequal normalizing factors for each term. As a result of the improper choice of the intensity ratio of the reference and object beams, differentiation of the zero order term is seen in the figure. Although the bleaching is effective for the reduction of this effect, the process was not adopted in this experiment because of the possibility of increased light scattering.

Fig. 7 shows the impulse response of another hologram of the random mask. Fairly good reconstruction without differentiation is obtained when the intensity ratio of the two beams is properly selected.

The patterns for training and classification are shown in Table 2. These patterns consists of 6 × 4 matrix of transparent and opaque zones. The matrix points of (2, 4), (3, 3), (4, 2), (5, 2) and (2, 2), (4, 1), (4, 3), (5, 4) are always transparent and opaque respectively for each pattern belonging to class α whereas (3, 2), (4, 4), (5, 1), (6, 2) and (1, 1), (1, 4), (3, 1), (6, 3) are so for each pattern belonging to class β. Other matrix points are randomly selected to be transparent and opaque to make the total number of transparent and opaque zones is equal for all patterns.

Fig. 6. Matched filtering of random mask.

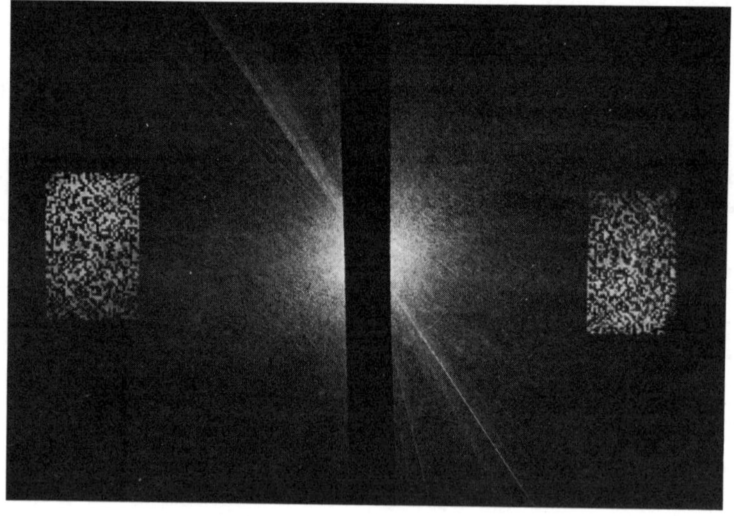

Fig. 7. Impulse response of the hologram of random mask.

Table 2. Classification of patterns.

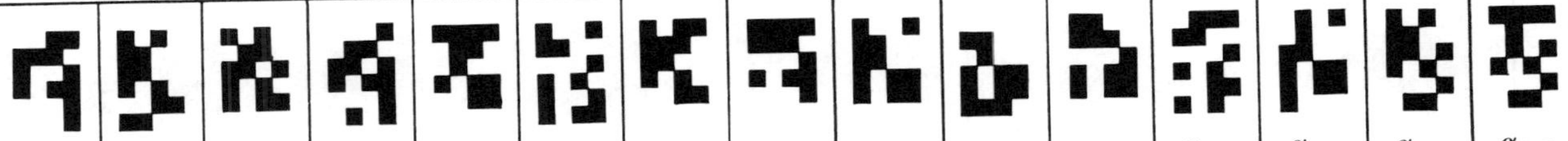

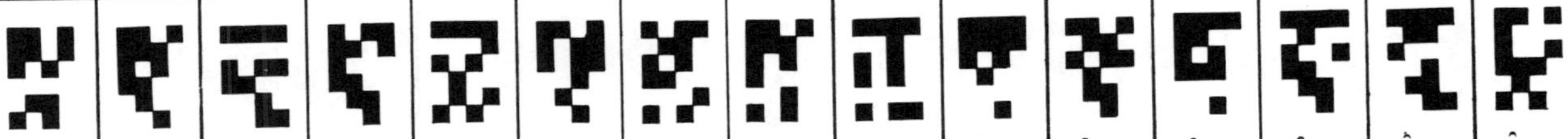

		Training patterns										Unknown patterns				
Class α		α_1	α_2	α_3	α_4	α_5	α_6	α_7	α_8	α_9	α_{10}	α_{11}	α_{12}	α_{13}	α_{14}	α_{15}
	y_α	1.03	1.02	1.02	1.03	0.65	1.03	1.12	0.98	1.10	1.00	1.00	0.91	1.03	1.00	1.01
	y_β	0.97	0.98	0.98	0.97	1.35	0.97	0.88	1.02	0.90	1.00	1.00	1.09	0.97	1.00	0.99
Class β		β_1	β_2	β_3	β_4	β_5	β_6	β_7	β_8	β_9	β_{10}	β_{11}	β_{12}	β_{13}	β_{14}	β_{15}
	y_α	0.89	0.97	0.92	0.81	0.92	0.89	0.84	0.76	0.91	0.89	0.92	0.93	0.94	0.95	0.91
	y_β	1.11	1.03	1.08	1.19	1.08	1.11	1.16	1.24	1.09	1.11	1.08	1.07	1.06	1.05	1.09

Table 3. Correlation filters.

T_n	$(1 - P_{\beta k})$	$(1 - P_{\alpha k})$
T_p	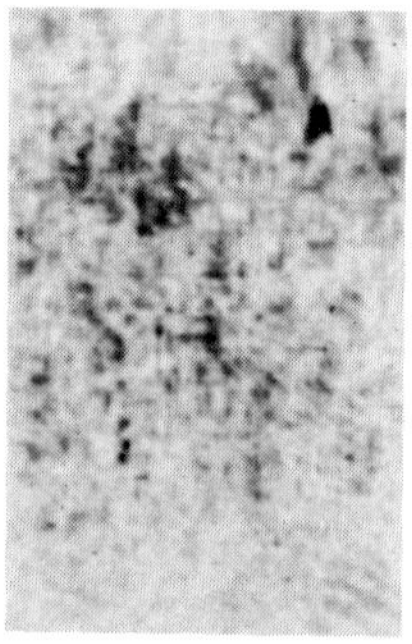 $P_{\alpha k}$	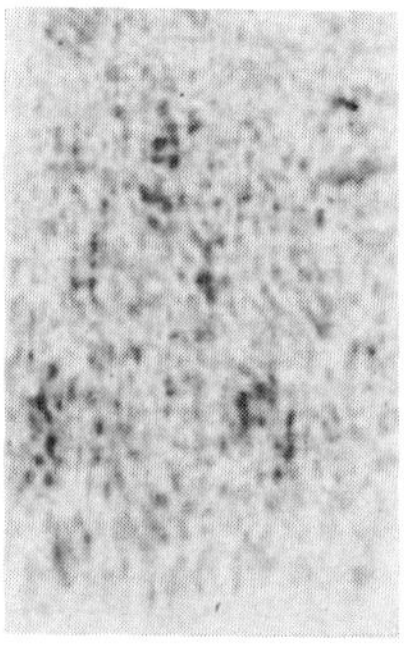 $P_{\beta k}$
$T_n T_p$	$P_{\alpha k}(1 - P_{\beta k})$	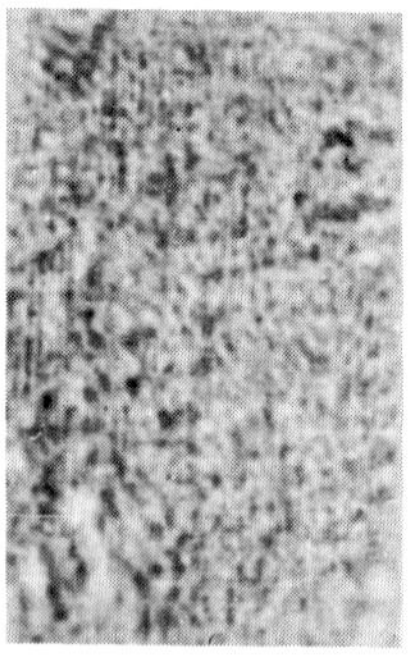 $P_{\beta k}(1 - P_{\alpha k})$

For training, successive exposures are made using the training patterns from $s_{\alpha 1}$ to $s_{\alpha 10}$ placed in plane P_1 with the shutter inserted before the half of the photographic film placed in plane P_3 of the optical system of Fig. 4. In the same way, successive exposures are made for the training patterns from $s_{\beta 1}$ to $s_{\beta 10}$ with the shutter inserted before the opposite half of the photographic film. The obtained correlation filter T_n is used to make the contact print T_p. Table 3 shows the correlation patterns of these filters. In this table, the trasnmittance of composite filters of T_n and T_p which are placed in plane P_3 of Fig. 4 for pattern classification is also shown.

The numerical values shown in Table 2 is the normalized results of classification. The correct classification was made for the patterns of class β as y_β is larger than y_α. For the patterns of class α, however, the discrimination ratio was very low and yet some misjudgements occurred which may be partly due to the incompleteness of the rotational symmetry of the random mask.

6. CONCLUSION

In this paper, an attempt has been made to describe optical information processing using holographic techniques. It has been shown that coherent optics is a very powerful and versatile tool for performing a number of operations on two-dimensional signals. Fourier transform, Fresnel transform, correlation, convolution and deconvolution are some examples.

As an example of holographic information processing, an optical PAPA device was made. The preliminary experiments are promising although the discrimination ratio is not very high in the present stage. To make an exact evaluation, the effect of the characteristic curve of a photographic emulsion and the setting error of both the hologram and correlation filters have to be considered.

REFERENCES

1. S. Tanaka: Oyo Buturi 37 (1968) 860.
2. A. K. Rigler and T. P. Vogl: Appl. Optics 5 (1966) 1086.
3. N. George and J. W. Matthew: Appl. Phys. Letters 9 (1966) 212.
4. T. A. Shankoff and R. K. Curran: Appl. Phys. Letters 13 (1968) 239.
5. A. Labeyrie and J. Flamand: Optics Commun. 1 (1969) 5.
6. S. Lu: Proc. Inst. Elect. Electronics Engrs 56 (1968) 116.
7. E. N. Leith and J. Upatnieks: J. Opt. Soc. Amer. 54 (1964) 1295.
8. G. W. Stroke: Appl. Phys. Letters 6 (1965) 201.

9. G. L. Turin: IRE Trans Inform. Theory IT-6 (1960) 311.
10. D. Huhn, E. Spiller and U. Wagner: Phys. Letters 27A (1968) 51.
11. A. Vander Lugt: IEEE Trans Inform. Theory IT-10 (1964) 139.
12. A. W. Lohmann and H. W. Werlich: Phys. Letters 25A (1967) 570.
13. A. Vander Lugt, F. B. Rotz and A. Klooster, Jr.: Optical and Electro-Optical Information Processing, ed. J. T. Tippett, D. A. Berkowitz, L. C. Clapp, C. J. Koester and A. Vanderburg, Jr.: (MIT Press, Cambridge, 1965) p.125.
14. A. Kozma and D. L. Kelly: Appl. Optics 4 (1965) 387.
15. K. Preston, Jr.: Electronics 38 (1965) No.18, 72.
16. J.-Ch. Viénot et J. Bulabois: Rev. Opt. 44 (1965) 621; C. R. Acad. Sci. (Paris) 263B (1966) 1300; Optica Acta 14 (1967) 57.
17. B. R. Brown and A. W. Lohmann: Appl. Optics 5 (1966) 967.
18. S. Lowenthal et Y. Belvaux: C. R. Acad. Sci. (Paris) 262B (1966) 413; Rev. Opt. 46 (1967) 1; Optica Acta 14 (1967) 245.
19. A. Vander Lugt: Appl. Optics 5 (1967) 1221; 1760; Optica Acta 15 (1968) 1.
20. V. V. Horvath, J. M. Holeman and C. Q. Lemmond: Laser Focus 3 (1967) No.11, 18.
21. D. J. Raso: J. Opt. Soc. Amer. 58 (1968) 432.
22. R. A. Binns, A. Dickinson and B. M. Watrasiewicz: Appl. Optics 7 (1968) 1047.
23. R. P. Grosso, S. J. Kisher and R. M. Vesper: Applications of Lasers to Photography and Information Handling, ed. R. D. Murray (The Society of Photographic Scientists and Engineers, New York, 1968) p.131.
24. G. W. Stroke and R. G. Zech: Phys. Letters 25A (1967) 89; Japan. J. Appl. Phys. 7 (1968) 764.
25. G. W. Stroke, G. Indebetouw and C.Puech: Phys. Letters 26A (1968) 443.
26. G. W. Stroke: Phys. Letters 27A (1968) 405.
27. C. S. Weaver and J. W. Goodman: Appl. Optics 5 (1966) 1248.
28. J. E. Rau: J. Opt. Soc. Amer. 56 (1966) 1490.
29. R. Hioki, H. Kawashima, H. Kato and K. Imai: Oyo Buturi 36 (1967) 713.
30. J. W. Goodman, W. H. Huntley, Jr., D. W. Jackson and M. Lehmann: Appl. Phys. Letters 8 (1966) 311.
31. J. D. Gaskill: J. Opt. Soc. Amer. 58 (1968) 600.
32. M. Noguchi and T. Ose: Seisan Kenkyu 20 (1968) 464.
33. G. W. Stroke, R. Restrick, A. Funkhouser and D. Brumm: Phys. Letters 18 (1965) 274; Appl. Phys. Letters 7 (1965) 178.
34. K. S. Pennington and R. J. Collier: Appl. Phys. Letters 8 (1966) 14.
35. D. Gabor: Nature 208 (1965) 422.
36. A. Dickinson: Marconi Rev. 30 (1967) 40.
37. B. M. Watrasiewicz: Nature 216 (1967) 302.
38. A. Gamba: Proc. Inst. Radio Engrs 49 (1961) 349; Suppl. Nuovo Cimento 26 (1962) 169; 26 (1962) 176.
39. A. Gamba, L. Gamberini, G. Palmieri and R. Sanna: Suppl. Nuovo

Cimento 20 (1961) 112.

40. A. Borsellino and A. Gamba: Suppl. Nuovo Cimento 20 (1961) 221.
41. E. Gagliado: Suppl. Nuovo Cimento 20 (1961) 232.
42. G. Palmieri and R. Sanna: Suppl. Nuovo Cimento 23 (1962) 266.
43. G. Palmieri: Suppl. Nuovo Cimento 23 (1962) 276.
44. A. Gamba, G. Palmieri and R. Sanna: Suppl. Nuovo Cimento 23 (1962) 280.
45. A. Gamba: Suppl. Nuovo Cimento 26 (1962) 371.
46. For example, G. Nagy: Proc. Inst. Elect. Electronics Engrs 56 (1968) 836.
47. H. Kato and S. Tanaka: to be published.

CORRELATION TECHNIQUES BY HOLOGRAPHY AND ITS APPLICATION TO FINGERPRINT IDENTIFICATION

Jumpei TSUJIUCHI*, Kiyofumi MATSUDA** and Naoya TAKEYA**

Tokyo Institute of Technology, Meguro-ku, Tokyo*

The Government Mechanical Laboratory, Suginami-ku, Tokyo**

I. INTRODUCTION

The optical correlation technique is a powerful means for optical pattern recognition [1][2], and holography brings a new and precise method known as optical matched filtering [3] using a Fourier-transform hologram.

As far as the realization of a holographic correlator for practical use is concerned, several fundamental problems such as the configuration of the optical system, aberrations due to the filter, and the influence of slight displacement of the filter have been discussed by Vander Lugt [4], and many useful results have been obtained.

We have made a holographic correlator to be used mainly for fingerprint identification. This correlator has a mechanism by which a slight difference in object dimension can be compensated because the object to be processed is prepared as a photographic transparency which can record the slight defference of magnification that may occur. The result of the processing can be obtained both by visual observation of the reconstructed image and by photoelectric measurement, and a special detecting system to obtain the quantitative result has been used.

This paper is concerned with these practical considerations of the holographic correlator, and some results of fingerprint identification experiments are also shown.

II. FUNDAMENTAL TREATMENT

As shown in Fig. 1, an object O_A prepared as a photographic transparency is placed in the converging beam from a lens L_1 at the distance z from the focal plane H, and the amplitude transmission of the object is given by

$$f_A(u,v) = e_A\, a(u,v) + (1 - e_A)\, g(u,v), \tag{1}$$

where a(u,v) is the relative amplitude transmission of the object, e_A the diffraction efficiency corresponding to the maximum height of the object transmission from the background, and g(u,v) the shape of the uniform aperture inside of which the object exists, the amplitude distribution of the signal wave impinging upon the plane H becomes [5]

$$W_A = A \exp(ik \frac{x^2 + y^2}{2z}) \iint f_A(u,v) \exp(-ik \frac{ux + vy}{z}) \times \exp(-ik \frac{(ux + vy)^2}{2z^3})\, du\, dv, \tag{2}$$

where A is a constant proportional to the amplitude of the source.

The last term in the integral in (2) is an additional phase term which makes the diffraction deviate from the Fraunhofer type. If p_{max} is the maximum spatial frequency of the object to be recorded in the hologram, we have

$$x_{max} = y_{max} = p_{max}\, \lambda z, \tag{3}$$

so the maximum phase change from the Fraunhofer condition becomes

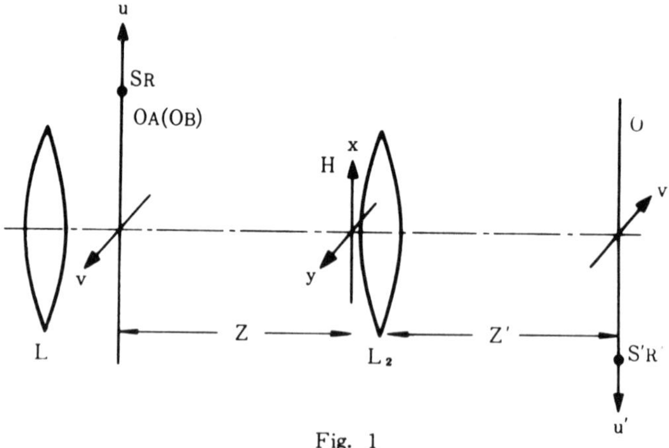

Fig. 1

$$\psi_{max} = \pi (u^2 + v^2)_{max} \, p^2_{max} \lambda / z. \tag{4}$$

The allowable limit of the Fraunhofer diffraction can be estimated by $\psi_{max} < \pi/2$, and if we put $\lambda = 6328$ A, $u_{max} = v_{max} = 10$ mm and $p_{max} = 30$ lines/mm, we have $z > 228$ mm, and (2) becomes under these conditions

$$W_A = A \exp(ik \frac{x^2 + y^2}{2z}) \left\{ e_A \, F[a(u,v)] + (1 - e_A) \, F[g(u,v)] \right\}, \tag{5}$$

where F[ ] means the Fourier transformation.

The reference wave emitted from a point source $S_R(u_o,0)$ in the same plane as the object has the amplitude

$$W_R = R \exp(ik \frac{x^2 + y^2}{2z}) \exp(-ik \frac{u_o x}{z}), \tag{6}$$

where R is also a constant.

Interference fringes due to these two waves in the plane H make a Fourier-transform hologram of O_A, and we have the component by which the direct image of S_R is to be reconstructed;

$$\begin{aligned} H_R &= \gamma W_A^* W_R \\ &= \gamma AR \exp(-ik \frac{u_o}{z}) \left\{ e_A \, F[a(u,v)] + (1 - e_A) \, F[g(u,v)] \right\}, \end{aligned} \tag{7}$$

where γ is a photographic gamma of the hologram.

If an object O_B with the amplitude transmission

$$f_B(u,v) = e_B \, b(u,v) + (1 - e_B) \, g(u,v) \tag{8}$$

is put in place of O_A, the wavefront illuminating the hologram becomes

$$W_B = B \exp(ik \frac{x^2 + y^2}{2z}) \left\{ e_B \, F[b(u,v)] + (1 - e_B) \, F[g(u,v)] \right\}, \tag{9}$$

where B, b(u,v) and e_B are correspond to A, a, e_A. The reconstructed image of S_R is formed in O' plane by the use of a lens L_2, and the amplitude distribution of the image S_R' becomes

$$r'(u',v') = F^{-1}[W_R H_R]$$
$$= \gamma ABR \left\{ e_A e_B E_A E_B \Phi_{AB} + e_A(1 - e_B)\, E_A E_G \Phi_{AG} + e_B(1 - e_B)\, E_B E_G \Phi_{BG} + (1 - e_A)(1 - e_B)\, E_G E_G \Phi_{GG} \right\}, \tag{10}$$

where Φ_{AB} is the cross correlation function of O_A and O_B;

$$\Phi_{AB}(u',v') = \frac{\iint a^*(u + u',\ v + v')\ b(u,v)\ du\ dv}{E_A\ E_B} \tag{11}$$

with

$$E_A = \left\{ \iint |a(u,v)|^2\ du\ dv \right\}^{1/2}$$
$$E_B = \left\{ \iint |b(u,v)|^2\ du\ dv \right\}^{1/2}. \tag{12}$$

When O_B is identical with O_A, it becomes

$$\Phi_{AB}(0,0) = \Phi_{AA}(0,0) = 1 \tag{13}$$

and if O_A has random structure

$$\Phi_{AB} = \Phi_{AA} = \delta\,(u',\ v'), \tag{14}$$

Φ_{AG} and Φ_{BG} are almost constant, and Φ_{GG} becomes a very broadly spreading function, then the image S_R' is seen as a small bright spot surrounded by nearly uniform background light, the amplitude distribution of which becomes

$$r'_{AA} = \gamma ABRe_A^{\,2}\ E_A^{\,2}\ \Phi_{AA} + \text{const.} \tag{15}$$

If O_B is not identical with O_A, Φ_{AB} spreads out very broadly and no spot image is formed. So, the identification of O_A with O_B can be made merely by the observation of the bright spot image.

When O_B has partially identical structure with O_A, b(u,v) in (8) may be written as

$$b(u,v) = h(u,v)\ a(u,v) + c(u,v)$$

$$= a'(u,v) + c(u,v), \tag{16}$$

where h(u,v) is a masking function which modifies the identical structure with O_A and c(u,v) the noise structure, the reconstructed image becomes

$$r'_{AB} = \gamma \, ABR \, e_A e_B E_A E_{A'} \, \Phi_{AA'} + \text{const.} \tag{17}$$

Let Q_A and Q_B be intensities of spot images corresponding respectively to r'_{AA} and r'_{AB} except for the background light, we have

$$Q_B / Q_A = e_B^2 E_{A'}^2 \Phi_{AA'}^2 / (e_A^2 E_A^2 \Phi_{AA}^2) \tag{18}$$

and supposing I_A and I_B are intensities diffracted by O_A and O_B respectively, we have

$$I_B / I_A = e_B^2 \, E_B^2 / (e_A^2 \, E_A^2), \tag{19}$$

then

$$\Phi = E_{A'} \, \Phi_{AA'} / (E_B \, \Phi_{AA}) = (Q_B I_A / Q_A I_B)^{1/2}. \tag{20}$$

This quantity Φ can be considered as the discrimination ratio [2] in this case. Similarly, if the original object O_A has partially identical structure with O_B, we have

$$\Phi = E_{B'} \, \Phi_{BB'} / (E_A \, \Phi_{AA}) = (Q_B I_A / Q_A I_B)^{1/2}. \tag{21}$$

In order to measure I_A or I_B, a negative photographic image D is taken at the focus without any object, and total intensities Q_o and Q_o' are measured with and without the stop D respectively, the intensity transmission of D becomes $T = Q_o'/Q_o$. After O_A is set, the total intensity at H without D becomes $P_A = I_o + I_A$, where I_o is the non-diffracted light and I_A the diffracted light due to O_A, and $P_A' = TI_o + I_A$ with D. We have then

$$I_A = \frac{P_A' - TP_A}{1 - T}, \tag{22}$$

and I_B is also obtained by a similar method.

The compensation of the object magnification can be made by the displacement of the object toward the hologram. If O_B is magnified by $m = 1 + \Delta m$, the compensation can be made by changing z into mz, the wavefront in this case is

$$W_B' = B \exp(ik \frac{x^2 + y^2}{2mz}) \iint f_B(mu,mv) \exp(-ik \frac{mux + mvy}{mz}) \, du \, dv, \tag{23}$$

so the final image becomes

$$r''(u',v') = F^{-1}[\exp(ik \frac{x^2 + y^2}{2z})(1 - \frac{1}{m})] * r'(u',v'). \tag{24}$$

The first term corresponds to the defocus of the output image which produces a slight spreading of the spot image. Supposing the effective aperture of the hologram is determined by (3), the radius of the defocused spot image becomes

$$\bar{u}' = (1 - \frac{1}{m}) \, p_{max} \lambda z'/2 = \Delta m \, p_{max} \lambda z'/2, \tag{25}$$

for example p_{max} = 30 lines/mm, z' = 300 mm and $\Delta m = 0.1$, we have $\bar{u}' = 0.3$ mm.

Lateral shift of the object $(\Delta u, \Delta v)$ makes an additional phase term $\exp(ik(\Delta ux + \Delta vy)/z)$ to W_B, and it produces only a lateral shift of S_R' which can be easily compensated by the shift of the observing system.

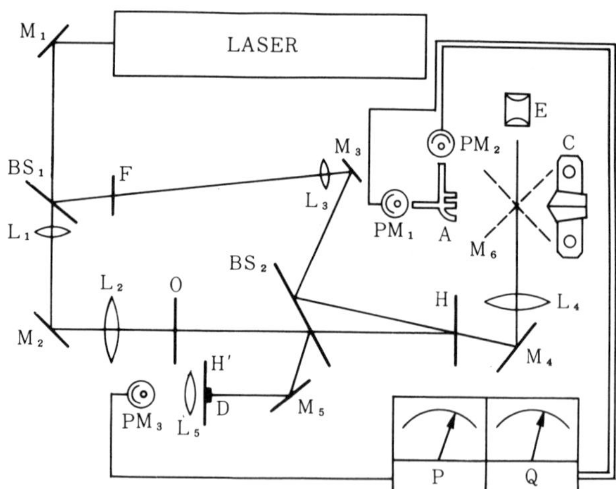

Fig. 2 Holographic correlator

III. HOLOGRAPHIC CORRELATOR

Fig. 2 shows the schematic diagram of the holographic correlator we have constructed for fingerprint identification. An object O_A is placed at the position O in the converging beam which is emitted from the lens L_2 and which can move along the optical axis of L_2, and the Fraunhofer diffraction wave can be obtained in the focal plane H. The reference source is realized at the focus of a lens L_3 so as to obtain a Fourier-transform hologram in the plane H.

The reconstruction is made by putting an object O_B at the position O in place of O_A, and the diffraction wave illuminates the hologram placed in H plane. The reconstructed wavefront forms the direct image of S_R by the use of a lens L_4, and this image can be observed by an ocular E and also by a camera C using a rotary mirror M_6. The compensation for lateral shift of the object can be made by small tilt of a mirror M_4.

A special detecting aperture A made by optical fiber bundles is used to eliminate the effect of the background light superposed on the output image S_R'. This aperture has, as shown in Fig. 3, two concentric apertures, and the difference of intensities measured by each aperture is used as output signal. A part of the signal beam is divided by a beam splitter BS_2 and the diffraction pattern is obtained in H' plane, where a photomultiplier PM_3 serves to measure the intensities of diffracted light I_A and I_B past a stop D.

Quantitative descriptions of the correlator are as follows; distance between the object and the hologram $z = 430 \pm 45$ mm, focal length of the observing system $z' = 300$ mm, effective aperture of the object to be used 20×30 mm, and the wave length $\lambda = 6328$ A.

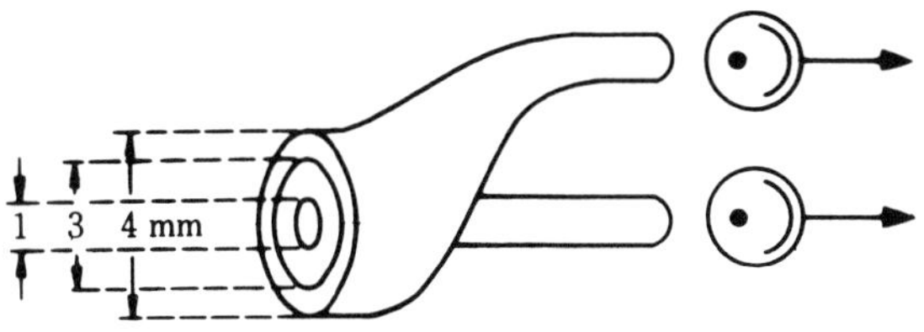

Fig.3 Detecting aperture of output image

IV. APPLICATION TO THE FINGERPRINT IDENTIFICATION

As an object of the holographic correlator, a fingerprint seems to be one of the most suitable object: a fingerprint is obtained always in natural size and has a sort of binary structure with fairly high and limited spatial frequency domain about 2 - 10 lines/mm ; with such an object, it is easy to produce a Fourier transform hologram of good quality for use as matched filters, because the distribution of diffracted light is not mainly concentrated in the central portion of the focus and the hologram can record easily these frequencies which will be essential to the identification.

Furthermore, this method has an excellent capacity of detecting the signal from the object submerged in noise this is a very important feature of this application because the fingerprint obtained in the crime scene (latent fingerprint) is often damaged by other unnecessary fingerprints and by the structure of the surface touched by the fingers. The application of the holographic correlation technique to fingerprint identification was reported before by Horvath, Holeman and Lemmond [6], and we have made a series of experiment for the same purpose.

To process the fingerprint by this correlator, it is photographed in constant magnification about natural size and made into a negative transparency which shows fingerprint ridges as bright lines (see Fig. 4). A latent fingerprint obtained in the scene is used as original object 0_A and recorded fingerprints, stored by the police, as objects to be examined 0_B, or vice versa, and the discrimination ratio Φ for each fingerprint 0_B is measured.

Immersion of the object is used to compensate for irregularity in the film surface using liquid gates. A film on which the image of the fingerprint is recorded is put between two well polished glass plates with a small amount of dibuthyl phtalate. If this liquid immersion is not used, the discrimination ratio between two identical fingerprints decreases if these two are recorded on different films. The table shows an example of the results obtained by Kodak 649 F 35 mm film.

0_A 0_B	immersion immersion	immersion no immersion	no immersion no immersion
Φ	1.00	0.81	0.61

The two photographs in Fig. 4 show the image obtained by this correlator when 0_A is and is not identical with 0_B, and we have Φ = 1.00 in the former case and Φ = 0 in the latter case.

Fig. 5 shows the result when 0_B is identical with 0_A but 0_B

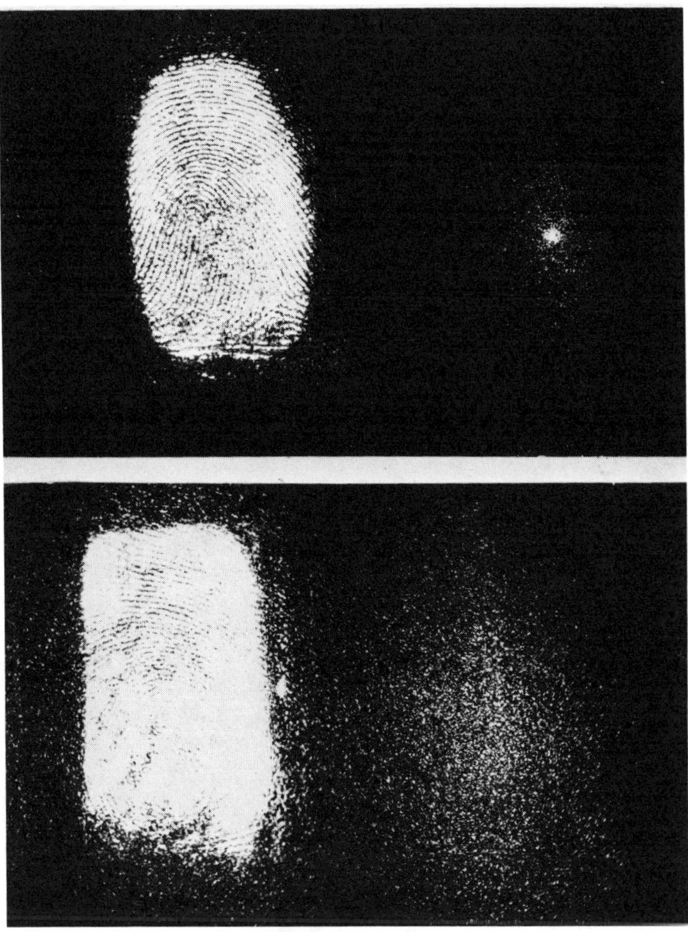

Fig. 4 Output images of the correlator. Each photograph containes the image of a fingerprint to be examined on the left and the reconstructed spot image on the right. The upper photograph corresponds to the case when O_A is identical with O_B, and the lower one corresponds to the case when O_A is not identical with O_B.

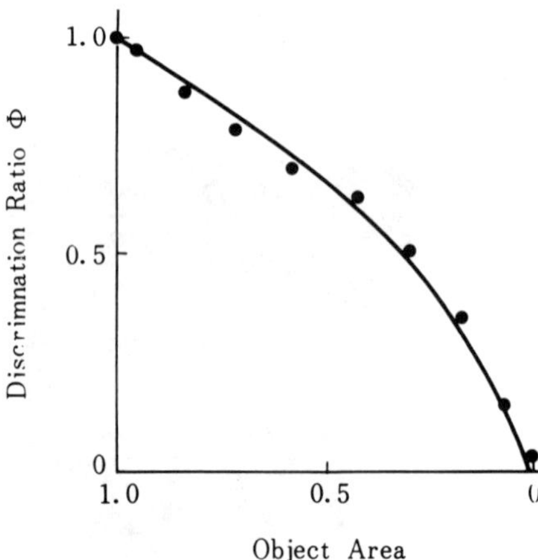

Fig. 5 Relations between discrimination ratio and object area

is masked down by a knife edge from one side to the other. From this result, it is understood that the discrimination ratio decreases according to the area of the fingerprint to be examined, but it may be concluded that the identification can be made with less than 15 % of the area where $\Phi^2 = 0.1$. Fig.6 shows the discrimination ratios about three different portions of the same area for the identical fingerprints, and the values of Φ are almost the same if the area of the object is the same.

The next experiment is made with two fingerprints of the same finger but with different printing pressures. The result shows that the printing pressure causes a slight deformation of the fingerprint, and the discrimination ratio varies from 1.00 to 0.36 according to the pressure.

This result is very important to the practical applications of this technique to criminal investigation. As an exhibit for use in court, fingerprint identification is made by 12 - 15 characteristic points such as ridge endings, ridge bifurcations, short ridges, etc.. But this correlation is made mainly by the whole shape of ridges but not by the detailed characteristic points. Even though the fingerprint is deformed by pressure or by the nature of the surface touched by the fingers, the detail is hardly deformed and the identification by the use of characteristic points is not affected. But the identification by correlation is very sensitive to the shape deformation and it must be used very carefully to avoid misunderstanding. To solve this difficulty, it may be useful to determine a suitable threshold of the discrimination

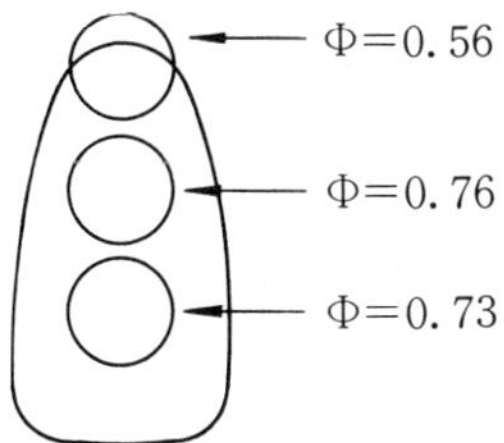

Fig. 6 Discrimination ratios about three different portions of the same area in a fingerprint

ratio as a result of a lot of experiments, and also to record two or three fingerprints with different pressure when they are prepared for storage.

Finally, the effect of noise superposed on the fingerprint are examined. An experiment shows $\Phi = 0.37$ for two superposed fingerprints none of them could be identified by visual observation. From this, although the value of the discrimination ratio is not so large, we believe that the identification of fingerprints damaged by noise can be made if a suitable threshold of Φ is determined.

V. CONCLUSION

The utility of the holographic correlation technique to fingerprint identification is very interesting, and it is important to evaluate the role of this technique in the practical application.

In order to investigate the fingerprint, the ridge pattern is classified into many items and a latent fingerprint picked on the criminal scene is investigated first according to this classification with a computer. This may be called the first step of the investigation. After picking out some fingerprints of the same item as the latent fingerprint from a great number of stored fingerprints, the second step of the investigation is carried out by visual observation of experts, and this work requires skill and is sometimes made difficult by insufficient area or by noise damage.

The holographic method can be used in two ways. The first way is to use it in the first step of the investigation; the latent fingerprint is processed by this correlator without any classfication and processing is made with a great number of stored fingerprints and so the time allowed for each processing must be extremely short. As liquid immersion of many stored fingerprints will be practically impossible, fingerprints must be stored as Fourier-transform holograms whose phase disturbing effect is much weaker than that of the object transparencies because of the smallness of hologram area. These holograms will be recorded for example on roll films and the film passed at the position given by the Fourier spectrum of the latent fingerprint. The positionning of each hologram must be made very precisely and the latent fingerprint placed at the object position is rotated when a new hologram is put in position to find the proper azimuth. If the correlator is used in the second step, the processing may be made more slowly and carefully and the stored fingerprint can be used both as the object in a liquid gate and as the hologram. In both cases, the first and second step, when photographic material without any phase disturbance is available, the processing will be, like other coherent optical processing, much more easy and precise than the present state with a liquid gate.

In conclusion, it may be said that the holographic correlation technique can be used as a unique method to investigate fingerprints once photographic materials without any phase disturbance are available and once a suitable threshold of discrimination ratio for each case is established.

REFERENCES

[1] D. McLachlan, Jr.: J. Opt. Soc. Amer. 52 (1962) 454

[2] J. D. Armitage and A. W. Lohmann: Appl. Opt. 4 (1964) 461

[3] A. Vander Lugt, F. B. Rotz and A. Klooster, Jr.: Optical and Electro-Optical Information Processing, Edited by J. T. Tippett et al. (MIT Press, 1965) 125

[4] A. Vander Lugt: Appl. Opt. 5 (1966) 1760, 6 (1967) 1221

[5] M. Born and E. Wolf: Principles of Optics (Pergamon Press, 1964) 382

[6] V. V. Horvath, J. M. Holeman and C. Q. Lemmond: Laser Focus (June 1967) 18

OPTICAL IMAGE DEBLURRING METHODS

Jumpei Tsujiuchi - Tokyo Institute of Technology
Tokyo, Japan

George W. Stroke - State University of New York
Stony Brook, New York

ABSTRACT

Greatly sharpened images may be extracted from photographs which have been blurred either by accident (motion, lack of focus, atmospheric turbulence, etc.) or deliberately, for instance when 'coded' in view of special image processing or synthesis applications. Four image-deblurring methods may be distinguished: 1. Coherent optical analogue processing, 2. Incoherent optical analogure processing (non-electronic) 3. Incoherent opto-electronic analogue processing and 4. Digital computer processing. Of these, the methods 1 and 3 are reviewed in some detail, and some previously unpublished experimental results are given for illustration. Particular emphasis is also given to the distinction between image-deblurring methods operating directly in the spatial (image) domain, very powerful but developed so far only in special cases (e.g. defocused and motion-blurred images), on the one hand, and, on the other the spatial-frequency filtering image-deblurring methods, also quite general, but still in an early state of perfection. Both methods may require considerable further work before widespread implementation

INTRODUCTION

Recent work has demonstrated that it is possible to extract greatly sharpened 'deblurred' images from photographs which have been blurred either by accident or, in some cases, deliberately, for instance when 'coded' in view of special image processing applications. In simple words, it has truly become possible to turn a bad photograph into a good image in a great number of situations, including cases when the photographs were blurred by motion, lack of focus and by atmospheric turbulence.

In a general way, optical image deblurring methods implement principles based on the 'spatial frequency filtering' formulation given by Maréchal and Croce in 1953, when the 'deblurring' is carried out in the spatial frequency (spectral or Fourier) domain.* The first experimental verifications of 'spatial frequency filtering' in optics using 'complex' (amplitude and phase filters were described by Tsujiuchi in 1958, for the cases when the deconvolving filter could be computed and its phase part realized by evaporation. The principles for optical "spatial frequency filtering" of blurred photographs in the general case was first given by G. W. Stroke and R. G. Zech in 1967, when they showed that the required complex spatial filter could be realized by holography, directly from the experimental 'spread function' (i.e. directly from the experimentally obtained blurred image of a point, or equivalently of an edge), without any computation! The principle of holographic 'Fourier-transform division' filtering has now been extensively verified, in a number of dramatic cases, some of which are shown as illustration for the first time here.

There exists another class of situation where optical analogue deblurring (i.e. deconvolution) of the blurred photographs may be carried out directly in the spatial domain, rather than in its spectral (Fourier-transform) domain. In these cases, the deblurring is carried out by literally again convolving the convolved (blurred) photograph with a suitable 'deblurring masking function', most readily in a scanning arrangement. Such a method was described already in 1961, by D. J. McLean, in view of the Wild (Culgoora) imaging radio-telescope application, and again, more recently, in 1968, by Tsujiuchi, and in 1969, by the University of Arizona researcher W. Swindell. The most recent of these spatial-domain

---

*For a general background see:

E. L. O'NEILL, Introduction to Statistical Optics (Addison-Wesley Publ. Co. Reading, Mass. 1963)

G. W. STROKE, An Introduction to Coherent Optics and Holography (Academic Press, New York, 2nd Edition, 1969)

J. W. GOODMAN, Introduction to Fourier Optics (McGraw Hill, New York, 1968)

scanning deconvolution methods is that of Stroke, Halioua and Indebetouw: it combines _incoherent_ use of laser scanning with the use of the general holographic image-deblurring filter of Stroke and Zech.

Finally, there exists still another class of cases which arises notably when accidental blurring of the photograph appears either unavoidable or is to be expected. In these cases, it has appeared to be advantageous to actually "pre-code" the photographs with the aid of a suitable 'coding spread function', which subsequently lends itself to much more effective optical analogue (holographic) decoding than the 'spread function' corresponding to a well-corrected optical system. This principle appears to have also first formally proposed and demonstrated by Stroke in 1968. It was later adopted for further theoretical extension by a number of authors, including J. W. Goodman.

In addition to the optical (analogue) image deblurring methods, there exist also direct electronic 'image enhancing' methods, using television scanning principles, as well as digital picture processing methods which start with suitable (e.g. flying-spot) scanning of the blurred photographs, and use digital electronic computers to carry out the mathematical operations required for the deconvolution of the blurred photographs.

The television 'image enhancing' methods are based on the principles first described by P. Goldmark and J. M. Hollywood already in 1951, and, in their most recent form, by R. McMann and A. Goldberg in 1968. Digital picture processing has proven to be successful in the hands of a number of researchers, notably in those of J. L. Harris since 1966.

In this paper, we propose to discuss principally the optical analogue image deblurring methods, notably because they have proven to be a particularly powerful tool in cases where a considerable number of image 'points' must be deblurred simultaneously and rapidly.

However, it should be clear, even at this time, that the various image deblurring methods tend naturally to combine features common to several classes. Accordingly, distinctions such as those which we make, in keeping with tradition, are to be considered principally as being convenient for classification. Four classes of image deblurring methods are singled out in TABLE I. They include several other image processing methods in addition to those already introduced above. Another possible way by which image-deblurring methods could be classified would use a distinction between processing in the _spatial_ (i.e. image) domain, on the one hand, and, on the other, processing in the _spatial frequency_ (spectral, Fourier) domain. We have used an appropriate sign

TABLE 1

IMAGE DEBLURRING METHODS

1. COHERENT OPTICAL ANALOGUE PROCESSING

1.1 Processing of ordinary blurred images

a. non-holographic spatial filtering [1,2,3]

b. holographic spatial filtering [4,5]

1.2 Processing of pre-coded images

a. holographic method [6,7]

1.3 Electro-Optical Methods

a. Ultra-sonic delay line modulation methods [25]

2. INCOHERENT OPTICAL ANALOGUE PROCESSING (NON ELECTRONIC)

2.1 Optical "masking" methods

→ a. photographic masking [8]

→ b. methods using quenching of flourescent screens [8]

2.2 Photo-chemical "masking" effects

→ a. Herschell effect [9]

→ b. Adjacency (Eberhard) effect [8,10]

3. INCOHERENT OPTO-ELECTRONIC ANALOGUE PROCESSING

→ a. logetronics [8]

→ b. spatial-domain deconvolution [including television, flying spot and other line-by-line scanning] [11-17]

→ c. Laser-light spatial-domain scanning deconvolution using the general Stroke and Zech holographic deblurring filter [22].

4. DIGITAL COMPUTER PROCESSING

4.1 Digital Filtering

→ a. deconvolution [18,19]

b. modulation of Fourier spectrum [19]

4.2 Analytical methods

a. image motion deblurring [20]

→ b. integral equation solution [21]

NOTE: → indicates processing in the spatial (image) domain

(arrow) to indicate the processing methods which operate directly in the spatial (image) domain. For completeness, the reader may wish to consult the outstanding group of papers in the proceedings of the 1964 Boston symposium on optical and electro-optical information processing*.

* J. T. Tippett et al. editors, Optical and Electro-Optical Information Processing (M.I.T. Press, Cambridge, Mass. 1965).

PART I

GENERAL PRINCIPLES OF SPATIAL-FREQUENCY FILTERING IN OPTICS AND NON HOLOGRAPHIC IMAGE DEBLURRING METHODS

1. Spatial Filtering Principles in Optics (Coherent Light) (general considerations*)

Let an optical system (Fig. 1) be used to image a plane field (x,y) into a plane field (x'y'). Let the function h(x',y') represent the intensity distribution in the image of a point, throughout the field: the function h(x',y') is known in optics as the "point-spread function" or as the "spread function" for short. It is nothing but the impulse-response function of the optical system. Let o (x,y) be the intensity distribution in the object and i(x',y') the intensity distribution in the geometrical-optics image of the object may be described by o(x,y), then we have the following well-known convolution integral relation:

$$i(x',y') = \iint_{-\infty}^{+\infty} o(x,y)\ h(x'-x,y'-y)\ dx\ dy \qquad [1]$$

We may write this equation in the spatial-frequency (Fourier transform domain) in the form

$$I(u,v) = O(u,v)\ H(u,v) \qquad [2]$$

where I(u,v) is the spatial Fourier transform of i(x',y'), and so on. The coordinates (u,v) are the spatial frequency coordinates. The functions I(u,v) and O(u,v)are known as the spectral representations of the image i(x',y') and of the object o(x,y) respectively. One can thus consider an optical system as being a linear filtering operator on the Fourier spectrum of the object. The optical system is then characterized by its spatial frequency transfer function H(u,v).

In a perfect system, the impulse-response function h(x',y') would be a δ-function. Its Fourier-transform H(u,v) would be a constant for all spatial frequencies. However, in the case of imperfections, such as imperfect focusing, the function H(u,v) is

*A More detailed presentation is given in Part II, within the context of holographic image deblurring. This Part I is to some extent based on the paper presented by J. Tsujiuchi on 23 August 1968 at the United Nation's Conference on the Exploration and Peaceful Uses of Outer Space, Vienna [Ref. 14].

not any more a constant for all frequencies, in the case of simple or simply-connected (non-obstructed) apertures. The general effect of a decreasing amplitude in the spatial frequency transfer function H(u,v) is both a decrease of "resolution" with increasing frequencies and a significant related decrease in the contrast of the images.

Among the most frequent imaging imperfection, it may be of interest to single out the following, as we do in TABLE II

TABLE II

SPECTRAL REPRESENTATION OF TYPICAL IMAGE IMPERFECTIONS

(a) Slightly blurred image

$$H(\psi) = 1 - a\psi^2/\psi^2_{max} \tag{3}$$

(b) Out of focus image

$$H(\psi) = 2\, J_1(b\psi)/(b\psi) \tag{4}$$

(c) Literally moved image

$$H(u) = \sin cu/cu \tag{5}$$

(d) Double image

$$H(u) = \cos du \tag{6}$$

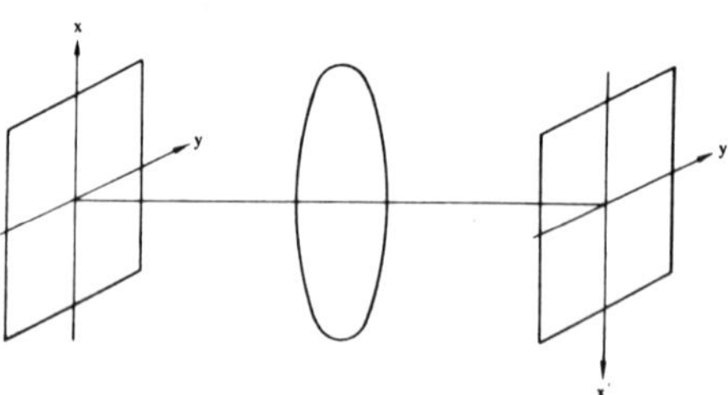

FIG. 1 An Optical System

where $\psi = (u^2+v^2)^{1/2}$, and where a, b, and c are constants, measuring the magnitude of the imperfections.

The principle of "spatial filtering" may now be readily understood with the aid of the following equations. Let the spectrum I(u,v) of the image i(x',y') be "filtered" (in amplitude and in phase) by a filter characterized by the function T(u,v). The spectrum of the image, upon traversal of the filter, becomes

$$I'(u,v) = I(u,v)\ T(u,v) = O(u,v)H(u,v)T(u,y) \tag{7}$$

We may now chose the filter function to be

$$T(u,v) = 1/\ H(u,v) \tag{8}$$

so that the filtered image becomes equal to the geometrical image o(x',y') of the object, except, to a first approximation, at or near the spatial frequencies for which H(u,v) = 0. This is the principle of "spatial frequency filtering".

Another effect of the use of spatial filtering is the reduction of "noise". For example, if we can chose the filtering function in such a way that it passes the main frequencies characterizing the image, while rejecting the "noise". In case that there exists no correlation between the image spectrum and the noise spectrum, the noise-rejecting filter function is

$$T(u,v) = \Phi(u,v)/[\Phi(u,v) + \Phi_n(u,v)] \tag{9}$$

where Φ and Φ_n are the Wiener spectra of the image and of the "noise", respectively [26]. In a typical image, Φ_n is a Gaussian distribution with comparatively large extent in the spatial frequency domain, while Φ tends to concentrate near the low frequencies. In such a case, H(u,v) may take on the form of a low-pass filter.

On the other hand, we may note that the filter for a slightly blurred picture has the general character of a high-pass filter. Accordingly, we find that the filtered image may tend to become "noisy" in the case of the presence of signigicant amount of "noise". In such a case, one would chose as a filtering function one described by the equation

$$T[u,v] = \frac{H^*(u,v)\ \Phi(u,v)}{|H(u,v)|^2\Phi(u,v) + \Phi_n(u,v)} \tag{10}$$

in order to suppress the "noise" as much as possible [27].

From the preceding remarks, the process of spatial-frequency filtering is a mathematical process which consists in a complex multiplication of the spectrum of the image by the filtering function. Significantly, the multiplication of the various spectral components may be considered, to a very good approximation, to be essentially a multiplication of each image frequency by a corresponding filtering factor (amplitude and phase), independently of the other frequencies. While this is of course not true in the strictest mathematical sense (and therefore results in Gibbs-like and similar effects), this point of view does permit one to understand why spatial-frequency filtering does in fact work so well in practice, in spite of the frequency presence of a number of isolated "zeros" (nulls) in the spatial frequency transfer function $H(u,v)$. The implementation of the spatial frequency filtering process may be carried out both with the aid of optical analogue methods (in general coherent) and with the aid of digital electronic computers. However, before discussing some of the optical methods, notably those using holographic filters [4], we introduce the important alternate principle, namely that of spatial deconvolution, which may be carried out in incoherent light, directly in the spatial (image) domain.

2. SPATIAL-DOMAIN DECONVOLUTION IN OPTICS (IN INCOHERENT LIGHT)

Let us consider the image $i(x',y')$ in the form given by equation (1):

$$i(x',y') = \iint o(x,y)h(x'-x,y'-y)\,dx\,dy \qquad (1)$$

For the purpose of further discussion, we write Eq. (1) in its operational form

$$i = o \otimes h \qquad (11)$$

Let us now conceive of a situation in which the image is further convolved with some function $s(x', y')$, for instance by flying-spot or similar scanning, using a beam of which the light-intensity distribution in the image-plane, upon incidence onto the film, is equal to $s(x',y')$. The amplitude of the light transmitted through the film is then

$$i \otimes s = o \otimes h \otimes s \qquad (12)$$

It is clear that the amplitude of the light transmitted through the film will be equal to the object intensity $o(x',y')$ if we can make

$$h \otimes s = \delta \qquad (13)$$

or, equivalently, in the spatial-frequency (Fourier) domain:

$$HS = 1 \tag{14}$$

We thus find that the appropriate function s(x',y'), also known as the "masking function" is such that its spatial Fourier transform is:

$$S(u,v) = 1/H(u,v) \tag{15}$$

By comparing eq. (15) with eq. (8), we note, first og all, that the desired "masking function" s(x'y') is nothing but the Fourier transform of the spatial-filtering "filtering function" T(u,v).

Stroke, Halioua and Indebetouw [22] recently showed the remarkable fact that the image-deblurring scanning function (the "masking" function) is always a real function (partly positive and partly negative). Indeed, since h(x',y') is a real function, it follows that its Fourier transform H(u,v) is hermitian, i.e.

$$H(u,v)^* = H(-u,-v) \tag{16}$$

Accordingly, the functions S(u,v) = [1/H(u,v) is also hermitian, i.e.

$$\frac{1}{H(u,v)^*} = \frac{1}{H(-u,-v)} \tag{17}$$

But eq. (17) may be written in the form

$$\frac{1}{H(u,v)}^* = \frac{1}{H(-u,-v)} \tag{17'}$$

and therefore, since S(u,v) is hermitian, one concludes that the deblurring scanning function s(x,y) must be real! Stroke et al. [22] further concluded that a general scanning deconvolution method would best be carried out in laser light, rather than with incoherent light, as had been done by Tsujiuchi [14] since 1966 and more recently by Swindell [17] for special cases. Indeed, bu using the general image-deblurring holographic Fourier-transform division filter of Stroke and Zech [4], first described in 1967 for holographic image deblurring applications (see part II.), and by illuminating the Fourier transform division filter [which has a transmittance in amplitude equal to 1/H(u,v)], one materializes in the focal plane of a lens, following the filter, the desired scanning function s(x,y), as may be readily concluded from the above, including eq. (15) The blurred photograph, described by eq. (1) is then scanned with s(x,y) according to eq. (12) and the image, reconstructed "point by point" is equal to o(x,y) with the usual diffraction limit of the original image-forming aperture. Significantly, as observed by

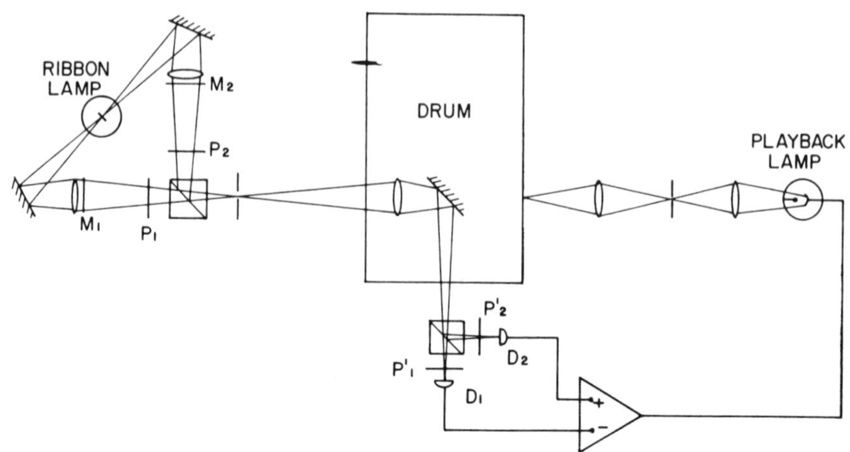

FIG. 2 Spatial domain scanning image deblurring arrangement (after W. Swindell, loc. cit.)

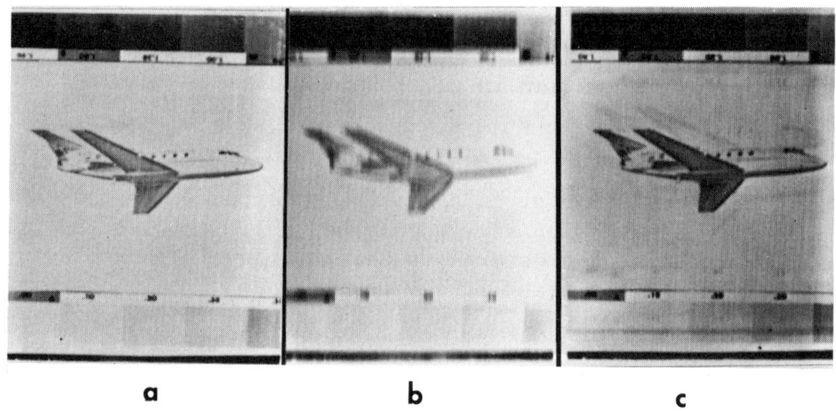

FIG. 3 Spatial domain scanning image deblurring: experimental results (using apparatus of FIG. 2) [after W. Swindell, private communication for publication of this work]: (a) shows a photograph of the object scene after being scanned on the processor and thus forms the basis for comparison with the processed image. (b) shows the object scene after having been scanned with a slit, thus simulating linear uniform image motion blur. (c) shows the results of processing (b) with a continuous tone mask designed to reduce this specific degradation.

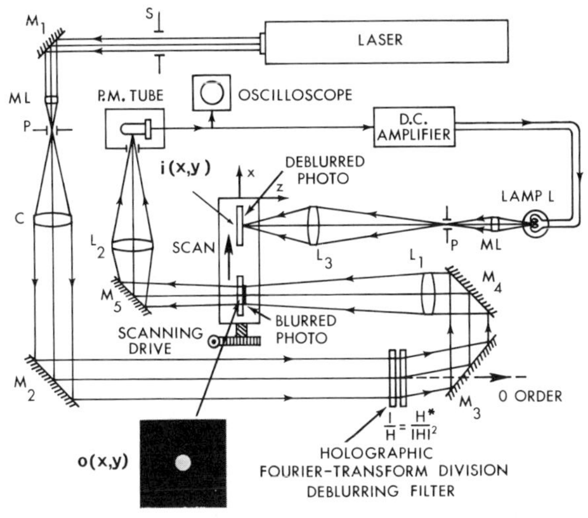

4a

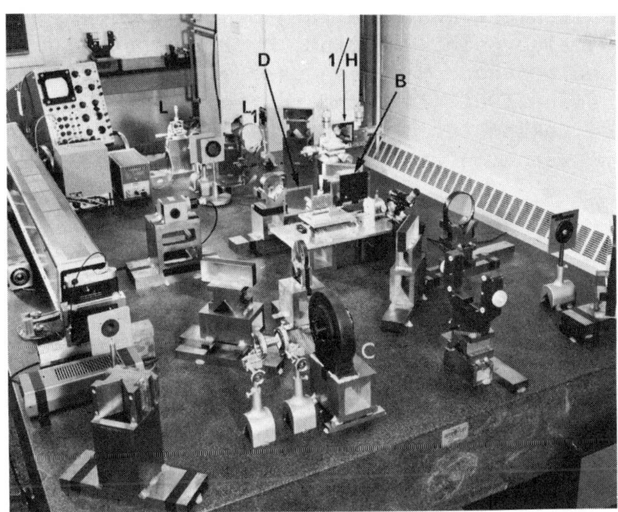

4b

FIG. 4.a Spatial-domain laser-scanning image deblurring using Stroke and Zech holographic Fourier-transform division filter (after G. W. Stroke, M. Halioua, G. Indebetouw and F. Poisson, Optics Vommun. Vol. 1, No. 8, pp. 355-358. March 1970). The inserts show: o(x,y) = badly blurred image point (diameter of blurr circle: about 2 mm with f-240 mm lens used); i(x,y)=deblurred sharpened image point. (b) Experimental arrangement used in spatial-domain laser scanning image deblurring work (after Stroke, Halioua, Indebetouw and Poisson, loc. cit.)

Stroke et al. [22], the image-forming part of their proposed scheme remains incoherent, while fully exploiting the brightness of the laser, because of the point-by-point scanning: the usual "coherent" image-processing difficulties (e.g. interference phenomena resulting from phase-transmission of the film, laser speckle, etc.) are thus suppressed!

The general laser-scanning spatial-domain image deconvolution scheme of Stroke et al [22] using the general holographic Fourier-transform division filter of Stroke and Zech [4] was preceded, as we mention by the incoherent-light flying-spot deconvolution scheme used by Tsujiuchi [14] since 1966 and by that, independently by Swindell [17], since 1968. Indeed, as shown by Tsujiuchi and by Swindell, the practical implementation of the spatial deconvolution becomes readily feasible, in special cases, by using two orthogonally polarized beams, one consisting of the positive part of s(x',y'), say S_+ and the other of the negative part, say s_-. Upon transmission through the blurred image point, the two beams are subtracted in a suitable arrangement [FIG. 2] and the resulting current, when transformed into a suitable light intensity, then represents the deblurred image point. The entire blurred image is thus deblurred, point by point, in a suitable scanning arrangement: the deblurred image is reconstructed, also point by point, by illuminating a photographic emulsion. Such a scheme has been used with television display, as by Tsujiuchi since 1966 [14], and the results are shown in FIG. 5-8 Another way of taking into account the positive and negative values of the masking function was deomonstrated by McLean [15].

The remarkable results obtained by Swindell are shown in FIG. 2 and 4 which indicate the great power of directly scanning spatial-domain image deconvolution methods. A schematic diagram, a photograph of the apparatus and an early experimental result of the holographic spatial-domain scanning image deblurring method of Storke et al [22] is shown in Fig. 4a. and 4b. The results of the Tsujiuchi method are discussed in greater detail below, and the results are shown in FIG. 5-8.

We may also recall the direct "aperture-correcting" method of D. J. McLean [15], who used a photographic spatial-domain "mask" directly with an annular-aperture lens, to correct for the well-known $J_o^2(\pi D\theta/\lambda)$ diffraction pattern of the (thin) annular aperture of the J. P. Wild radio telescope, as originally conceived [23]. Indeed, Wild had shown (1961) that this spread function, having the well-known most undesirable rings surrounding the bright central point, could be readily converted into the spread function corresponding to a fully-filled circular aperture, having the same diameter as the annulus. Wild showed that the correction function would have the form shown in FIG. 5. It is seen that this function is, for the most part, a positive function, but it does have some negative parts. In this particular case, as shown by McLean, the simplest method of taking into account the positive and the negative parts

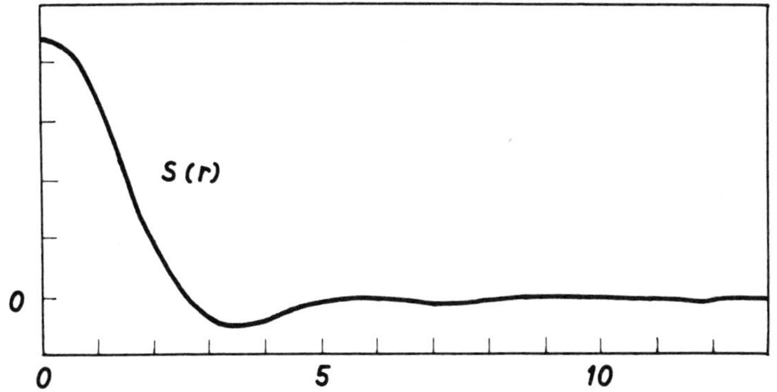

FIG. 5 Aperture correction function for an annulus

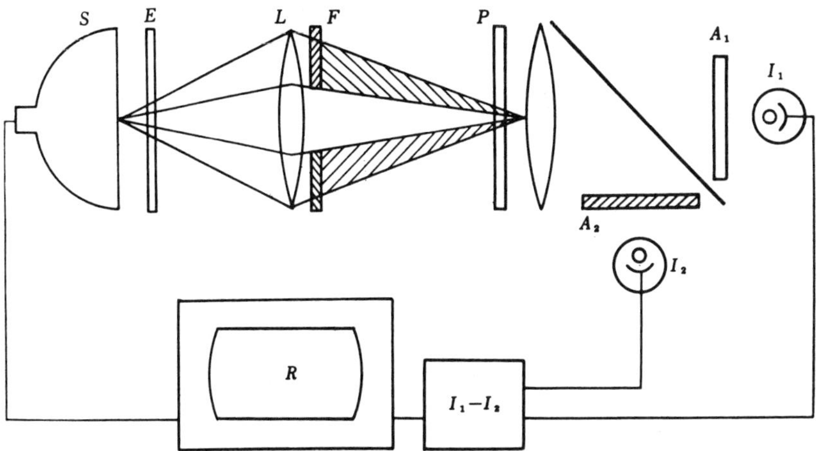

FIG. 6 Incoherent spatial filtering system

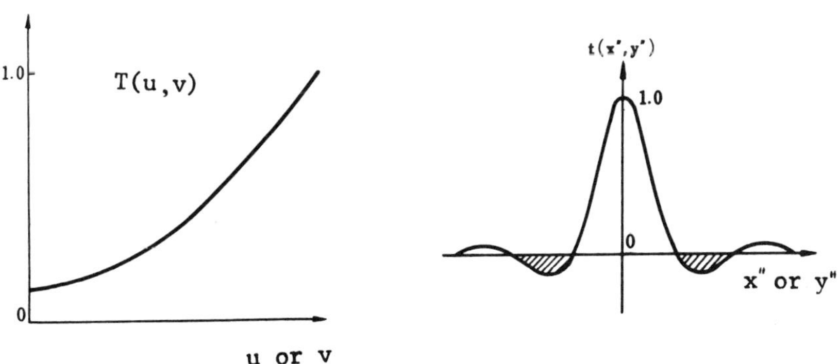

FIG. 7 A high pass filter T(u,v) for slightly blurred pictures and its Fourier transform t(x",y").

of s(r) was to add to the modified correction (masking) function a suitable constant background, which in turn was truncated at such large distance from the center that its image was essentially uniform over the limited field of interest. The masking aperture was made photographically by a long exposure of a black and white generating pattern while the latter was spinning about its center. In the final version of his Culgoora "radio-heliograph', (24), Wild has chosen to use direct electronic analogue computation, rather than the optical processing scheme described. The use of coherent-light electro-optical signal processing using ultrasonic delay-line modulators, for this and other image processing applications should also be noted [25].

In order to complete the illustration of the possibilities offered by the spatial deconvolution methods in incoherent light, we give some further details of the method shown in FIG. 6, after Tsujiuchi [14]. In a general way, the spatial-filtering deconvolution in incoherent light is performed in a system which has a transfer function equal to T(u,v). However, in the special case of slightly blurred pictures, T(u,v) is a high-pass filter shown in FIG. 7a. Its Fourier transform, the impulse response of the system, i.e. t(x", y") has a negative part, shown shaded in FIG. 7b. The negative part, having a negative intensity, cannot be realized with a conventional incoherent imaging system. Its realization is achieved as follows.

As described by Tsujiuchi [14] a flying-spot scanning system is used, as shown in FIG. 6. The screen of a cathode ray tube is imaged onto the blurred transparency with the aid of an objective L. The light beam transmitted through the transparency is recorded by two photomultiplier tubes I_1 and I_2 which give two signals corresponding to the two beams transmitted through P. A polarizer E is placed next to the lens L and two analyzers A_1 and A_2, respectively in front of I_1 and I_2. If the two axis of A_1 and A_2 are orthogonal with respect to each other, and the axis of A_1 is parallel to E, and the axis of F is at 45° from E, and if the lens L is slightly out of focus, then the scanning spot on P is formed of both the central and of the marginal parts, as shown, in such a way that the central part of the beam is recorded by I_1 and the marginal part (the shaded part of FIG. 6) is recorded by I_2. If the signals are subtracted in such a way that the receiver R receives the signal I_1-I_2, together with the synchronisation signal from S, then one may indeed obtain the impulse response function t(x",y") with its "negative" intensity in this arrangement. The shape of t(x",y") is controlled by the amount of defocusing of L, the aportioning of the external beam through F relative to the central beam, and by the gains of I_1 and I_2. An illustration of an experimental result obtained with this system is shown in FIG. 8 (a, b, and c). A photograph of the apparatus is shown in FIG. 9

FIG. 8 (a)

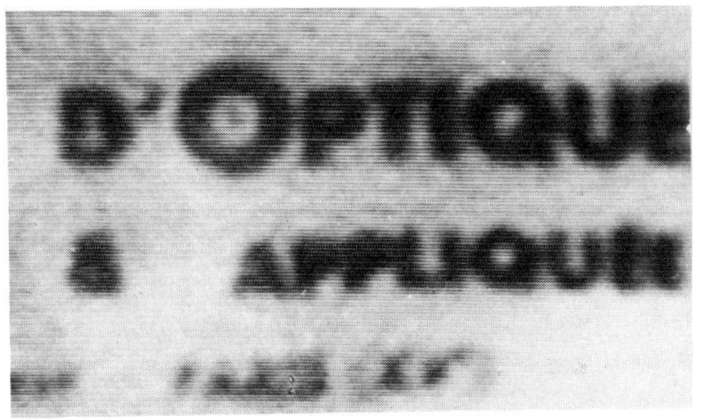

FIG. 8 (b)

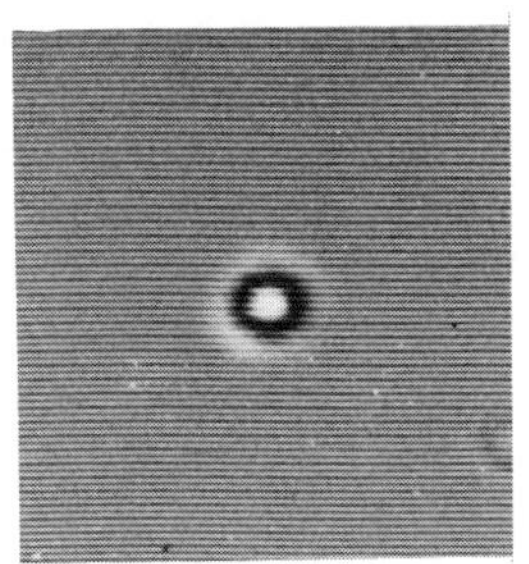

FIG. 8 (c)

FIG. 9

3. SPATIAL-FREQUENCY (FOURIER) DOMAIN NON-HOLOGRAPHIC COHERENT DEBLURRING

Spatial-frequency domain filtering, according to A. Maréchal and P. Croce [1] may be carried out in an arrangement such as that shown in Fig. 10. In this arrangement, let P be the image i(x',y') to be "deblurred" (i.e. filtered). The image P is illuminated by a beam of coherent light, originating from the point S and focused in S' by the converging lens L_1. Upon traversal of the transparency P, the light distribution in the (s,t) plane is the Fourier spectrum I(s,t) of the amplitude transmittance of the image i(x',y'). By placing a spatial filter F with a transmittance T(s,t) into the (s,t) plane, one obtains in the focal plane of the lens L_2 a "restored" image which is "deblurred" to a very satisfactory degree, as discussed in detail in PART II. Here we shall limit ourselves to the discussion of some of the experimental problems of their solution. [2]

a. Preparation of the transparency to be deblurred

Because of the illumination of the transparency with coherent light, the achievement of an amplitude transmission equal to the intensity distribution requires certain precautions (see also PART II).

First of all, it is necessary to process the transparency in such a way that the gamma is equal to -2. Alternately, one tries to have a low contrast in the image.

Next, one must avoid the introduction of spurious phase variations by transmission through the transparency. Such phase variations result in very damaging "noise-like" intensity variations by Fourier transformation in the final image (recall the fact that only the phase variations alone, in bleached holograms, are sufficient to display full-contrast range intensity images, by Fourier transformation in the eye!). The spurious phase variations are suppressed by immersion of the transparency between two very good glass plates, using a liquid of an index similar to that of the emulsion. Tsujiuchi [14] has successfully used buthylphtalate between two carefully polished glass plates.

b. Preparation of the filter

The problem is to realize a filter such that its amplitude transmission function is equal to the "restoring" function. In the general case [see PART II] the required Fourier-transform division filter is prepared holographically according to Stroke and Zech [4]. In a number of special cases, the filters may be prepared according to the method first described by Tsujiuchi [2], using the evaporation method for the phase part, as discussed below.

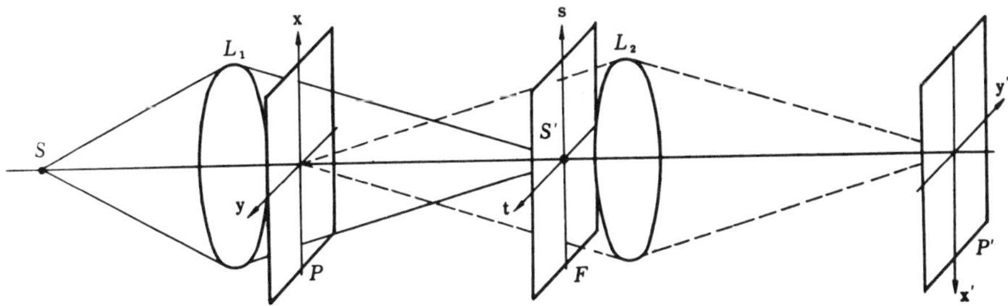

FIG. 10 Spatial filtering system

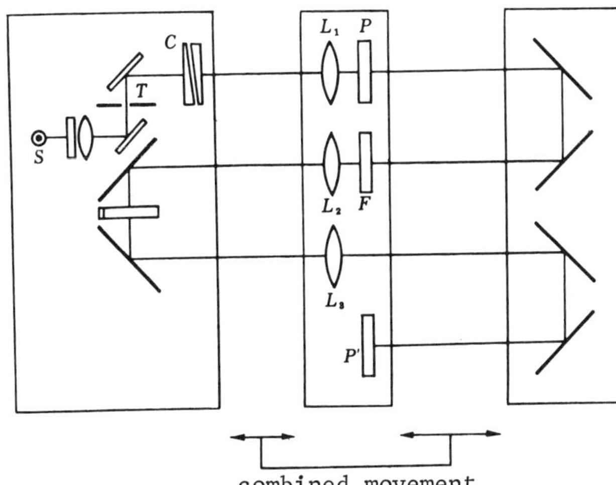

FIG. 11 Variable frequency filtering apparatus schematic

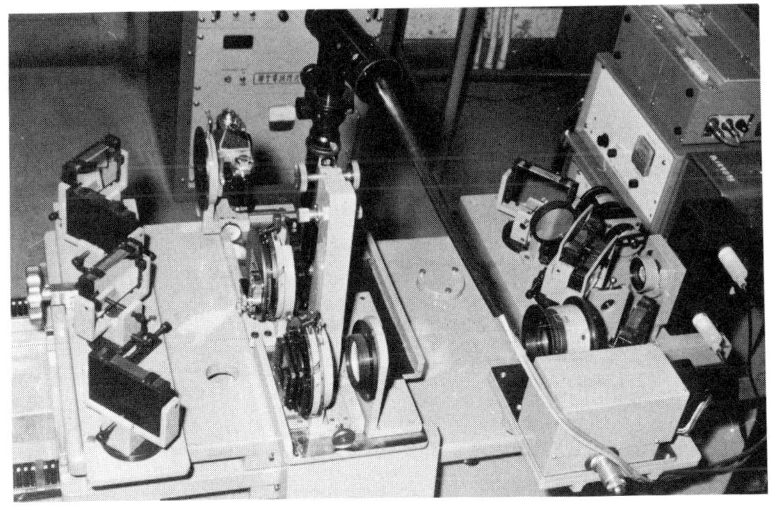

FIG. 12 Variable frequency filtering apparatus

For slightly blurred images, and for images with slight amount of "noise", an amplitude filter alone may be sufficient. However, in more general cases, it is necessary to have a combination of amplitude and phase filter. The amplitude filter may be prepared photographically, by taking a photograph of an object having an intensity distribution proportional to the transfer function of the system used in the recording of the 'blurred' photograph. The phase part of the filter is realized by depositing a half-wave thin layer of MgF_2, introducing a phase delay of π. The two parts of the filter (amplitude and phase) are carefully aligned, close to each other, with their centers in as perfect coincidence as possible, and by using between them a layer of immersion liquid, so as to avoid spurious phase delays from the amplitude filter. A new filter must be prepared for each different transfer function $H(s,t)$: this makes the method rather tedious in practice. In order to circumvent a part of this problem, Tsujiuchi has introduced a variable-frequency arrangement. This permits one to continuously vary the magnification of the Fourier spectrum, while maintaining a constant magnification for the final image. The arrangement thus permits to adapt any desired spatial frequency (within the adjustments) in the filter to the spectrum of the blurred image: the "optimum" filtering is determined by inspection of the final image. The Tsujiuchi variable-frequency adapted filtering apparatus is shown in FIG. 11. A monochromatic light source S is used to illuminate a small hole T, which, in turn, is imaged in F by the lens L_1. The blurred image P to be filtered is placed immediately after the lens L_1 as shown, and its Fourier spectrum is produced in the F-plane. The spatial-frequency "deblurring" filter F is placed in F, and the deblurred image is produced in P' by means of the optical system L_2-L_3. It is this system which is the varying frequency system, arranged also so as to maintain a constant magnification between P and the final image P'. The filtering apparatus of FIG.11 also incorporates the possibility of sweeping the intensity distribution in the F-plane in order to permit one to measure the Wiener spectrum both in the blurred as well as in the deblurred image. The intensity sweeping and measurement are obtained with the aid of a rotating wedge C and a microphotometer which may be inserted between F and L_2.

A photograph of the apparatus of FIG. 11 is shown in FIG. 12 and experimental results obtained with it in FIG. 13.

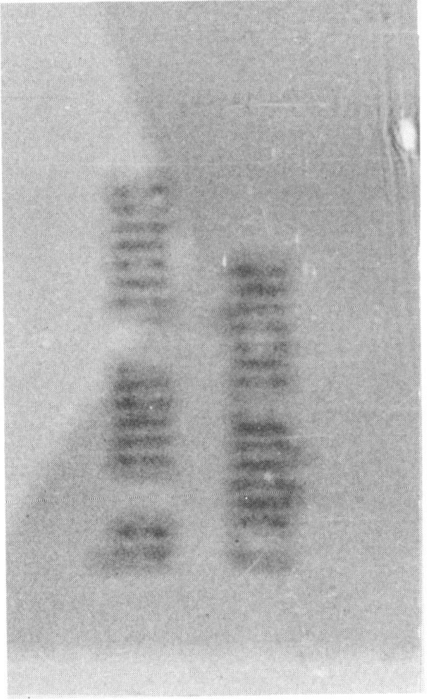

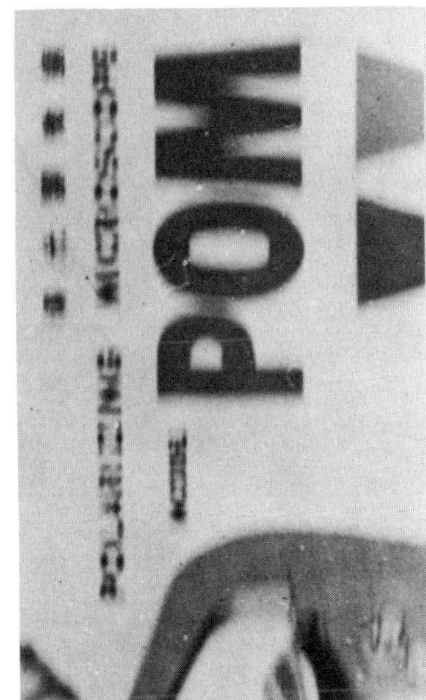

FIG. 13

REFERENCES FOR PART I

[1] A. Maréchal et P. Croce: C. R. Acad. Sci. Paris 237 (1953) 607
[2] J. Tsujiuchi: Progress in Optics Vol. II, edited by E. Wolf North Holland Pub. Co. (1963) 131
[3] T. M. Holladay and J. D. Gallatin: J. Opt. Soc. Am. 56 (1966) 869
[4] G. W. Stroke and R. G. Zech: Phys. Letters 25A (1967) 89
[5] J. Tsujiuchi, T. Honda and T. Fukaya: Opt. Communications to be published
[6] G. W. Stroke: Phys. Letters, 28A (1968) 252
[7] J. W. Goodman and W. T. Rhodes: U. S. - Japan Seminar on "Information Processing by Holography" (1969)
[8] A. W. Lohmann: Opt. Act. 6 (1959) 319
[9] D. H. Kelly: J. Opt. Soc. Am. 51 (1961) 1095
[10] K. Sayanagi: J. Appl. Phys. Japan 29 (1960) 155
[11] P. Goldmark and J. M. Hollywood, Proc. IRE, 39 (1951) 1314
[12] R. McMann and A. Goldberg, J. Soc. Mot. Pict. Telev. Engrs. 77(1968)221
[13] T. P. Cheatham and A. Kohlenberg, National Convent. IRE (1954)
[14] J. Tsujiuchi, ICO-7 Preprint (1966) 90, United Nation's Conference on the Exploration and Peaceful Use of Outer Space (1968)
[15] D. J. McLean, Proc. Roy. Soc. (London), 263A(1961)545-551
[16] J. K. Hawkins and C. J. Munsey, J. Opt. Soc. Am. 57(1967)914
[17] W. Swindell, Un. of Arizona, Optical Sciences Newsletter, Vol. 4, No. 1 (January 1970), p. 24.
[18] F. C. Billingsley: Proceeding of Computerized Image Techniques Seminar (1967) S.P.I.E. II - 1
[19] T. S. Huang and H. L. Kasnitz; ibid. XVII -1
[20] J. L. Harris, Sr.: J. Opt. Soc. Am. 56 (1966) 569
[21] R. W. Harris and C. K. Rushforth: Scientific Report No. 8 Air Force Cambridge Research Laboratories (1966)
[22] G. W. Stroke, M. Halioua and G. Indebetouw, Physics Letters 31A, No. 6(23 March 1970) p. 341.
[23] G. W. Stroke, M. Halioua, G. Indebetouw and F. Poisson, Optics Commum. 1 (1970) 355-358.
[24] J. P. Wild, Proc. Roy. Soc. (London) 262A(1961)84-99
[25] J. P. Wild, Proc. Roy. Soc. (London) 286A(1965)499-509
[26] L. B. Lambert, M. Arm and A. Aimette, in Optical and Electro-Optical Information Processing, J. T. Tippett at al. ed. (MIT Press, 1965)
[27] H. Saito, Seisan Kenkyu (Monthly Journal of the Institute of Industrial Science, University of Tokyo), 100(1958)8 (in Japanese
[28] C. W. Helstrom, J. Optc. Soc. Am., 57(1967)2937

PART II

HOLOGRAPHIC IMAGE DEBLURRING METHODS*

Introduction

Two forms of holographic spatial filtering may be used for the deconvolution of accidentally blurred images and of images deliberately 'coded' for aperture-synthesis purposes. The first is generally applicable and uses a Fourier-transform division filter realized by holography from the intrumental 'spread function' of the system. The second, even more straightforward, is particulary suited for the special case when the autocorrelation function of the spread function is or may suitably be made to be sharply peaked. This form should permit one to realize a new class of optical systems in cases where conventional (lens and mirror) systems may not be readily realized (e.g. for x-ray and ultrasonic imaging, for space astronomy and photography, among others).

Recent work has indeed demonstrated that it is possible to use holographic methods to extract greatly sharpened 'deblurred' images from photographs which have been blurred by accident or deliberately coded, for instance in view of 'aperture synthesis' applications [1-8]. The blurring or coding of the blurred photograph $g(x,y)$ may be expressed as the spatial convolution:

$$g(x',y') = \iint_{-\infty}^{+\infty} f(x,y)h(x'-x,\ y'-y)dxdy, \tag{1}$$

between the desired image $f(x,y)$ and the instrumental impulse response function (spread function, i.e. image of a point) $h(x,y)$. It is well known that equation (1) takes the form of the product:

$$\overline{G}(u,v) = \overline{F}(u,v)\overline{H}(u,v) \tag{2}$$

in the spatial Fourier-transform domain, where we have

$$F(u,v) = \iint_{-\infty}^{+\infty} f(x,y)\exp[2\pi i(ux+vy)]dxdy \tag{3}$$

---

* This part is based to some extent on the invited paper presented by G. W. Stroke on 24 September 1968 at the Symposium on Applications of Coherent Light, International Commission for Optics, Florence, Italy. The paper appeared in Optica Acta, (1969) Vol. 16, No. 4, pp. 401-422. Because of its integrated content, it appeared desirable to use the format notations and equation numbering as given in the original paper. The parts from G. W. Stroke's OPTICA ACTA paper are reprinted by kind permission from Taylor & Francis Ltd.

and similarly for G and H, with suitable normalization*. It is also known that a Fourier-transform relation such as that of equation (3) exists between the spatial part of the complex electric field amplitude E(x,y) in the pupil of a perfectly corrected (ideal) lens system and the field E(u,v) in the image plane (u,v)**, i.e.

$$\bar{E}(u,v) = \iint_{-\infty}^{+\infty} \bar{E}(x,y)\exp[2\pi i(ux+vy)]dxdy \tag{4}$$

It is also known that one may readily realize by transmission through a photographic transparency an electric field vectro distribution $E_T(x,y)$ equal to the exposure g(x,y) of a photographic transparency i.e.

$$\bar{E}_T(x,y)=g(x,y) \tag{5}$$

by suitable processing of the photographic negative exposed to the 'intensity' g(x,y), where we have

$$g(x,y)=\langle\tilde{\bar{E}}(x,y,t)\tilde{\bar{E}}^*(x,y,t)\rangle \tag{6}$$

---

* For a general background, notations and approximation used see, e.g.

Stroke, G. W., 1966, An Introduction to Coherent Optics and Holography (New York: Academic Press); second, revised and enlarged edition, December 1968.

Stroke, G. W., 1967, "Diffraction Gratings" in Handbuch der Physik, Vol. 29, edited by S. Flugge (Berlin: Springer Verlag), pp. 426-754, notably pp. 504-570.

** Another way of expressing the relation between the transmittance of the photographic plate and the exposure (kindly suggested by Prof. D. Gabor) is as follows: One may readily realize an amplitude transmission t(x,y) of a photographic emulsion proportional to the exposure E(x,y) by suitable processing. The exposure is proportional to the intensity:

$$I=\langle E(x,y,t).\ E^*(x,y,t)\rangle$$

In the linerar range of the Hurter and Driffield curve, where the slope is $\tan^{-1}\gamma$ and where we have the quation log (1/T=γlogE(x,y) the intensity tramsmission is

$$T(x,y)=\text{const}.I^{-\gamma}$$

and the amplitude transmission:

$$t(x,y)=[T(x,y)]^{1/2}=\text{const}.I^{-1/2\gamma}$$

and by making $\gamma=-2$ the amplitude transmission t(x,y) becomes proportional to E. Another condition which meets these requirements, according to Tsujiuchi [ref 14, Part I] is that obtained when the image is recorded with low contrast.

under the usual assumptions [see Stroke, loc. cit. 1966, 1967, 1968], and similarly for the spread function h(x,y). For instance, in the linear range of the Hurter and Driffield curve, where the slope is $\tan^{-1}\gamma_n$, and where we have

$$\log \frac{1}{|\overline{E}_T|^2} = \gamma_N \log g(x,y) \tag{7}$$

the 'amplitude' transmission of the negative exposed to g(x,y) is

$$|\overline{E}_T| = g(x,y)^{-\gamma_N/2} \tag{8}$$

By taking suitable care, it is possible to make the phase transmission of the negative to be a constant, so that we may write:

$$\overline{E}_T(x,y) = g(x,y)^{-\gamma_N/2} \tag{9}$$

It should be clear (as first pointed out by Gabor [14]) that we may

$$\overline{E}_T(x,y) = g(x,y) \quad [\gamma_N = -2] \tag{10}$$

with the condition $\gamma_N = -2$. In practice [1] this condition may be realized by making a contact print of the negative (N), and by processing this positive (P) in such a way that the condition:

$$\gamma_N \gamma_P = 2 \tag{11}$$

be satisfied. In other words, the exposure of the positive P is:

$$[g(x,y)^{-\gamma_N/2}]^2 = g(x,y)^{-\gamma_N} \tag{12}$$

and the transmission of this positive (P) is:

$$\overline{E}_T = [g(x,y)^{-\gamma_N}]^{-\gamma_P/2} = g(x,y), \quad [\gamma_n \gamma_p = 2] \tag{13}$$

if the condition of equation (11) is satisfied.

Finally, as we recall below, with particular reference to this work, it is now well known (see, e.g. [8]), that the three angularly separated waves, produced by holograms, have electric field-vector amplitudes which can readily be made to be independent of the photographic recording and processing conditions under some specified conditions. For instance, let $E_o(u,v)$ describe the field produced by the 'object' field in the plane (u,v) of the hologram, and let $E_R(u,v)$ be the field similarly produced by the 'reference' source in the hologram plane. The exposure of the hologram is (with the time-averaging understood):

$$I(u,v) = (\overline{E}_o + \overline{E}_R)(\overline{E}_o + \overline{E}_R)^*, \tag{14}$$

i.e. in the well-known form:

$$I(u,v)=[|\bar{E}_O|^2+|\bar{E}_R|^2]+[\bar{E}_O\bar{E}_R^*]+[\bar{E}_O^*\bar{E}_R], \tag{15}$$

where the three separated waves have been separately bracketed. The electric field vector distribution of the field transmitted through the hologram, processed with a γ, may be made to be proportional to $I(u,v)$, with the γ appearing only as a scale factor, provided only that one respects the simple condition:

$$|\bar{E}_R|^2 >> |\bar{E}_O|^2 \tag{16}$$

In practice, the condition >> may mean approximately:

$$|\bar{E}_R|^2 \cong \ (4 \text{ to } 10)\,|\bar{E}_O|^2$$

as an example. Indeed, if we write equation (15) in the form:

$$I(u,v)=|\bar{E}_R|^2\left\{1+\frac{|\bar{E}_O|^2}{|\bar{E}_R|^2}+\frac{\bar{E}_O}{\bar{E}_R}+\frac{\bar{E}_O^*}{\bar{E}_R^*}\right\} \tag{17}$$

and if we recall that the electric field transmitted through the hologram (negative!) is (upon illumination with a wave of unit amplitude):

$$\bar{E}_T = I(u,v)^{-\gamma_N/2} \tag{18}$$

we readily see, by a binomial expansion of equation (18), that the form $(1 + \varepsilon)$, with ε small, of equation (17) gives:

$$\bar{E}_T \cong |\bar{E}_R|^{-\gamma_N}\{1-\frac{\gamma_N}{2}[\frac{|\bar{E}_O|^2}{|\bar{E}_R|^2}+\frac{\bar{E}_O}{\bar{E}_R}+\frac{\bar{E}_O^*}{\bar{E}_R^*}]\}, \tag{19}$$

i.e.

$$\begin{aligned}\bar{E}_T \cong\ & |\bar{E}_R|^{-\gamma_N-2}\,[|\bar{E}_R|^2-\frac{\gamma_N}{2}|\bar{E}_O|^2] \\ & -\frac{\gamma N}{2}|\bar{E}_R|^{-\gamma_N-2}\,[\bar{E}_O\bar{E}_R^*] \\ & -\frac{\gamma_N}{2}\bar{E}_R^{-\gamma_N-2}\,[\bar{E}_O^*\bar{E}_R]\end{aligned} \tag{20}$$

This equation must be carefully considered for the purposes in this work. First, we may note that the field $\bar{E}_T$ transmitted through the hologram (upon illumination with a wave amplitude) will indeed have its imaging waves proportional to $[\bar{E}_O\bar{E}_R^*]$ and $[\bar{E}_O^*\bar{E}_R]$, i.e.

$$\frac{\gamma_N}{2}\,|\bar{E}_R|^{-\gamma_N-2}[\bar{E}_o\bar{E}_R^*]\propto\bar{E}_o\bar{E}_R^* \qquad (21)$$

and

$$\frac{\gamma_N}{2}\,|\bar{E}_R|^{-\gamma_N-2}[\bar{E}_o^*\bar{E}_R]\propto\bar{E}_o^*\bar{E}_R \qquad (22)$$

to the extent that the amplitude $|\bar{E}_R(u,v)|$ of the reference wave may be considered as constant in the hologram plane. The condition $[|\bar{E}_R|=\text{const.}]$ may, in practice be readily realized for a number of unfocused waves. This will be the case for a plane wave (originating from a point source 'at infinity'), a spherical wave (e.g. as originating from a point source near the object). In fact, as may be readily verified by experiment, most fields in which the object is not focused on the plate encountered in optics (as distinguished from a field near a source, in a 'focused' image region, in shadows, etc.) happens to have an amplitude which is uniform, i.e. constant, over very extended regions! In particular, the far-field region of an 'extended' reference source, as we use it in holography, as well as the field produced by it by Fourier transformation, both have amplitudes which may be considered as constant, within the requirements of equations (21) and (22). Under these conditions, we may henceforth write that the field transmitted by the hologram of equation (15) upon illumination with a wave $\bar{E}_R(u,v)$ as:

$$\bar{E}_T(u,v)\propto\bar{E}_{R'}(u,v)I(u,v) \qquad (23)$$

i.e.

$$\begin{aligned}\bar{E}_T(u,v)\propto\bar{E}_{R'}&[|\bar{E}_o|^2+|\bar{E}_R|^2]\\&+\bar{E}_o\bar{E}_R^*\bar{E}_{R'}\\&+\bar{E}_o^*\bar{E}_R\bar{E}_{R'}\end{aligned} \qquad (24)$$

Because of the Fourier-transforming property of lenses, as expressed in equation (4) and because of the possibilities of obtaining linear relations between the fields transmitted by a photographic transparency and its exposure (equation (13)) on the one hand, and on the other, between the field transmitted by a hologram and its exposure (equation (24)), we shall show that Fourier-transform holography [9], and notably 'extended-source Fourier-transform holography' [11,12] offer a particularly straight-forward solution to the 'inversion' of the integral equation (1).

Indeed, if we write the integral equation (1) in the symbolic form*:

$$g(x,y)=f(x,y)\circledast h(x,y) \qquad (25)$$

*The notation used is that introduced by: Bracewell, R.N., 1965, The Fourier Transform and its Applications (New York: McGraw Hill Book Company). See also[8].

where ⊗ indicates a spatial convolution [8], and if we recall the Fourier-transform equivalent of equation (25):

$$G(u,v)=F(u,v)H(u,v) \tag{2}$$

then it should be clear that the desired image f(x,y) may be extracted from the photograph g(x,y), provided that we can 'extract' the function F(u,v) from the function G(u,v) and take the Fourier transform of F(u,v), for instance with another lens. That this should be possible, in principle, was first suggested by Maréchal and Croce [12] in 1953, and again, at that time for the purposes of improving on the spectra produced by insufficiently perfect diffraction gratings, by the author [7] in 1955, who also proposed a solution in the form of a Fourier-transform 'division' as we discuss below. However, it was only when we realized [1] that holograms, as first described by Gabor [13-15] in 1948, could be readily used to help in the realization (materialization) of the required spatially-complex filters, that the practical solutions, such as those which we present with new theoretical and experimental detail below, became possible.

In brief, two methods permit one to extract the function F(u,v) from the function G(u,v) of equation (2), and form it, by a second Fourier transformation, the desired image function f(x,y)

One method [1] consists in dividing G(u,v) by the function H(u,v), i.e. by performing the operation:

$$\overline{G}\overline{H}^{-1}=\overline{F} \tag{26}$$

This method, as may be readily shown, is quite general, and permits one to retrieve the desired function to the limit of 'diffraction' of the original focusing system, already in the first step. The method does require, however, considerable care in the photographic steps involved, as we show below.

Another method, recently originated by the author [5,6] appears to be much more powerful, as well as considerably simpler. Its application to the special case of image 'deblurring', for images formed by initially 'well-corrected' systems, which were out-of-focus or moved during the exposure, is less general than the Fourier-transform division method, but much simpler to implement, at the present stage of development. Moreover, we show below that the second method has already formed the basis for the design of a new class of optical systems which, by themselves are, paradoxically rather poor (deliberately poor) image-forming systems, as far as the formation of the first-step photograph g(x,y) is concerned. This second method of image deconvolution (decoding) consists in onne form, of multiplying the function G(u,v) by the complex conjugate

$\bar{H}^*(u,v)$ of the function $\bar{H}(u,v)$, where we recall that $\bar{H}(u,v)$ is the spatial Fourier transform of the spread function $h(x,y)$, according to equation (3). The second method of image deconvolution is described by the equation [5,6]:

$$\bar{G}\,\bar{H}^*=\bar{F}\,\bar{H}\,\bar{H}^*=\bar{F} \tag{27a}$$

$$\bar{H}\,\bar{H}^*=1$$

According to a well-known theorem (see, e.g. [8]), the 'decoding' condition:

$$\bar{H}\,\bar{H}^*=1 \tag{27b}$$

may be written in the (x,y) domain in the form:

$$\iint_{-\infty}^{+\infty} h(x,y)h^*(x+x',\ y+y')\,dxdy=\delta(x',y') \tag{28}$$

where (x',y') is the Dirac delta function. We may write equation (28) in the equivalent symbolic form (see [8]):

$$h \star h^*=\delta \tag{29}$$

where the symbol $\star$ indicates (here) the spatial autocorrelation operation described by equation (28). The conditions (27b) and (29) which must be satisfied in order that the operation $\bar{G}\,\bar{H}^*$ may indeed yield $\bar{F}$ as desired, according to equation (27a), may be formally related by the Fourier-transform theorem:

$$\bar{H}\,\bar{H}^*=1 \rightleftarrows h \star h^*=\delta \tag{30}$$

referred to above, where the symbol $\rightleftarrows$ indicates 'by Fourier transformation'. It may be helpful to recall that the 'decoding' conditions $\bar{H}\,\bar{H}^*=1$ (equation (27a)), i.e. $h \star h^*=\delta$ (equation (29)) are in fact extensions of the method of 'extended-source Fourier-transform holography', (see [10,00] and [8, pp. 127-137]). A particularly remarkable example of a function described by equation (29), suggested by Gabor [40], is that of a simple diffusing (ground) glass. This is further illustrated in FIG. 3.

Further details with regard to the second method which we may call the 'correlative holographic decoding method' have been published in [5,6,16]. Similarly, we may call the first method the 'holographic Fourier-transform division method', for short. Before proceeding to the discussion of these methods, with their experimental illustrations, it may be in order to maks some reference to the advantages in speed which these optical computing methods have in comparison with digital-computer solutions of the deconvolution.

problem, e.g. as demonstrated by Harris [17] since 1966. According to Ansley [18], the Fourier transform an n x m matrix from n^2m^2 to $n \log_2 m \times m \log_2 n$. This is to be compared to minutes required for the same operation when it is performed holographically. In fact, at present it seems that a maximum of 250 x 250 lines (i.e. image portions of about 2•5 x 2•5 mm with 100 lines/mm resolution) may be reasonably processed with current methods. However, when successful, the computer-decoded images may in the end present a perfection which may be in many ways more readily superior to that obtained by our first method. Accordingly, it may well be that the holographic deconvolution methods may in some case also serve as powerful first-step search methods, for the improvement of images, to determine the image or image sets for which further digital computer processing for image refinement may be justified. It may also be of some interest to note analogies between these methods and some of the forms of apodization [20], and other forms of non-holographic coding and decoding, e.g. as used in spectroscopy [21-23] as well as image-sharpening methods of purely electronic nature, e.g. as used in television [24-25], to which we have made extensive reference in the past as well as in PART I. We also recall that optical filtering methods using filters generated in part in this manner and in part photographically have been described by a number of authors, notably by Tsujiuchi [26].

1. Holographic Fourier-transform Division Method

Our method [1] consists in recognizing that the equation:

$$g(x,y)=f(x,y)\otimes h(x,y) \tag{25}$$

i.e. its equivalent Fourier-transform equation:

$$\overline{G}(u,v)=\overline{F}(u,v)\overline{H}(u,v) \tag{2}$$

may be readily solved by the Fourier-transform division:

$$\overline{G}\,\overline{H}^{-1}=\overline{F} \tag{26}$$

provided that the required Fourier-transform division filter can indeed be realized. We showed in [1,3] that such a filter could indeed be realizes, notably in the form of a two-part 'sandwich'

$$\overline{H}^{-1} = \frac{\overline{H}^*}{|H|^2} \tag{31}$$

Each of the two filter components, $\overline{H}^*$ and $|\overline{H}|^2$ can be separately realized, both from the Fourier-transform H(u,v) of the spread function h(x,y), i.e. of the image of a point, as formed by the system under the conditions of use. It may be of some interest to note, for instance in connection with the diffraction patterns formed by imperfect gratings (see, e.g. [7]), that the spread function $h(x,y)$ may in fact also be readily obtained by computation (e.g. using a digital computer), when a direct measurement with adequate precision may not be readily achieved. The computation of $h(x,y)$ is based on the fact that

$$h(x,y) = \bar{E}_H(x,y)\bar{E}_H(x,y)^* \tag{32}$$

where $\bar{E}_H(x,y)$ is the spatial Fourier transform of the electric field 'amplitude' $\bar{E}_H(u,v)$ in the wavefront, when the system is illuminated by a wave from a point in the object domain. The wavefront $\bar{E}_H(u,v)$ may be readily measured interferometrically, or indeed holographically. In general, however, for most image-forming applications, the function $h(x,y)$ may in fact be determined with adequate precision simply from its photograph.

To realize in the focal plane of a Fourier-transforming lens the field $\bar{H}(u,v)$, we must produce in the pupil of the lens the field $h(x,y)$. This may be readily achieved according to the steps involved in the negative-positive procedure leading to equation (13) Using the field $\bar{H}(u,v)$, we obtain the $\bar{H}^*$ component of the filter simply by recording the corresponding hologram, i.e. by making the $\bar{H}(u,v)$ interfere with a suitable reference field, e.g. a plane wave. We obtain a hologram in the form:

$$I(u,v)=1+|\bar{H}|^2+\bar{H}+\bar{H}^* \tag{33}$$

Under the conditions discussed in the first section, notably in equations (15) to (24), illumination of this hologram with a wave of unit amplitude will indeed produce an imaging wave $\bar{H}^*$ as desired.

The $|\bar{H}|^{-2}$ may be obtained as follows. We first record the exposure $|\bar{H}|^2$ produced by the field $\bar{H}(u,v)$. If we now process the photographic negative with a γ equal to 2, the field transmitted through the negative will indeed be equal to $|\bar{H}|^{-2}$. We have for this field (with $\gamma=2$):

$$\bar{E}_T=[|\bar{H}|^2]^{-\gamma/2}=(|\bar{H}|^2)^{-1} \tag{34}$$

Accordingly, if this field, produced by this component of the 'sandwich' filter is used to illuminate the hologram component, described by equation (33), the field transmitted in the imaging-wave side band will be:

$$|\bar{H}|^{-2}\bar{H}^*= \frac{\bar{H}^*}{|\bar{H}|^2} \tag{35}$$

upon illumination with a unit-amplitude wave. It follows immediately that illumination of the sandwich filter with the field $\bar{G}(u,v)$ will produce an imaging wave equal to:

$$\overline{G}\,\frac{\overline{H}^*}{|\overline{H}|^2} = \overline{F}_{BL} \cong \overline{F} \tag{36}$$

as desired. We now distinguish between the field $\overline{F}_{BL}$ actually obtained by this filtering within the spatial bandwidth in which the spatial frequencies are being filtered, on the one hand, and on the other, the field $\overline{F}$ which would, correspond to a field which, by Fourier transformation would give a 'super-resolved' image f(x,y) i.e. an image without any limitations such as that which would characterize even a perfectly corrected lens system, notably the 'diffraction' limitations. It may be shown [27,28], at least mathematically, that the filtering step of equation (36) may in fact be considered as a possible first step in obtaining a certain amount of 'super-resolution' improvement, notably in cases when a suitable absence of image-degrading noise might be experimentally achievable. We hasten to note, in this context, that the situation for obtaining 'super-resolution' in one-step imaging e.g. in radar and radio-astronomy [29,30]. The possibility of thus investigating 'super-resolution' by two-step imaging, using holographic processing, as an example, may be of particular interest also in further progress in the application of holographic image-computing methods to the solution of the 'phase problem' in x-ray crystallography [8,31,32]. These preliminary considerations may be further clarified as follows. It can be shown that the Fourier-transform division indicated in equation (36) permits one to restore essentially all the spatial frequencies within the bandwidth corresponding to the image-forming aperture used to record g(x,y) respectively h(x,y). This is true, except for the isolated nulls of the $\overline{H}(u,v)$ function, for which the frequencies are not properly restored.

Indeed by noting that

$$h(x,y) = \overline{E}_H(x,y)\overline{E}_H(x,y)^* \tag{32}$$

and that

$$\overline{E}_H(x,y) = \iint_{-\infty}^{+\infty} \overline{E}_H(u,v)\exp[2\pi i(ux+vy)\,du\,dv, \tag{37}$$

as we mentioned before, we immediately have (see, e.g. equation (30)):

$$\overline{H}(u,v) = \overline{E}_H(u,v) \star \overline{E}_H(u,v)^* \tag{38}$$

which we may also write (see, e.g. [8, Chap. III. equation (23b)]) in the form:

$$\overline{H}(u,v) = \overline{E}_H(u,v) \circledast \overline{E}_H{}^*(-u,-v) \tag{39}$$

The filter function $\overline{H}(u,v)$ is thus in effect nothing but the 'convolution of the wavefront $\overline{E}_H(u,v)$ with itself according to equation (39). Since the wavefronts $\overline{E}_p(u,v)$, perfect, and $\overline{E}_H(u,v)$ imperfect ('blurred') occupy the same aperture both for the system 'in focus', as well as when the system is 'out of focus' (or when it 'blurrs' the image by motion) we may conclude, at least to this approximation, that the blurred spread function $h(x,y)$, as well as that $h_p(x,y)$ which would correspond to the system 'in focus' contain spatial frequencies up to the same limit, in the (u,v) domain. The difference between the function $\overline{H}_p(u,v)$ corresponding to the 'in focus' system, and the function $\overline{H}(u,v)$ corresponding to the 'blurred' spread function $h(x,y)$ is that the spatial frequencies in $\overline{H}(u,v)$ are generally smaller in amplitude, and shifted in phase, relative to the spatial frequencies in $\overline{H}_p(u,v)$. It is in fact the restoration of the spatial frequencies of $\overline{H}(u,v)$ to the values in $\overline{H}_p(u,v)$ which may be considered as the interpretation of the 'spatial filtering' restoration process. In other words, an 'in focus' system, have a 'diffraction-limited' spread function $h_p(x,y)$ produces an image:

$$g_p(x,y)=f(x,y)\circledast h_p(x,y) \tag{40}$$

and also not a 'super-resolved' image $f(x,y)$! The field corresponding to the 'in-focus' image in the Fourier-transform domain is:

$$\overline{G}_p(u,v)=\overline{F}(u,v)\overline{H}_p(u,v) \tag{41}$$

which gives, by a second Fourier transformation (except for possible additional diffraction limitations), the field $g_p(x,y)$, identical to that described by equation (40), and again not a 'super-resolved' image $f(x,y)$! Naturally, all of our previous papers implicitly made this assumption, as we explicitly stated in [1]. By comparing equations (41) and (2), it should indeed be clear that the image-deconvolution methods aim, in this first approximation, in 'restoring' the function $\overline{H}_p(u,v)$ from the function $\overline{H}(u,v)$. Because the filtering restoration methods operate, in effect, on each spatial frequency separately, both in amplitude and in phase, it should be clear that the existence of a number of isolated zeros in the $\overline{H}(u,v)$ function, as used here for division, will not, in general, significantly affect the success of the filtering*. The existence of a number of isolated zeroes in the function $\overline{H}(u,v)$ used for Fourier-transform division filtering, does not invalidate the validity of equation (26), within the limitations discussed here above. In other words, even though some of the spatial frequencies may not be restored by means of the Fourier-transform division methods, a

---

*This is essentially true in cases where there are a limited number of zero's in $H(u,v)$ and to the extent that corresponding 'Gibbs'-like effects may be neglected in the first approximation.

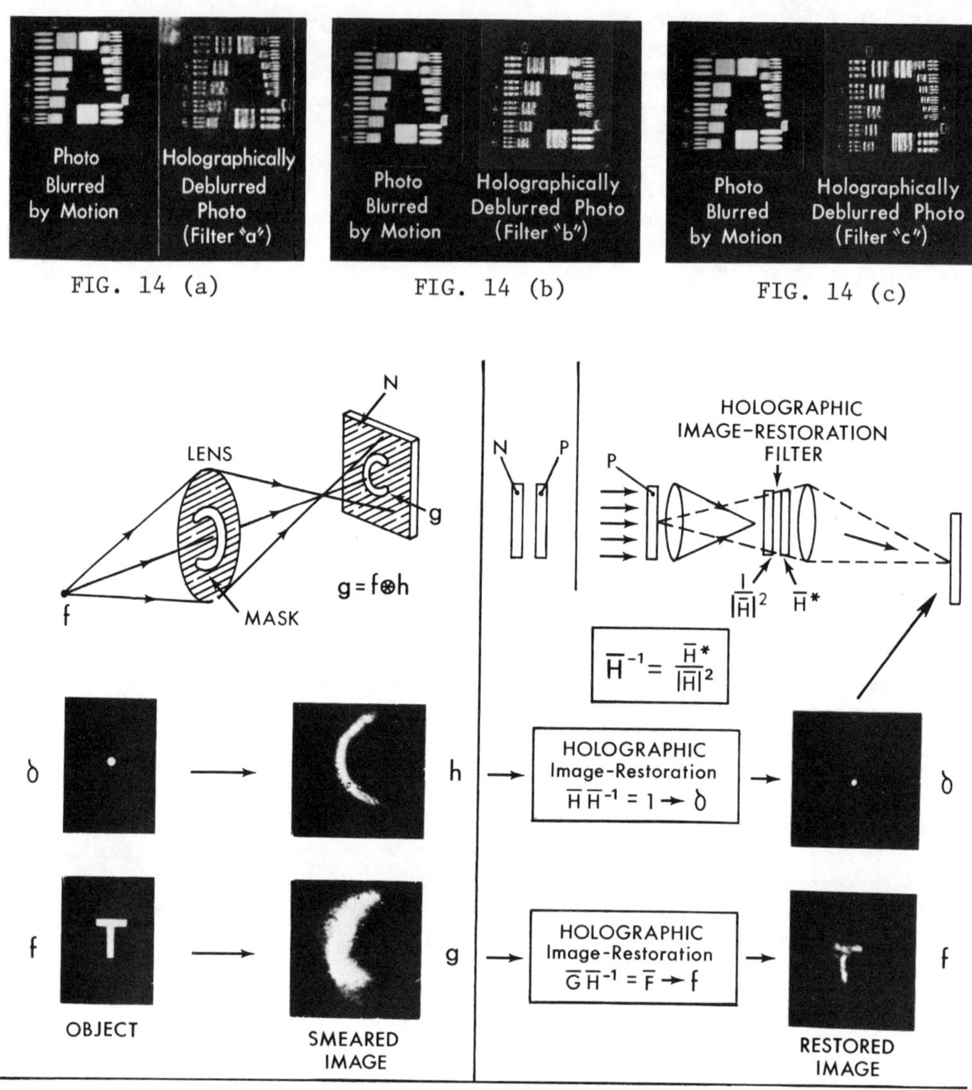

FIG. 14 (d) Holographic image restoration using holographic Fourier-transform division method of Stroke and Zech (loc. cit.) This remarkable early experimental verification was obtained by Lohmann and Werlich (loc. cit.).

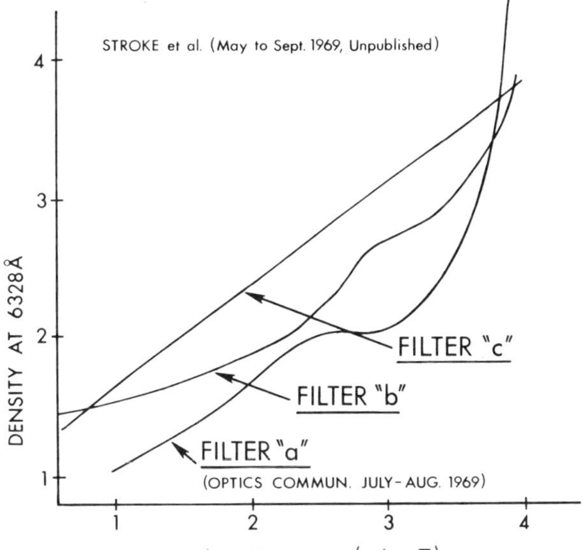
STROKE et al. (May to Sept. 1969, Unpublished)
DENSITY AT 6328Å
FILTER "c"
FILTER "b"
FILTER "a"
(OPTICS COMMUN. JULY-AUG. 1969)
log Exposure (= log I)
THREE FILTER-FUNCTIONS
USED FOR DIFFERENT-EMPHASIS DEBLURRING
FROM SAME BLURRED PHOTO

FIG. 15

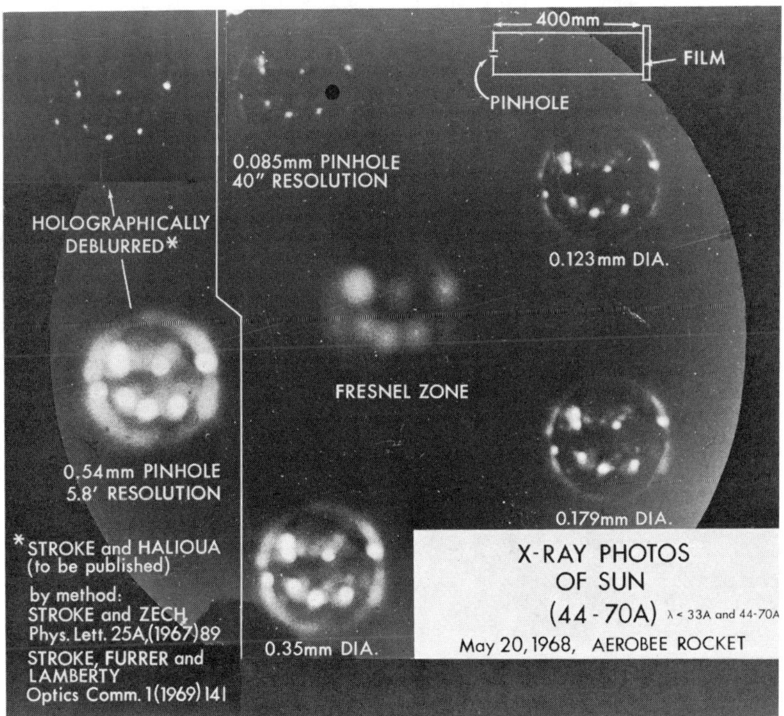
400mm
FILM
PINHOLE
0.085mm PINHOLE
40" RESOLUTION
HOLOGRAPHICALLY
DEBLURRED*
0.123mm DIA.
FRESNEL ZONE
0.54mm PINHOLE
5.8' RESOLUTION
0.179mm DIA.
*STROKE and HALIOUA
(to be published)
by method:
STROKE and ZECH
Phys. Lett. 25A,(1967)89
STROKE, FURRER and
LAMBERTY
Optics Comm. 1(1969)141
0.35mm DIA.
X-RAY PHOTOS
OF SUN
(44-70A)
λ < 33A and 44-70A
May 20, 1968, AEROBEE ROCKET

FIG. 16

noticeable improvement in the 'deblurred' image will in general be observed, provided that a significant part of the spatial frequency function $H_p(u,v)$ is indeed properly restored, in amplitude and in phase, especially, in many practical cases, in the lower frequency regions, inasmuch as the higher frequency regions are in general already quite attenuated in amplitude, even in perfectly corrected image-forming systems. For instance, in a perfectly corrected image-forming system, using a rectangular aperture, the spatial frequency amplitudes decrease linearly from the lowest to the highest frequency transmitted, along a direction parallel to a side of the aperture. The 'super-resolutions' which we mention above, could conceivably be obtained in a number of ways from the function $H_p(u,v)$. One form of 'super-resolution' which may be readily achieved, and which appears to involve few conceptual difficulties or objections, would simply consist in raising the spatial frequency amplitudes above the line of linear decrease, discussed above. Clearly, we may add, the preceding discussion refers to the formation of images, and to the recording of $g(x,y)$ in incoherent light, while the spatial filtering operations are, on the contrary, carried out in coherent light. It may be shown (see, e.g., [8]) that the spatial frequency response function, analogous to the $H_p(u,v)$ functions discussed above, for perfectly-corrected systems operating in coherent light are perfectly flat in amplitude up to a frequency equal to one-half of the cut-off frequency of incoherent-light imaging systems: at that frequency, the spatial frequency of response of the coherent-light imaging systems abruptly drops to zero. It should be clear, also that the function $H(u,v)$, $H_p(u,v)$ are nothing but the 'spatial frequency response functions' of the optical systems, in analogy with the terminology used for temporal frequency response functions in electrical circuit theory and in electronics, among others (for a general background, see, e.g., [8, Chap. III]).

Experimental verifications of the Stroke and Zech holographic Fourier-transform division method [1] were first described by Lohmann and Werlich [2] for the deconvolution of a badly blurred image of a single 'black-on-white' point as well as for an equally badly blurred 'black-on-white' letter 'T'. Their results are reproduced in FIG. 14.d. We have ourselves [3,4] given the first experimental verifications for a 'continuous-tone' three-dimensional macroscopic object using the theory given above. Further refinements of these methods have now been obtained and are shown here for the first time (see FIGURES 14-16) as obtained in unpublished work by Stroke and his students.

In conclusion, we may perhaps note that the complexity of the problem of 'realizing' (i.e. of synthesizing) such optical filters as the H^{-1} filter, equal to $H^*/|H|^2$, as discussed above, may perhaps be illustrated by, to some extent, comparing it to the mathematical difficulty of synthesizing electrical filters, even passive, in electronics, even when the frequency-transfer function is known. For instance, systematic methods for synthesizing RLC filters (R = resistance, L = inductance, C = capacitance), are still not known in a general way! Similarly, many years sometimes elapse before a method, such as that which we describe, is discovered in electronics for systematically synthesizing certain filter types, e.g. RL, RC or LC, especially if constraints such as 'equal values' for all resistances and all capacitances in a ladder-network may be desirable, as one example. In addition to the comparable theoretical difficulties in optics, there exists the additional difficulty resulting from the need of actually 'manufacturing' (photographically and holographically) the individual filter components, with exacting precisions, and for maintaining critical alignments of the optical-system components in the processing, as we discuss in the next section.

2. Correlative holographic decoding (deconvolution) method

It may be of interest to stress again the generality and power of the basic equations of holography first introduced by Gabor [13-15]. All the various forms of imaging and image processing, using holograms, may be shown to make use of the same fundamental 'Gabor' equations (see, e.g. [8]). However, there exists a basic difference between the image decorrelation (image deconvolution) methods, which we describe here, and which result in a decoded (deblurred) image, on the one hand, and on the other, the image correlation methods, first introduced by Vander Lugt [41], and which result in a coded pattern, for instance a bright dot or a cross, indicating the location of the centre of gravity of the correlation of the area [41] or of the contour [42][43], of the 'known' pattern with the areas or contours of the patterns in the image. The Vander Lugt 'matched' filtering methods consists in correlating a function $f_1(x,y)$ with the best possible 'in focus' image function $f(x,y)$. The correlation may be carried out in the spatial-frequency Fourier-transform domain with the aid of a Fourier-transform hologram $[1 + |\bar{F}_1|^2 + \bar{F}_1 + \bar{F}_1^*]$ of the coding function f_1. When the image function $f(x,y)$ is illuminated by a plane wave of coherent light, the spatial Fourier transform of $f(x,y)$ obtained in the focal plane of a lens is $\bar{F}(u,v)$ and the field transmitted through the hologram is $\bar{F}[1 + |\bar{F}_1|^2 + \bar{F}_1 + \bar{F}_1^*]$. One of the holographic 'side-band' waves is $\bar{F}\,\bar{F}_1^*$ and its Fourier transform is thus equal to the correlation $f \star f_1^*$, rather than to the decorrelation, as it is being sought in the image-deblurring methods! In this sense, it may be said that the image-deblurring methods aim at achieving the solution of an integral equation (e.g. the solution of equation (1) above), while the 'matched' filtering pattern-recognition methods aim at realizing the (in practice) much more straightforward correlation integral

$$\iint_{-\infty}^{+\infty} f(x,y)f_1(x + x', y + y')dxdy = f \star f_1.$$

For further details, see also, e.g. [8], notably also for diagrams and photographs of experimental arrangements required to materialize equations given in this paper and in references cited.

The basic principles of our new methods [5,6] were given above in equations (27) to (30). The realization of equation (27) may be accomplished by means of two holographic schemes.

Let g(x,y) be the 'blurred' photograph, such that

$$g(x,y) = f(x,y) \circledast h(x,y), \tag{25}$$

with the special requirement that the autocorrelation function of the spread function h(x,y) be sharply 'peaked', i.e. that we have

$$h \star h^* = \delta. \tag{29}$$

To this corresponds the equivalent condition:

$$\overline{H}.\overline{H}^* = 1 \tag{27a}$$

and we recall that

$$h \star h^* \rightleftarrows \overline{H}\,\overline{H}^*. \tag{30}$$

2.1 Extended-source holographic scheme A

Let us record the Fourier-transform hologram [8] of the function g(x,y), using the photographic precautions discussed in the Introduction. The hologram is:

$$I(u,v) = 1 + |\overline{G}|^2 + \overline{G} + \overline{G}^*. \tag{42}$$

By recalling the equation:

$$\overline{G}(u,v) = \overline{F}(u,v)\overline{H}(u,v), \tag{2}$$

we may write equation (42) in the form:

$$I(u,v) = 1 + \overline{G}^2 + \overline{F}\,\overline{H} + \overline{F}^*\overline{G}^*. \tag{43}$$

If we now replace the hologram into its recording position and illuminate it by the wave H, originating from a suitably realized copy (see Introduction) of a photograph of the spread function h(x,y), placed into the position of the point source, used to

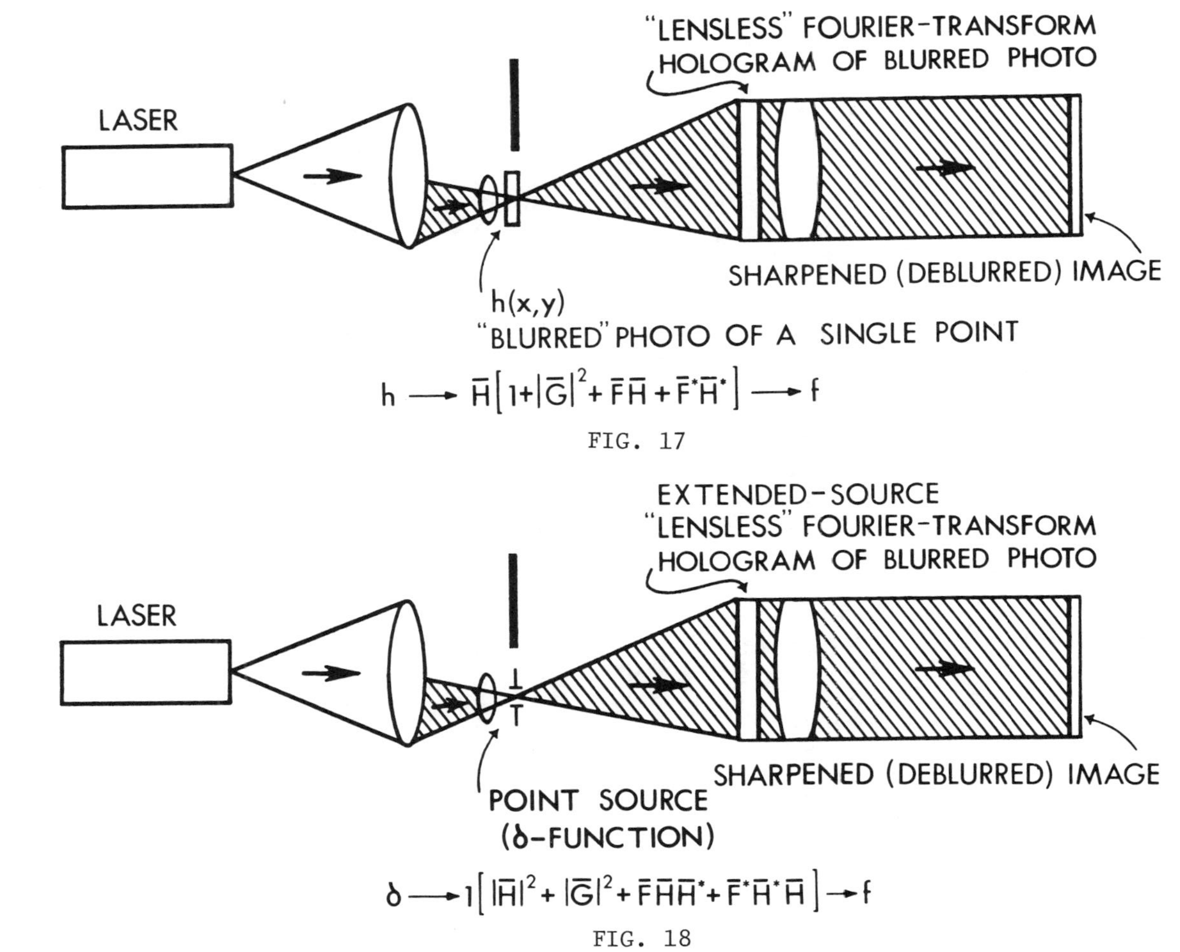
LASER
"LENSLESS" FOURIER-TRANSFORM
HOLOGRAM OF BLURRED PHOTO
SHARPENED (DEBLURRED) IMAGE
h(x,y)
"BLURRED" PHOTO OF A SINGLE POINT
h ⟶ H̄[1+|Ḡ|² + F̄H̄ + F̄*H̄*] ⟶ f
LASER
EXTENDED-SOURCE
"LENSLESS" FOURIER-TRANSFORM
HOLOGRAM OF BLURRED PHOTO
SHARPENED (DEBLURRED) IMAGE
POINT SOURCE
(δ-FUNCTION)
δ ⟶ 1[|H̄|² + |Ḡ|² + F̄H̄H̄* + F̄*H̄*H̄] ⟶ f

FIG. 17

FIG. 18

record the Fourier-transform hologram of equation (42) then the wave transmitted in the last imaging term of equation (43) (see figure 17) will be equal to

$$\overline{F}^*\overline{H}^*\overline{H} = F^* \qquad (\text{if } \overline{H}\,\overline{H}^* = 1). \tag{44}$$

Since $\overline{F}^*$ is nothing but the complex conjugate of the imaging wave F, the corresponding image, obtained by Fourier transformation in the focal plane of a lens 'looking' through the hologram is indeed equal to the desired image f(x,y), upon processing of the photograph with a γ equal to -2, i.e. by printing the positive according to the condition of equation (11)†.

2.2. Extended-source holographic scheme B

In this case, we record the Fourier-transform hologram of g(x,y) by using h(x,y) as the 'extended reference source' (e.g. as in figure 27, Chap. VI of [8]). The hologram is

$$\begin{aligned} I(u,v) &= (\overline{G} + \overline{H})(\overline{G} + \overline{H})^* \\ &= |\overline{G}|^2 + |\overline{H}|^2 + \overline{G}\,\overline{H}^* + \overline{G}^*\overline{H}. \end{aligned} \tag{45}$$

By again recalling that $\overline{G} = \overline{F}\,\overline{H}$, according to equation (2), we may write equation (45) in the form:

$$I(u,v) = |G|^2 + |\overline{H}|^2 + \overline{F}\,\overline{H}\,\overline{H}^* + \overline{F}^*\overline{H}^*\overline{H}. \tag{46}$$

Because we are considering, in particular, the case when $\overline{H}\,\overline{H}^* = 1$, we note that the image-forming wave of interest, say

$$\overline{F}\,\overline{H}\,\overline{H}^* = \overline{F} \qquad (\text{if } \overline{H}\,\overline{H}^* = 1), \tag{47}$$

is indeed equal to $\overline{F}$ as desired. Accordingly, in this case it is sufficient to replace the hologram into its recording position, and illuminate it with a wave of unit amplitude (see figure 18), as originating from a point source, situated 'in place' of the extended source h(x,y) used in the recording.

It may at first sight appear that less experimental care may be required in the realization of scheme B than in that required for the realization of scheme A. In fact, as we show below, the success of both these schemes depends on very strict respect on certain alignment conditions of the 'extended source', i.e. of the corresponding wave H relative to the hologram in the reconstruction (scheme A), respectively relative to the wave G in the recording (scheme B). Certain additional experimental conclusions will be given, following the discussion in the next section.

---

†In practice, this condition need not be necessarily more rigorously satisfied in this last step than in direct photography of the image g(x,y), using a well-corrected system!

2.3. Extended-source holographic scheme C.

The two methods A. and B. presented above have the disadvantage that an individual hologram has to be recorded for each coded image. In a recent paper [47] G. Groh and G. W. Stroke proposed to avoid this drawback by recording once and for all the Fourier-transform hologram of the point-spread function h using a δ-function (i.e. a suitably illuminated pinhole) as the reference source. The essential terms stored in the hologram are

$$|H+1|^2 = HH^* + 1 + H^* + H$$

where we assume that the Fourier-transform of the δ-function may be written as 1, and where H is the Fourier transform of h. Using this hologram, it becomes quite simple to construct a decoding apparatus for all coded images obtained by the corresponding generalized system. One just has to place that hologram into the filter plane of an ordinary Fourier-transform arrangement. The field transmitted through the holographic filter, upon illumination with the 'coded' transparency $g = f \circledast h$ is equal to

$$GH^* = FHH^*$$

in the third term of the preceding equation, and the corresponding 'decoded' image is

$$f \circledast [h \star h^*] \cong f$$

to the extent that $h \star h^* \cong \delta$-function. A further advantage noted by Groh and Stroke [47] is that the zero-order terms in the image are not broadened, since they correspond to -functions, when the peaked auto-correlation function condition $[h \star h^* \cong \delta]$ is fulfilled.

3. Experimental alignment requirements

Equation (24) indicates the effect which differences between $\overline{E}_R$, the 'reference' wave used in the recording, and $E_{R'}$, the illuminating wave used in the reconstruction, will have on extracting the desired wave $\overline{E}_o$ from the hologram. The analysis applies equally to differences created by lack of corresponding alignment between the $\overline{H}$ and $\overline{H}^*$ waves, as they appear in the two hologram types described by equations (43) and (46) (for further details, see [8, pp. 127-137]).

In the simplest analysis, let the amplitudes of both $\overline{E}_R$ and $\overline{E}_{R'}$, be unity, so that we may write:

$$\overline{E}_R = |\overline{E}_R| \exp(i\phi_R) = \exp(i\phi_R), \qquad (48a)$$

$$\overline{E}_{R'} = |\overline{E}_o| \exp(i\phi_{R'}) = \exp(i\phi_{R'}). \qquad (48b)$$

In any event, the imaging wave $\bar{E}_o\bar{E}_R{}^*\bar{E}_{R'}$, produces by Fourier transformation an image described by the equation:

$$\bar{E}_o\bar{E}_R{}^*\bar{E}_{R'} \rightarrow T[\bar{E}_o] \circledast T[\bar{E}_R{}^*] \bigstar T[\bar{E}_{R'}] \quad , \tag{49}$$

where → indicates 'by Fourier transformation', and 'T' [ ] indicates 'Fourier transform of [ ]'. Since $T[\bar{E}_o]$ is in fact nothing but the desired image, we conclude from equation (49) that the image obtained, according to equation (49) is equal to the desired image convolved with the correlation of the conjugate of the recording source with the illuminating source used in the reconstruction. Clearly, if the recording and reconstructing sources are the same, and if their correlation is equal to the Dirac delta, function, then equation (49) becomes:

$$T[\bar{E}_o] \circledast \delta = T[\bar{E}_o] , \tag{50}$$

in agreement with the analysis of the preceding sections.

Let us now consider the simplest case of misalignment, namely the case described by equation (48). This would also be the case of interest in which $\bar{E}_R$ and $\bar{E}_{R'}$, are in fact produced by the same source, but when the source, in the reconstruction, is misaligned, so that the phase $\phi_{R'}$ is different from the phase ϕ_R, in the plane of the hologram.

In this case the imaging term of interest becomes:

$$\bar{E}_o \exp(-i\phi_R)\exp(i\phi_{R'}) = \bar{E}_o \exp[i\phi_{R'} - \phi_R)] , \tag{51}$$

from which we immediately recognize that the imaging wave is now not any more the wave $\bar{E}$, but rather a wave having a possibly very considerable wavefront aberration $(\phi_{R'} - \phi_R)$. We may recall (see, e.g. [7]) that wavefront aberrations as small as one-tenth of a wavelength (i.e. $2\pi/10$ radians$\cong 6/10$) become noticeable for high-resolution imaging, when the aberration extends smoothly over large regions of the wavefront. The tolerances for more localized aberrations is about ten times smaller still, and their effect is proportional to the area of the wavefront 'covered' by the aberrations! Aberrations of several radians certainly lead to intolerably large image degradation, in the case of the 'extended' aberrations.

It is possible to assess the magnitude of the misalignment phase errors for the two types of holographic experiments discussed in the previous section.

Let us consider, for simplicity, the case of the 'lensless Fourier-transform hologram' [8, pp. 127-137]. Let one of the waves from the reference source, located in the x plane, produce a field:

$$[\bar{E}_R(u)]_1 = \exp\left(i \frac{2\pi}{\lambda} \frac{u^2}{2f}\right) \tag{52}$$

in the hologram u plane, located at a distance f from the x plane, and parallel to it. We are assuming that the x extent of the source is small compared to the u extent of the hologram. Now let the corresponding wave from the source in the reconstruction originate from a position of the source displaced by a small distance x_o from the position used in the recording. The reconstructing wave may then be written in the form:

$$[\bar{E}_{R'}(u)]_1 = \exp\left(i \frac{2\pi}{\lambda} \frac{(u-x_o)^2}{2f}\right) \tag{53}$$

Since we are interested in the product $E_R{}^*E_{R'}$, and because x_o is small compared to u, we may write:

$$\bar{E}_R{}^*\bar{E}_R \cong \exp\left(-i \frac{2\pi}{\lambda} \frac{u^2}{2f}\right) \exp\left(i \frac{2\pi}{\lambda} \frac{u^2}{2f}\right) \exp\left(-i \frac{2\pi}{\lambda} \frac{2ux_o}{2f}\right) \tag{54}$$

We may conclude that the corresponding wavefront aberration is:

$$(\phi_{R'}-\phi_R)_1 \cong \frac{2\pi}{\lambda} \frac{ux_o}{f}. \tag{55}$$

In particular, equation (55) gives the aberration at the outer 'edge' of the hologram (or of the part of the hologram used in the reconstruction).

As an example, we may consider the magnitudes involved in the experiment illustrated in figure 19.

The wavelength λ used, 6328A is about 1/40 000 in. The distance f used was about 33 in. The images were reconstructed using an aperture having a diameter about 1 in. (giving u = 1/2 at the edge). With a displacement x_o = 0.001 in., the aberration is:

$$\frac{2\pi}{\lambda} \frac{ux_o}{f} = 2\pi \times 40 \times 10^3 \times \frac{1}{2} \times \frac{1}{10^3} \times \frac{1}{33} \tag{56}$$

i.e.

$$\phi_{R'} - \phi_R \cong 4 \text{ radians}$$

or about two-thirds of a wavelength. By recalling the Rayleigh tolerance of about one-quarter of a wavelength for 'perfect' imaging,

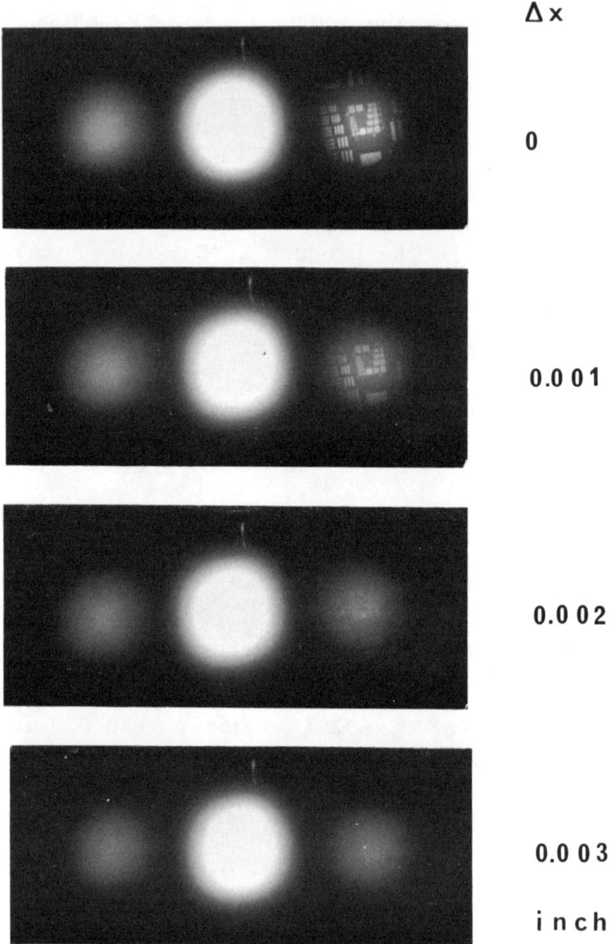

FIG. 19

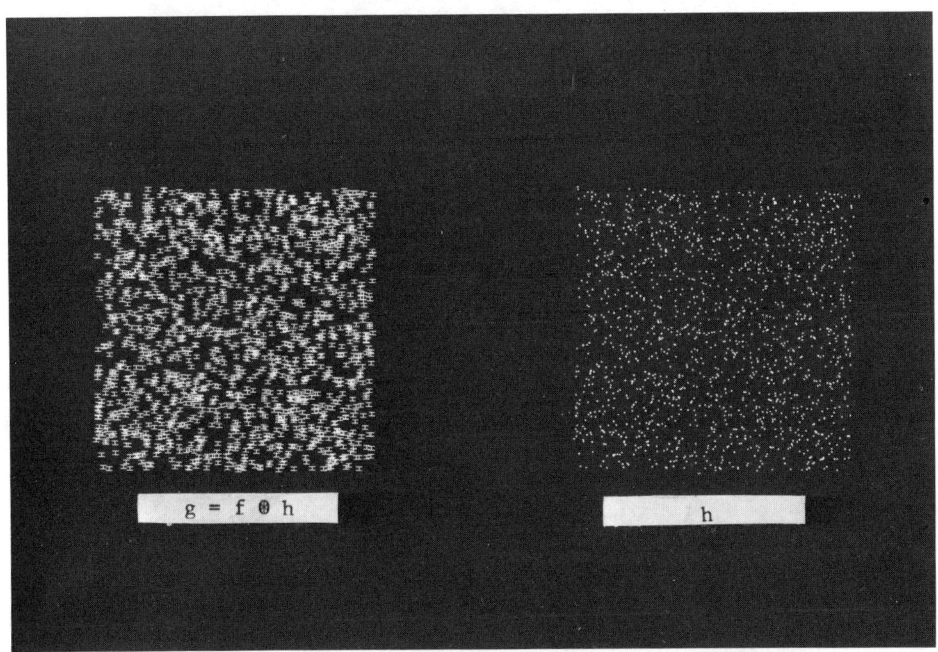

FIG. 20 Functions g(x,y) and h(x,y) used to record the Fourier-transform hologram for thepurpose of synthesizing the single image f(x,y) according to text. Dimensions of h(x,y): 2500 50-micron diameter pinholes, randomly disposed in a 18 x 18 mm square. Dimensions of g9x,y): 2500 "horizontal" letters "H" disposed in the same arrangement as h(x,y) within an 18 x 18 mm square. Height of "H": 300 microns, width of "bars": 50 microns (same as pinhole diameter); separation of "bars": 200 microns. Distance between center of g and center of h: 31 mm. Focal length of Fourier-transform hologram recording lens: 1069 mm. No diffusors were used in the hologram recording, with g and h placed in the front focal plane of the lens, but a density of about 3.02 was placed in front of g to help in approximately equalizing the two scattered fields. (After Stroke et al [16])

FIG. 21 Synthesized single image f(x,y) of the "horizontal" letter "H" reconstructed by holographic Fourier transformation from the hologram recorded according to FIG. 1 and text. The image shown is an enlargement, recorded in the best focus for the "horizontal" bars of the "horizontal" letter "H" obtainable with the 1 069mm lens used. (The "vertical" bar of the "horizontal" letter "H" clearly appears in other photos, focused at a distance where the "horizontal" bars shown here are less sharply imaged!) A positive contact print of the hologram recorded according to FIG. 1 was used in the reconstruction. The missing vertical bar in the letter "H" is due to incorrect weighting of the corresponding part in the hologram! (After Stroke et al [16]).

we should expect a displacement as small as x_o = 0.001 in. at a distnace f = 33 in. to produce a very noticeable deterioration in the image. This has in fact been fully borne out by our experiments, illustrated in figure 19. These results also help in explaining the results obtained in [33] carried out to verify our original method of [10, 11].

4. Image deblurring experiments using correlative holographic deconvolution.

In view of verifying the theory given above, we have actually carried out a deconvolution experiment, according to scheme B (equations (45) to (47), using for the object g(x,y) a computer-generated convolution (of the letter 'H') and for the spread function h(x,y) a computer generated random array of 2,500 pinholes (0.046 mm diameter) randomly disposed in a 17 x 17 mm area. The image g(x,y) had its 2,500 'images' disposed in the same area, and an attempt was made to have the two arrays as similar as possible, according to the tolerances of the previous section. This work, carried out in view of achieving a new method of x-ray imaging, using a multiple-pinhole camera, according to a suggestion by Dicke [34] is fully described in our ref. [16]. We did in fact obtain a perfectly convincing deconvolution of the hologram, using the method illustrated in figure 1. The results are shown in Fig. 20 and Fig. 21.

PART III

HOLOGRAPHIC COMPENSATION FOR DISTORTIONS CAUSED BY TRANSMITTING OR BY RECORDING MEDIUM

Because of their direct relevance to the subject discussed here, we now give some results of work on three-dimensional imaging through distorting media, using the Stroke method of 'lensless Fourier-transform holography' [9]. This method (see also [8, pp. 127-137]) permits one to achieve compensation for distortions when imaging through 'turbulent' and distorting media, as was first demonstrated by Goodman et al. [35], and again verified by Gaskill [36], in both cases for two-dimensional objects. Because of the approximations involved in the Fourier-transform compensation (see, e.g., equation (54) above), it appeared necessary to verify to which extent tolerances, such as those corresponding to our equations (55) to (57) could be achieved in the imaging of three-dimensional objects through a 'distorting' medium. The experiments and theory* are thus immediately applicable to assessing the effects of 'distortions' or deformations in the holograms, or corresponding misalignments, in terms of the applications considered here.

---

*(first presented by G. W. Stroke, by invitation, on 23 May 1968 at the 'Holography' Seminar of the Society of Photo-Optical Instrumentation Engineers in San Francisco, California). Also based on G. W. Stroke, Optica Acta (1969), Vol. 16, pp. 401-422.

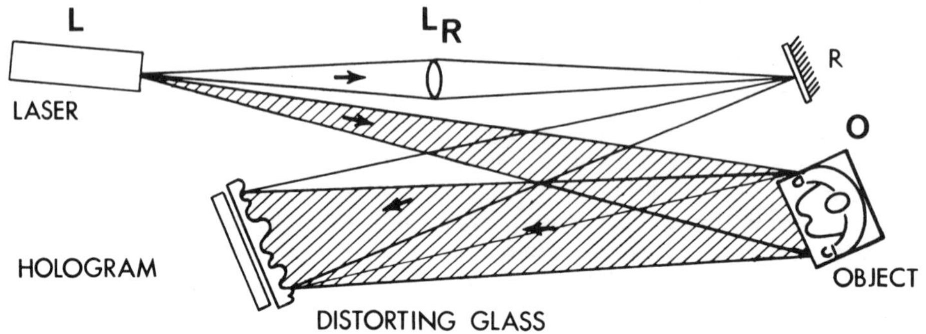

FIG. 22

DIRECT PHOTOGRAPH
THROUGH
DISTORTING GLASS

HOLOGRAPHIC PHOTOGRAPH
RECORDED THROUGH
DISTORTING GLASS

FIG. 23

Figure 22 shows a diagram of the experimental arrangement used. Let $\phi_D(u,v)$ be the distortion introduced onto a wave $E_O(u,v)$ originating from a point of the object, as shown, where (u,v) describes the hologram plane, as before. Let $\phi_{D'}(u,v)$ be the distortion introduced on the 'reference' wave originating from the reference point R. The imaging components in the hologram recorded according to this arrangement are thus:

$$E_{oi}E_R{}^*\exp[i(\phi_{Di} - \phi_D)], \qquad (59)$$

for each point of the object! Comparable terms describe the imaging components for the other (i-1) points of the object. It should thus be clear that 'compensation' for the distortions introduced on each object wave, and equal to $\exp[i\phi_{Di}]$, may be compensated by the distortion $\exp[i\phi_{D'}]$ introduced on the reference wave, provided that the value of the phase distortion of each wave, is compensated to the degree discussed above, i.e. say

$$\phi_{Di} - \phi_{D'} < \text{Rayleigh limit } (\pi/2 \text{ radians}) \qquad (60)$$

or, more rigorously to the limits given by Marechal [37], Toraldo di Francia [39], as well as those of the author [7], mentioned above. In particular, the Marechal limit for good imaging would require

$$\phi_{Di} - \phi_{D'} < \text{Maréchal limit } (\pi/4 \text{ radians}) \qquad (61)$$

throughout the entire wavefront (not just locally, as mentioned above).

It should be clear also, according to this analysis, that the image obtained through the 'compensated' lensless Fourier-transform hologram will be a considerably better image than that obtained by direct photography of the object, as viewed through the distorting medium!

Our experiments, illustrated in figure 23 have fully borne out this prediction, and indeed the orders of magnitude of distortion which may be compensated according to equations (60) and (61). The uncompensated 'distorted' image, obtained by direct photography through the distorting glass is shown in figure 23a. It shows the effects of a wavefront 'distortion' estimated to be about 4 radians in a direct (laser-light) photography of the object. The image may be compared to the third image from the top in figure 19 which corresponds to a wavefront 'distortion' of about 4 radians.

The remarkable improvement obtained by the method of holographic image compensation is shown in figure 23b.

It may be of interest to relate the phase distortions $\Delta\phi_D = [\phi_{Di} - \phi_{D'}]$ to thickness variations in the distorting medium, under given recording conditions. For example, if we call i_o the mean direction of incidence of an object wave with respect to the local normal of the distorting glass, and i_R the corresponding mean angle of incidence of the reference wave, elementary geometrical considerations give the relation:

$$\Delta\phi_D = [\phi_{Di} - \phi_{D'}] \ \frac{2\pi}{\lambda} \frac{1}{2} \frac{\Delta t}{n} (n-1)(i_o{}^2 - i_R{}^2), \quad (62)$$

where n is the refractive index of the distorting medium, and i_o, i_R, are respectively, small angles measured in radians. It is apparent that $\Delta\phi_D$ is zero when the two waves traverse the distorting medium along equal paths.

In the example illustrated in figure 22 and figure 23, the distance of the centre of the object from the reference point was about 100 mm, and the reference point was at about 1000 mm from the hologram plate, so that we may take, in this case $i_o = 0$ and i_R 1/10 radian. A path difference variation Δt of about 0.24 mm between the paths traversed by the reference wave and the object wave would thus create an aberration of about 4 radians. The distortion was obtained in this case by a deliberately irregular spraying of a transparent coating of some plastic on to an otherwise reasonably flat piece of plate glass. It is difficult to estimate, in this case, the exact amount of the thickness variations thus obtained, but the order of magnitude computed appears in reasonably good accord with the thickness variations on the plate. Even though the actual results in this case appear only qualitative, to some degree, the comparison between the direct photograph, figure 23(a), and the greatly improved image, figure 23(b), reconstructed from the hologram, when both are recorded through the same distorting medium, are of considerable interest within the context of this work.

REFERENCES

[1] STROKE, G. W., and ZECH, R. G., 1967, Physics Lett. A, 25, 89.
[2] LOHMANN, A. W., and WERLICH, H. W., 1967, Physics Lett. A, 25, 570.
[3] STROKE, G. W., INDEBETOUW, G., and PUECH, C., 1968, Physics Lett. A, 26, 443.
[4] STROKE, G. W., 1968, Photogr. Korr., 104, 82.
[5] STROKE, G. W. 1968, Physics Lett. A, 27, 405.
[6] STROKE, G. W. 1968, Physics Lett. A, 27, 252.
[7] STROKE, G. W., 1967, Handbuch der Physik, Vol. 29, edited by S. Flugge (Berlin: Springer Verlag), pp. 426-754.

[8] (a) STROKE, G. W. 1966, An Introduction to Coherent Optics and Holography (London, New York: Academic Press); (b) 1969, Ibid., second edition (April).
[9] (a) STROKE, G. W., 1965, Appl. Phys. Lett., 6, 201.
(b) STROKE, G. W., BRUMM, D., and FUNKHOUSER, A., 1965, J. Opt. Soc. Am., 55, 1327.
[10] STROKE, G. W., RESTRICK, R., FUNKHOUSER, A., and BRUMM, D., 1965, Appl. Phys. Lett., 6, 178.
[11] STROKE, G. W., RESTRICK, R., FUNKHOUSER, A., and BRUMM, D., 1965, Appl. Phys. Lett., 6, 178.
[12] MARECHAL, A., and CROCE, P., 1953, C. r. hebd. Seanc. Acad. Sci., Paris, 237, 607.
[13] GABOR, D., 1948, Nature, Lond., 161, 777.
[14] GABOR, D., 1949, Proc. R. Soc. A. 197, 454.
[15] GABOR, D., 1951, Proc. Phys. Soc., 64, 449.
[16] STROKE, G. W., HAYAT, G. S., HOOVER, R. B., and UNDERWOOD, J. H., Optics Communic. 1, No. 3, (1969) pp. 138-140.
[17] HARRIS, J. L., 1966, J. Opt. Soc. Am., 56, 569.
[18] ANSLEY, D., 1968, private communication to G. W. Stroke (May) based on a computation by J. W. Goodman.
[19] COOLEY, J. W., and TUKEY, J. W., 1965, Math. Comput., 19, 296.
[20] JACQUINOT, P., and ROIZEN-DOSSIER, B., 1964, Progress in Optics, Vol. III, edited by Emil Wolf (Amsterdam: North-Holland Publishing Co.), pp. 31-186.
[21] GIRARD, A., 1963, APPLIED OPTICS 2, 79.
[22] STROKE, G. W. and FUNKHOUSER, A., 1965, Physics Lett. 16, 272.
[23] BOUCHAREINE, P., and JACQUINOT, P., 1967, J. Phys., Paris, 28, C2-183.
[24] GOLDMARK, P., and HOLLYWOOD, J. M., 1951, Proc. Inst. Radio Engrs, 39, 1314.
[25] McMANN, R. and GOLDBERG, A., 1968, J. Soc. Motion Picture Telev. Engrs, 77, 221.
[26] TSUJIUCHI, J., 1963, Progress in Optics, Vol. II, edited by Emil Wolf (Amsterdam: North Holland Publishing Co.), pp. 133-182.
[27] PERINA, J., 1968, private communication to G. W. Stroke (September).
[28] PERINA, J. and STROKE, G. W., 1969 Optica Acta, (to be submitted).
[29] TORALDO DI FRANCIA, G., 1952, Suppl. to nuovo Cim., 9, 426.
[30] SIMON, J. C., BROUSSAUD, G., and SPITZ, E. 1959, C. r. hebd. Seance. Acad. Sci., Paris, 248, 2309.
[31] STROKE, G. W. and FALCONER, D. G., 1964, Physics Lett., 13, 306.
[32] TOLLIN, P., MAIN, P., ROSSMAN, M. G., STROKE, G. W., and RESTRICK, R. C., 1966, Nature, Lond. 209, 603.
[33] LANZL, F., MAGER, H. J., and WAIDELICH, W., 1968, Z. angew. Phys., 24, 156.

[34] DICKE, R. H., 1968, Astrophys. J., 153, L-101.
[35] GOODMAN, J. W., HUNTLEY, W. H., JACKSON, D. W., and LEHMANN, M., 1966, Appl. Phys. Lett., 8, 311.
[36] GASKILL, J. D., 1968, J. Opt. Soc. Am., 58, 600.
[37] MARECHAL, A., 1947, Thesis, Faculte des Sciences, University of Paris.
[38] TORALDO DI FRANCIA, G., 1958, La Diffrazione della Luce, Edizione Scientifiche Einaudi (Torino: Paolo Boringhieri).
[39] STROKE, G. W., 1967, Modern Optics (Proceedings of the 22-24 March 1967 Symposium)(Polytechnic Press of the Polytechnic Institute of Brooklyn), pp. 589-603.
[40] GABOR, D., 1968, private communication to the author (October).
[41] VANDER LUGT, A., 1964, I.E.E.E. Trans. Inf. Theory, 10, 139.
[42] LOWENTHAL, S., and BELVAUX, Y., 1966, C. r. hebd. Seanc. Acad. Sci., Paris, 262, 294; see also: 1967, Optica Acta, 14, 245.
[43] VIENOT, J. Ch., and BULABOIS, J., 1967, Optica Acta, 14, 57.
[44] STROKE, G. W., FURRER, F. and LAMBERTY, D. 1969, Optics Communications, 1, 141 This paper introduces holographic Fourier-transform division filtering using 'extended' dynamic range in the photographic filters.
[45] UPATNIEKS, J., VANDERLUGT, A., and LEITH, E. N., 1966, Appl. Optics, 5, 589.
[46] TSURUTA, T., and ITOH, Y., 1968, Appl. Optics, 7, 2139.
[47] GROH, G., and STROKE, G. W., Optics Commun. 8, No.9 (April 1970).

REAL TIME OPTICAL INFORMATION PROCESSING

Euval S. Barrekette

IBM-Thomas J. Watson Research Center

Yorktown Heights, New York 10598

Introduction

Coherent optical processing systems are particularly well-suited for handling the very large amount of information contained in two-dimensional images, since they can perform in parallel any of several operations on all the information in the image. This inherent advantage of such systems over the conventional serial approach of digital computing systems is well known and has been exploited in the past [1] to great advantage for such applications as contrast enhancement, filtering of correlated noise in telemetered images, recognition, etc., and more recently techniques have been developed for image deblurring [2].

Customarily, the filters are prepared in analog fashion and inserted mechanically into the appropriate plane in the optical information processor. This technique yields eminently satisfactory results in the laboratory where laborious and time consuming care can be taken to ensure that the filters are of proper size and properly aligned. However, as the amount of information to be processed is increasing at an accelerating rate, it is becoming clear that automatic systems for creating, changing and aligning filters in real time can play an important role in data reduction and processing.

The purpose of this paper is to describe an approach to real time spatial filtering, as well as to the problem of real-time conversion of incoherently illuminated subjects into coherent images for further processing. The feasibility of the techniques described herein has been demonstrated. Widespread utilization of such systems, however, must still await the solution of materials and engineering problems which in principal do not seem to pose insurmountable obstacles.

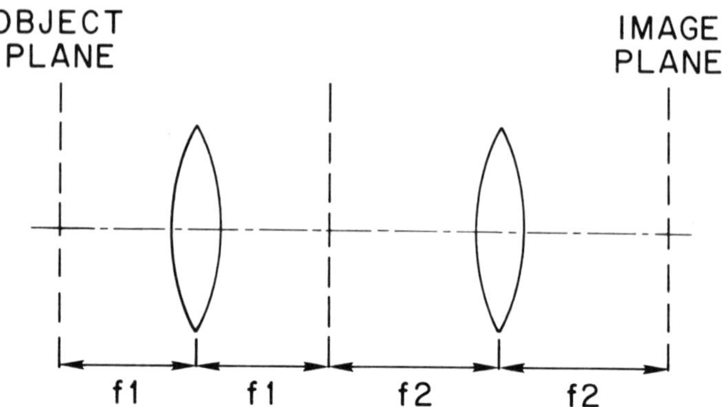

FIG. 1 Basic Unit of an Optical Processor

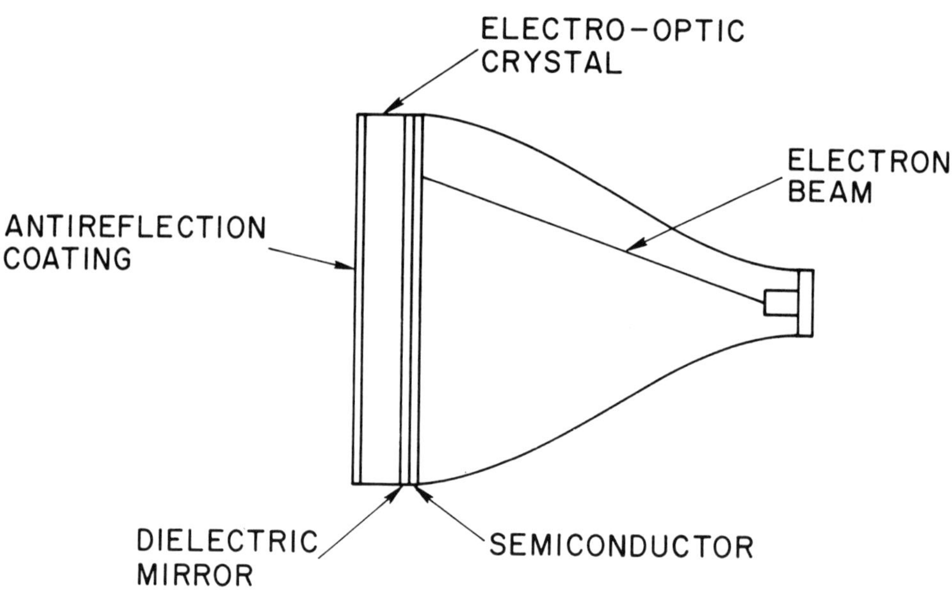

FIG. 2 Schematic of an Electron Beam Controlled Spatial Filtering Device

Electro Optical Filter

The basic structure of an optical processor is shown in Fig. 1. It is an imaging system consisting of two lenses separated by a distance equal to the sum of their focal lengths. If the object plane is illuminated by a coherent plane wave, the amplitude distribution in the common focal plane is approximated by the fourier transform of the amplitude distribution through the object [3].

Spatial filtering is accomplished by means of filters in the common focal plane. Extensive demonstrations of various filtering operations have been published by many authors over a period of many years [1,4], even predating Toepler's work with the schlieren system in 1864.

For real time filtering it is useful to have a filter which operates on the amplitude and/or phase of the light by means of physical effects which are electrically controllable. Electro-, magneto-, and acousto-optic devices can be used. Acousto-optic filtering has the shortcoming that it is inherently one dimensional. An acoustic column will effectively modulate the phase of a light beam transmitted through it in one dimension. To accomplish two-dimensional filtering a stack of acoustic columns must be used. Although the stack does not have equal resolution in two dimensions it is quite useful for some applications as for example in the processing of signals from phased array radars [5]. Magneto-optic filters, although feasible, are not as easily controllable as electro-optic ones. Accordingly, this paper is devoted to the description of truly two-dimensional filters relying on the longitudinal electro-optic effect.

A convenient embodiment of an electrically controllable electro-optic device is shown in Fig. 2. It is based on the von Ardenne tube [6] and in varying forms has appeared in the Electron Beam Scanlaser [7], as the Titus tube [8] and in an optical processor [9]. The device consists of a cathode ray tube whose face is a sandwich consisting of an electro-optic crystal such as KDP or $LiNbO_3$ with an antireflection coating and possibly a transparent conductor on one side and a dielectric mirror and a semiconductor of appropriate conductivity* on the other side facing the electron gun. A pattern of changes placed on the filter gives rise to a corresponding pattern of variations in birefringence in the crystal.

---

*The conductivity is chosen on the basis of the desired persistance of frame rate which can be set over a range from microseconds to days. A variable charge dwell time under external control is also conceivable using either a flood gun to remove charges from a dielectric surface or optical flooding of a photoconductor coated over a conductor placed on the dielectric mirror.

This electro-optic spatial filter can operate either on the phase or the amplitude of an impinging plane polarized beam of light. Thus if the plane of polarization of the incident light is along one of the axes of the index ellipsoid of the crystal, the light reflected from the filter will remain plane polarized but will be modulated in phase in relation to the pattern of changes in index induced in the crystal and thus to the pattern of charges placed on the filter. Alternately if the plane of polarization of the incident light is between two unequal axes of the index ellipsoid, the light reflected from the filter will be elliptically polarized. Thus after passing through an analyzer the reflected beam will be modulated in amplitude in relation to the charge pattern placed on the crystal.

Fig. 3 shows the electro-optic spatial filter placed within a folded basic processor. Results of several experiments [9] using the system are shown in Figs. 4,5,6.

In Fig. 4a is the image of a slit imaged through the system with no charges on the filter, and with the analyzer and polarizer aligned, while Fig. 4b is an image of the same slit with a circular charge pattern at the center of the Fourier spectrum of the slit. Sufficient charge was used to yield a 90° rotation in the plane of polarization of the beam at the location where the charge was deposited. Thus the low frequency components of the object where blocked out and edge enhancement resulted.

In Fig. 5a is shown the image of a square wave grating which was imaged through the system with the polarizer and analyzer aligned. In Fig. 5b is the image of the same grating imaged through the system with the analyzer and polarizer crossed. A two line raster of charges was placed on the crystal adjusted so as to allow only the two first orders of the grating to be transmitted thus yielding an image with twice the fundamental frequency of the object.

Fig. 6 is the image of a phase object which was created by bleaching a transparency of an intensity modulated object in contact with a grating. Thus the information in the object appears as phase modulation superimposed on a carrier frequency corresponding to the spatial frequency of the grating. The image was rendered visible using the system with crossed analyzer and polarizer; the electron beam was centered on one of the first orders of the grating and scanned an area large enough to encompass the band of frequencies with which the object modulates the carrier. Thus only the first order is passed and a visible image appears as in schlieren type projection.

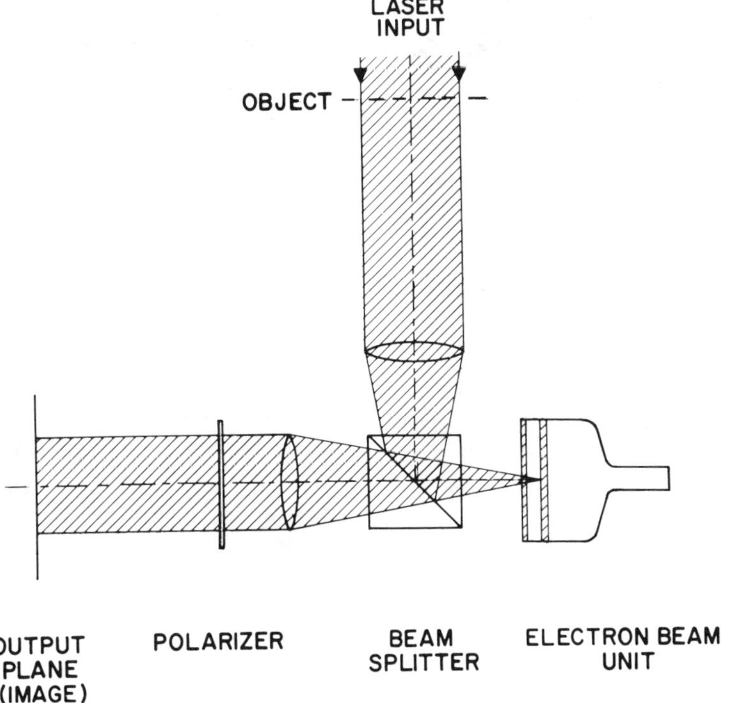

FIG. 3 Schematic of an Electro Optical Information Processor

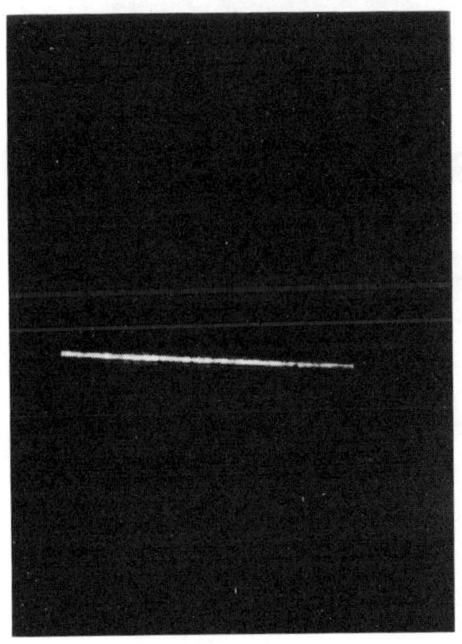

FIG. 4a Reproduction of a Single Slit through the System of FIG. 3

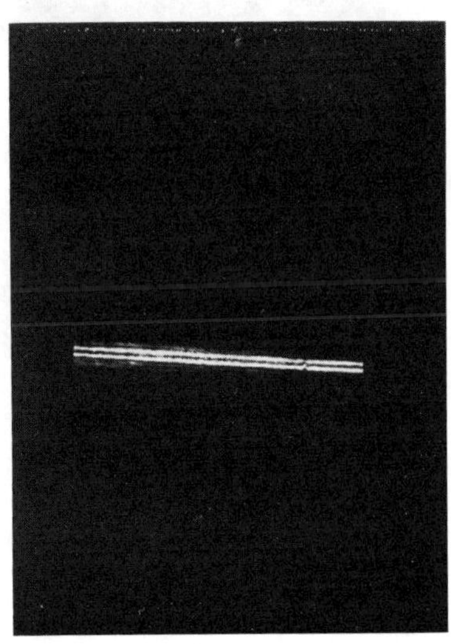

FIG. 4b Altered Image of the Slit of FIG. 4a Produced by Point Spatial Filter

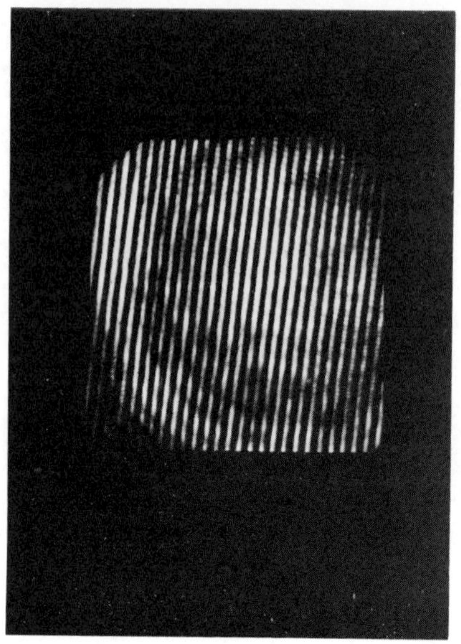

FIG. 5a Reproduction of a Square Wave Grating through the System of FIG. 3

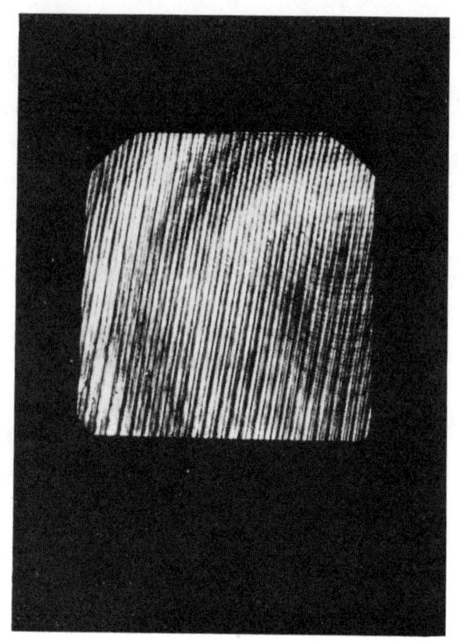

FIG. 5b Doubling of the Grating Frequency Produced by a Two-line Spatial Filter

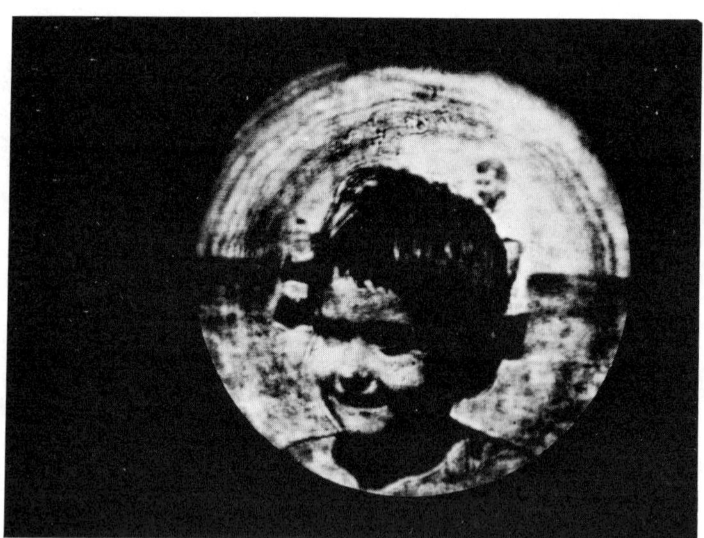

FIG. 6 Intensity Modulated Image of a Phase Modulated Object Produced by a Spatial Filter that Passes Only the First Order of the Phase Grating

FIG. 7a Electron Beam Controlled Spatial Filter with Colinear Propagation of Light and Electrons

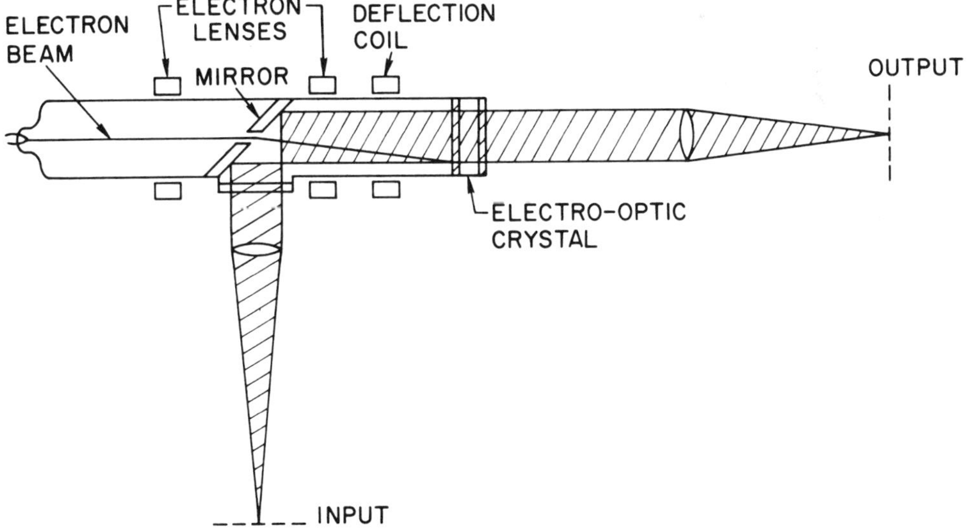

FIG. 7b Schematic of System Shown in FIG. 7a

A shortcoming of the system shown in Fig. 3 is that 75% of the light incident on the beam splitter is automatically lost before all other energy losses due to the filtering process itself. This problem is overcome with a cathode ray tube of the type reproduced in Fig. 7a and shown schematically in Fig. 7b. Here a lens focuses the electron beam onto a small aperture in a mirror and a second lens and deflection yokes place the transmitted beam on the crystal which in this geometry does not have a dielectric mirror. The filter works in transmission on the light entering the tube through the crystal and reflected out to the side. This geometry has the further advantage of colinear propagation of light and electrons; thus the problems of off axis optics or electron optics which limit the projection Titus Tube [8] or the Eidophor projector [10], are not encountered here.

Reactive Processing

For some applications, particularly when the object to be processed has low contrast, or if only low power is available, reactive processing, i.e., processing within a laser cavity becomes attractive [11,12], since the inherent nonlinearities of the laser and the large energy densities available within the cavity can then be used effectively.

A schematic of a reactive optical processor is shown in Fig. 8 as it would be used for conventional schlieren projection of phase objects. The right hand mirror of the resonator acts as a schlieren stop. It can be as large in extent as the zero order, but it must not obscure any of the side orders. In this configuration, light, which ordinarily is lost in a conventional schlieren projector, is fed back into the cavity where it can be stored without disturbing the modes of the resonator provided the depth of modulation of the phase object is sufficiently low. It has been demonstrated [12] that the power gain of reactive over passive processing has a theoretical upper limit which is the inverse of the fractional energy diffracted by the object into the first order when passively illuminated. Experiments with a phase grating have yielded a gain in excess of 5. Fig. 9 shows a reactive image of a jet of air blown into the laser cavity [12].

Clearly a real time filter can be placed in contact with the right-hand mirror of the resonator of Fig. 8 to yield a reactive system for operating on the side orders of the spatial spectrum of the object. Furthermore if the resonator contains several lenses with each pair of lenses separated by the sum of their focal lengths it becomes possible to conceive of a reactive processor* (Fig. 10) which contains several controllable elements of the type in Fig. 7; it will permit simultaneous operation on phase and amplitude as well as the insertion of the object into the cavity in the form of a pattern of birefringence on one of the filters.

---

*The stacked multi-element processor described herein could also be used in a mode wherein it is passively coherently illuminated.

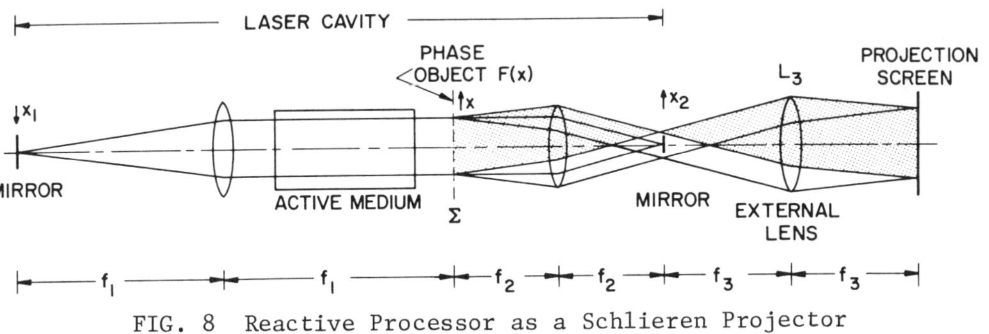

FIG. 8 Reactive Processor as a Schlieren Projector

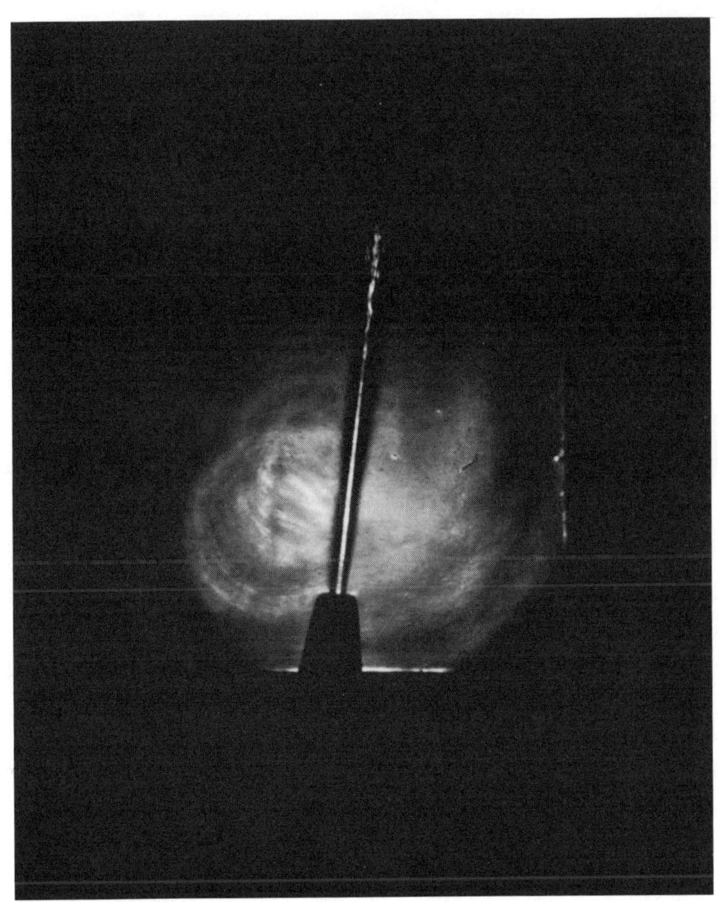

FIG. 9 Reactive Image of a Jet of Air Blown into Processor of

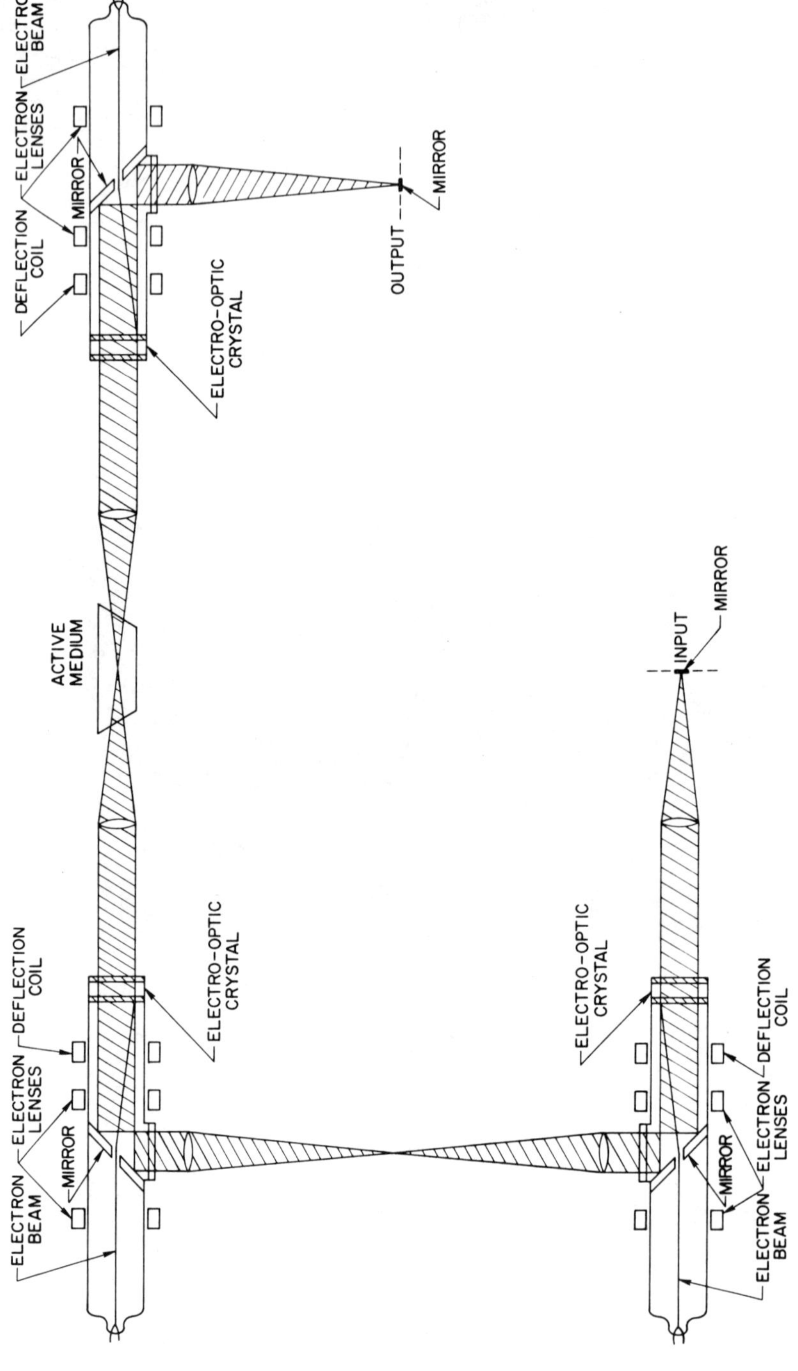

FIG. 10 Reactive Processor for Simultaneous Multiple Operations

Although one must anticipate some degradation due to the transfer functions of the various elements inserted within the light path in such a system, it is clear that operations will be performed in parallel over a large number of points in the object and that contrast enhancement will result from the nonlinearities arising from operations within the laser cavity. It must be borne in mind, when considering this approach, that digital processing of scanned data, involving numerous iterations in a basically serial processor, also suffer degradation from round-off errors and from the convolution of the object with the point spread function of the scanner in the initial data reduction.

Image Conversion

It is a straight-forward matter to employ electro-optic (or any other) real time two-dimensional spatial filter as an input element for inserting an object into the processor. However, if the object is an actual scene which is to be processed in real time and which cannot be conveniently illuminated coherently an incoherent to coherent image conversion system must be used.

One such device [13] relies on photoconductivity. It is shown in Fig. 11. The scene to be processed, is imaged onto an electro optic sandwich which has been charged by a corona discharge. The charged surface is a photoconductor deposited on a dielectric mirror which is on an electro-optic crystal. The other side of the crystal on which coherent light is incident is coated with a transparent conductor.

As in the electro-optic spatial filter, depending on the orientation of polarization of the incident coherent beam, the light reflected off the sandwich will be modulated either in phase or amplitude (in the latter case after passing through an analyzer). The modulation will correspond to the field distribution across the crystal which is related to the charge distribution and thus to the image on the crystal. The coherent beam can then be processed as described in the foregoing sections.

Image conversion can be passive or reactive. In the example illustrated in Fig. 11 the device is a reactive one, and a spatial filter placed at the right-hand mirror of the laser cavity will be analogous to that described in the foregoing for the reactive case (Fig. 8). With the active medium removed a conventional passive processor results.

An alternate device [14] for image conversion relying on photoemission is shown in Fig. 12. The coherent scene is imaged on an array of photoemitters, which replace the photoconductor in the device in Fig. 11. Here a wire grid in vacuum is used to draw the charges from the initially uncharged mosaic, whereas in the previous case the wire grid was necessary for forming the corona discharge. At any rate the charge distribution on the array is converted into modulation of the coherent beam incident from the opposite side of the crystal.

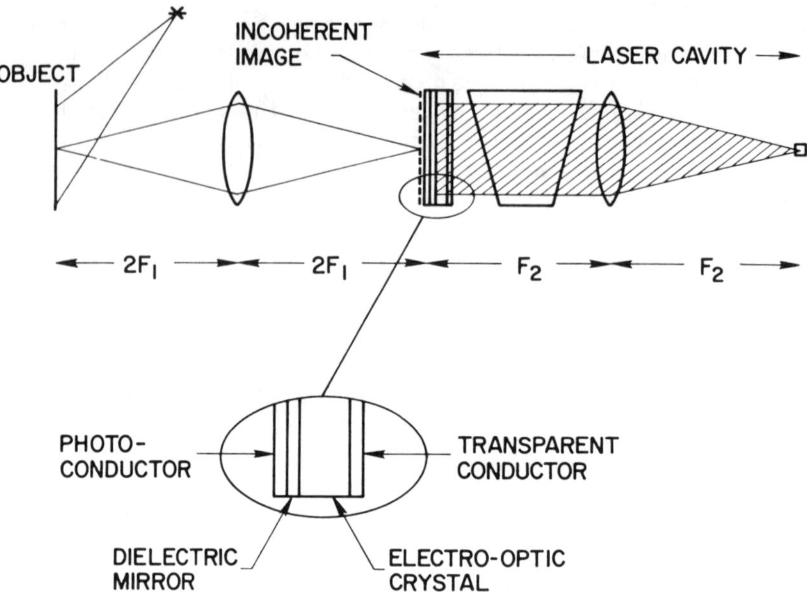

FIG. 11 Incoherent to Coherent Image Converter Relying on Photoconductivity

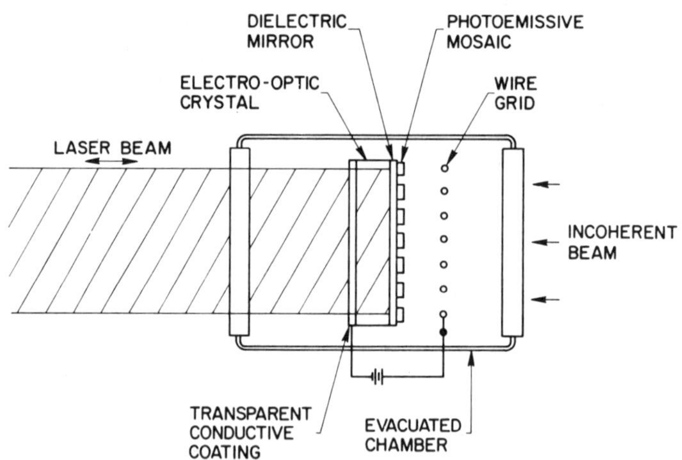

FIG. 12 Incoherent to Coherent Image Converter Relying on Photoemissivity

Conclusions

Several electro-optic devices were described which indicate a possible approach to real time image conversion and processing. It is clear that these are by no means the only possible configurations, nor are the effects they rely on the only ones available for these applications. The purpose of this paper is simply to show that real time processing is possible and should be considered as a contender for handling the wealth of image data being collected today.

It must be noted that these devices though feasible, are not a commercial reality today. The major obstacle that must be surmounted if real time processing is to become a reality are in the materials area. Crystals and coatings of reproducibly uniform characteristics and quality must be fabricated at competitive prices before this new technology can penetrate the marketplace. If these problems can be solved the full potential of optical parallel processors will be realized.

References

[1] see for example Joseph W. Goodman, Introduction to Fourier Optics, McGraw Hill, New York (1968), Edward L. O'Neill, Introduction to Statistical Optics, Addison-Wesley, Reading Massachusetts (1963), and George W. Stroke An Introduction to Coherent Optics and Holography, Academic Press, New York (1966) where many examples and extensive references are provided.

[2] see for example Jumpei Tsujiuchi and George W. Stroke, "Optical Deblurring Methods" Chapter 16 in this volume where an extensive bibliography is also available.

[3] J. Elmer Rhodes Jr. "Analysis and Synthesis of Optical Images" Am. J. Phys. 21(1953)337-343.

[4] see for example A. Vander Lugt "A review of optical data-processing techniques" Optica Acta, 15, (1968) 1-33 for an extensive review and bibliography.

[5] see for example Louis B. Lambert, Moses Arm and Alexander Aimette, "Electro-Optical Signal Processors for Phased Array Antennas" Chapter 38 of J. T. Tippett et al Optical and Electro-Optical Information Processing, MIT Press, Cambridge Massachusetts (1965)715-748. and Louis B. Lambert, "Optical Correlation", chapter in Raymond S. Berkowitz, Modern Radar John Wiley & Sons, Inc. N.Y. (1966)245-271.

[6] M. von Ardenne, US Patent 2,276,359 (1939).

[7] R. V. Pole and R. A. Myers, "Electron Beam Scanlaser" IEEE J. of Quant. Elect., QE-2, (1966)182-184.

[8] G. Marie "Un Nouveau Dispositif de Restitution D'Images Utilisant Un Effet Electro-Optique: Le Tube Titus" Philips Res. Repts., 22(1967)110-132.

[9] H. Wieder, R. V. Pole, P. Heidrich, "Electron Beam Writing of Spatial Filters" IBM J. of Res. & Dev. 13(1969)169-171.

[10] F. Fischer and H. Thiemann, "Theoretical Considerations on a New Method for Large-Projector Television", Schweiz. Arch. angeu. Wiss. Techn., 7(1944)305-318, 337-344, Ibid. 8(1942) 15-28, 135-143, 169-178, 199-212, 299-307.

[11] R. V. Pole, H. Wieder and E. S. Barrekette, "Reactive Optical Information Processing I: Theory of Information Recovery and Resonator Mode Structure", App. Opt. 6(1967)1571-1575.

[12] H. Wieder and R. V. Pole, "Reactive Optical Information Processing II: Factors Affecting the Applicability and Efficiency of the Method," App. Opt. 6(1967)1761-1765.

[13] R. A. Myers, R. V. Pole and H. Wieder, "Real-Time Incoherent to Coherent Image Conversion", IBM Tech. Discl. Bull., 11 (1969) 1314-1316 and R. V. Pole, "A Study of Automatic Target Recognition Techniques for Night Vision Devices", Final Report, Contract DAAK02-68-C-0118, July 5, 1968.

[14] L. Kuhn, Personal Communication, December 5, 1968.

Radar and Microwave Applications of Holography

Winston E. Kock

Vice President and Chief Scientist

The Bendix Corporation

Abstract

The similarity between side-looking radar records and true holograms, noted by the author at the first U. S. - Japan seminar in 1967, is briefly reviewed. Extension of the side-looking radar concept to stationary (doppler-free) coherent radars is discussed, including both monostatic and bistatic versions. The concept of a passive (cooperative) hologram radar system is put forward.

Because holography and side-looking radar both possess the hitherto unachievable ability to provide excellent resolution (focus) for objects at any point in the near field, one is tempted to dismiss as unimportant consideration of the far field performance of stationary coherent radar since ordinary radars perform this task quite well. However, coherent radar may achieve an equivalent performance more simply. It is shown that the array gain achieved by combining the outputs of all array elements of a radar is not lost in a coherent array radar even though the outputs of individual elements are photographically recorded.

Microwave holograms which employ liquid crystals in forming the hologram are discussed; also, a recently described procedure for using the hologram process to record both range and range rate of targets in a coherent radar. A method is described for achieving pulse compression in radars without using the frequency shift or "chirp" process. In this procedure the processing of the received signals is accomplished by photographically recording them and subsequently using laser illumination to obtain accurate range information. Extending the use of such pulse compression techniques to synthetic aperture radars is also discussed.

Introduction

At the first U. S. - Japan Seminar on Holography in 1967, the author reported on numerous applications of holography in the microwave field. Several publications have since appeared which were based on that presentation (1, 2, 3), one of which (2) provided numerous references to the activities in various laboratories. In this paper we first review briefly some of the earlier microwave hologram efforts and then report on more recent achievements in this area.

Microwave Holograms and Coherent Radar

In a coherent (side-looking or "synthetic aperture") radar system, (Fig. 1) an aircraft moving along a very straight path continually emits successive microwave pulses. The frequency of the microwave signal is held very constant, i. e., the signal remains coherent with itself for very long periods. During these periods the aircraft may have travelled several thousand feet, but because the signals are coherent, the many echoes which return during this period can be processed as though a single antenna as long as this flight path had been used. The effective antenna length is thus quite large. This large "synthetic" aperture has a very high resolving power, even at microwaves, enabling the radar to present extremely fine detail in its final pictures (4).

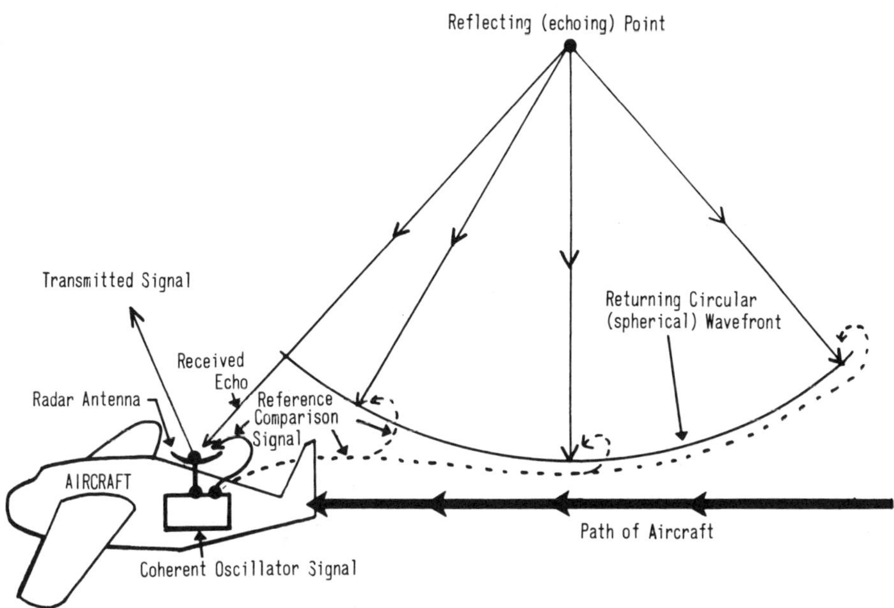

FIG. 1 In a side-looking radar, the circular wavefronts returning from a reflecting point are permitted to interfere with a coherent reference signal as the plane flies along a straight path; the recorded patterns are one-dimensional Fresnel zone plates.

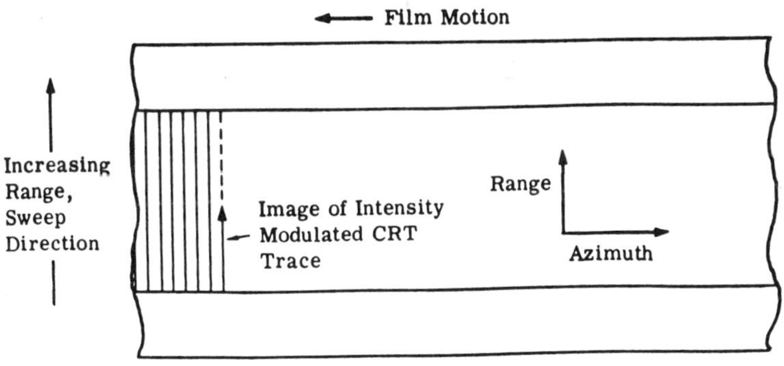

FIG. 2 A hologram record is formed in a synthetic aperture radar by photographing intensity-modulated CRT trace.

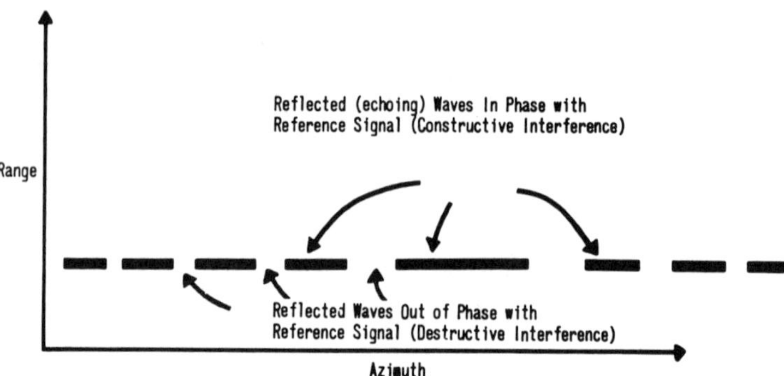

FIG. 3 The photographically stored record of a single reflecting point is a one-dimensional zone plate.

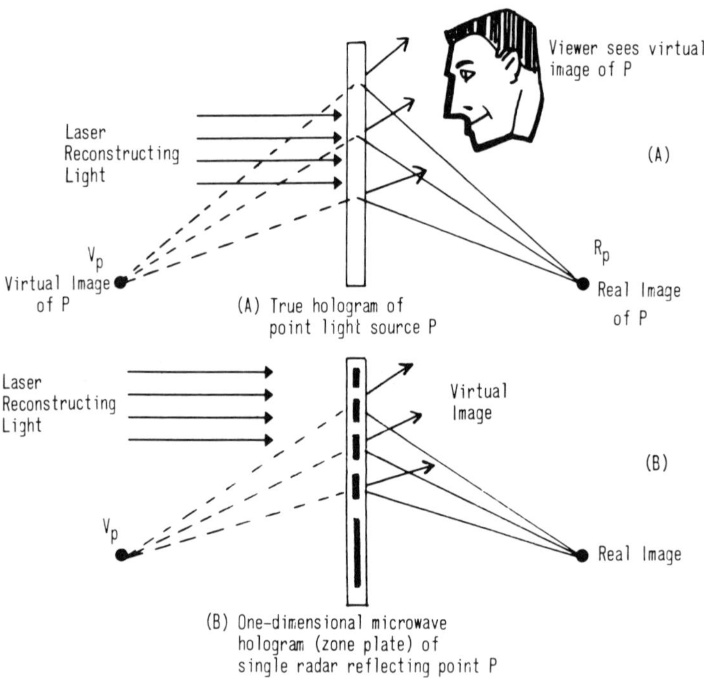

FIG. 4 A similarity exists between a hologram of a point source and the one-dimensional microwave zone plate holograms of a coherent radar.

The photographic record of the echoes received by such a coherent radar is actually a form of hologram (1, 2, 3). The microwave generator which produces the illuminating signal provides the hologram reference wave, and the reflected signals received along the flight path are made to interfere with this reference signal so that the complex interference pattern thereby generated and photographically recorded is completely equivalent to a hologram.

In figure 1 only one reflecting point is shown. The waves returning from this point have spherical wave fronts, whereas the oscillator reference signal is equivalent to a set of plane waves arriving in a direction which is perpendicular to the path of the craft. The received signal is combined with the coherent reference signal, amplified and made to intensity modulate a cathode - ray tube trace, as shown in Figure 2. Each vertical line in that figure plots signals received from all range points, with the points at greater range being recorded near the top of the vertical trace. As the airplane moves along and new pulses are emitted, the film is indexed (i. e. it is moved slightly to the left) to permit a new record of a new set of returns (a new vertical line) to be made.

For the case of only one reflecting point at a given range, as in Figure 1, the upward moving cathode ray beam would, for every pulse, be brightened only at that one point in range -- i. e. at only one spot on each vertical trace. The final result would thus be only one horizontal line of recorded echoes. But this line, as sketched in Figure 3, is not continuous. Because the returning waves are circular, the slant azimuth range, i. e. the distance from the aircraft to the reflecting point, changes, so that the combination of return and reference waves successively produces areas of constructive and destructive interference. At the greater slant angles where the distance from the plane to the point changes rapidly, this succession of in-phase and out-of-phase conditions occurs rapidly. When, on the other hand, the aircraft is practically abreast of the reflecting point, the range does not change rapidly, and the changes occur more slowly. The resulting record is a one-dimensional zone-plate hologram. If it is later illuminated by laser light, as sketched in Figure 4, it can reconstruct the reflecting point, just as the hologram in that figure reconstructs an image of the scene recorded by the hologram.

Both the range and azimuthal position of reflecting points which have generated zone plates can be determined by illuminating the hologram with laser light, as shown in Figure 5. In this figure two reflecting points are indicated which are displaced appreciably in range and slightly in azimuth. All such reconstructed images fall on a plane, and the tilt of this plane is determined by the original vertical tilt of the radar.

When synthetic aperture radar was first described, many could not understand how its high resolution could be maintained for nearby objects. Because of the great length of the synthetic aperture antenna, the majority of reflecting objects of interest are in the near field, not in the far field, or Fraunhofer region, where most radar antennas operate. For a near-field reflecting object, ordinary radars can only achieve maximum efficiency (gain) and maximum resolution when the phases of their receiving elements are adjusted so as to correspond to the arc of a circle centered on that object. Yet for other near-field objects, at other ranges and azimuth directions, these phases have to be different, so as to correspond to the arcs of other circles which are centered on these other points.

When it is recognized that the radar record is a hologram, the reason for this focussing ability is readily understood. In holography, each small light-reflecting point generates its own zone plate, and each of these recorded zone plates later causes coherent laser light to be reconstructed exactly at the equivalent point in space from whence the light emanated. Similarly, a side-looking radar captures photographically the curved wave fronts emanating from a reflecting point by combining them with a reference wave, thus generating a one-dimensional zone plate. Later, as in an optical hologram, coherent reconstructing laser light acquires, through diffraction of it by the zone plate, the properly curved wave fronts, causing the light to be focussed at points corresponding to the reflecting points in the original landscape. Thus, just as an optical hologram causes each point of a three-dimensional scene to be brought into sharp focus no matter what its distance from the photographic plate, so the microwave hologram of the radar record can impart good focus to all of its reconstructed points.

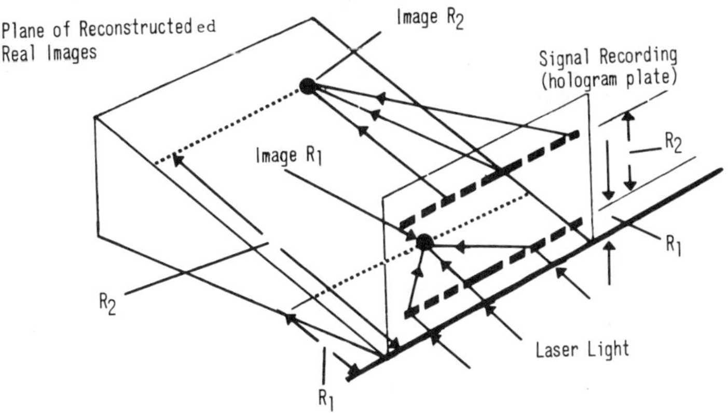

FIG. 5 Photographically recorded one-dimensional zone plates permit laser reconstruction of real images of two reflecting points.

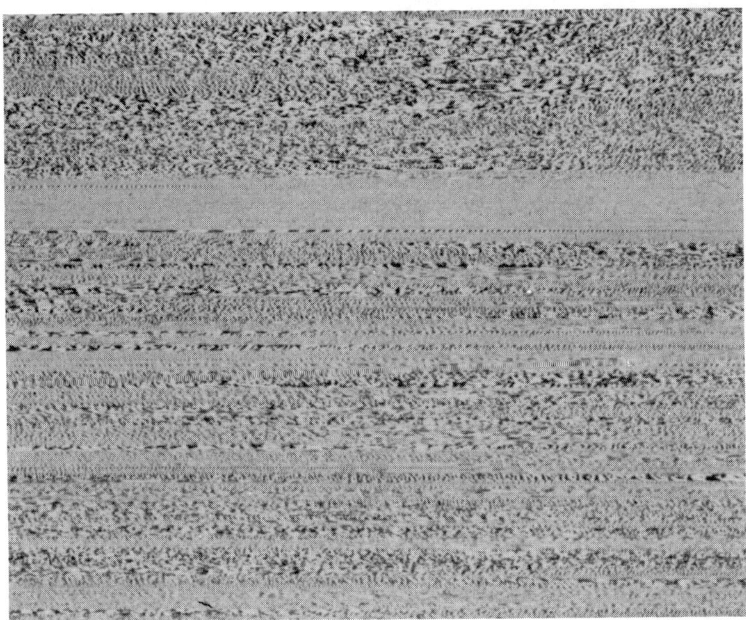

FIG. 6 An actual photographic record generated in a synthetic aperture radar.

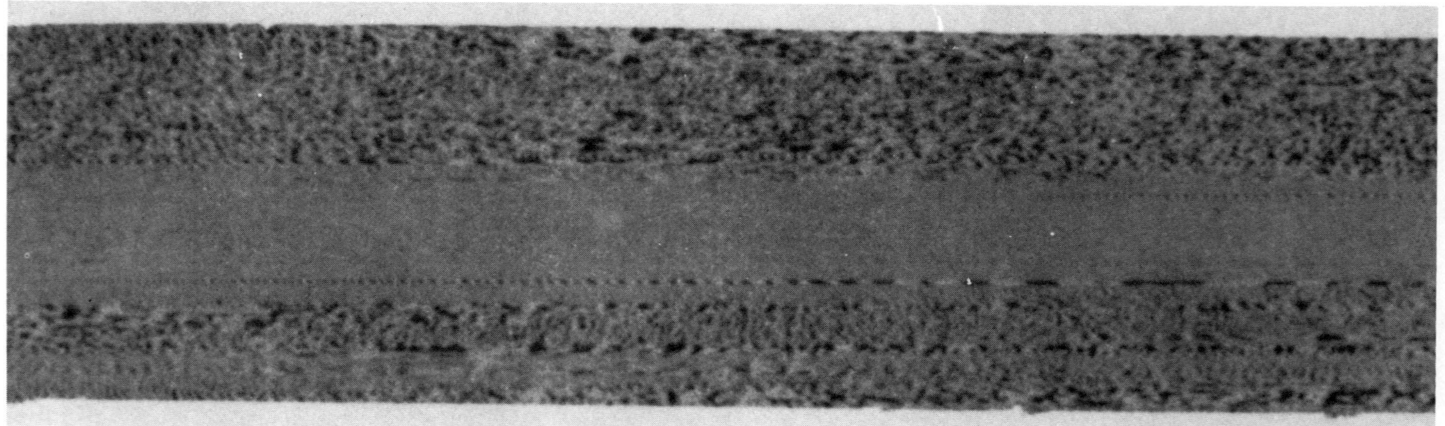

FIG. 7 Side-looking coherent radar records, like holograms, consist of many superimposed photographically recorded zone plates from which the original scene is reconstructed. In this radar record, which is an enlarged portion of FIG. 6, an isolated reflecting point has generated (near the bottom of the central blank area) an easily recognized one-dimensional zone plate which, with monochromatic processing, will yield azimuth information with a precision corresponding to a tiny fraction of the zone plate length. The zone plate can be looked upon as a long, broadside synthetic antenna with its concurrent high gain and beam sharpness.

FIG. 8 Proper optical processing of synthetic aperture radar holograms provide high resolution detail of the terrain flown over by the aircraft.

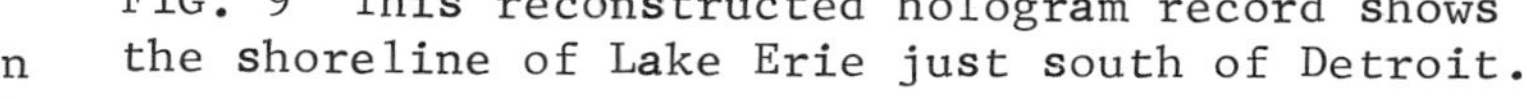

FIG. 9 This reconstructed hologram record shows the shoreline of Lake Erie just south of Detroit.

A typical microwave hologram, as generated by a side-looking radar, is shown in Figure 6. In Figure 7, which is an enlargement of a portion of Figure 6, a prominent one-dimensional zone plate, corresponding to the zone plate of Figure 3, is seen near the bottom of the blank area. When these microwave holograms are processed, photos of extremely good detail result, as shown in Figures 8 and 9.

Because of the similarity between coherent radar and holography, one is tempted to suggest that holography was re-invented during the 1950s by the University of Michigan group that developed side-looking radar under the leadership of Professor L. J. Cutrona. They generated zone plates, they employed offset procedures to separate the diffracted components, and they used coherent optics to regenerate the radar image. Fortunately for holography, one member of the group, Professor Emmett Leith, extended his coherent radar know-how to optical holograms, and in 1963 the first laser hologram came into existence.

Radar Records as Diffraction Devices

Certain advantages accrue when side-looking radar records are looked upon as holograms, i. e. as diffraction devices. At the 1967 U. S. - Japan seminar, the use, in this form of radar of the offset process to separate the real image, the virtual image, and the undeviated component was referred to. Use of the offset principle in purely optical holograms was first described by Leith and Upatneiks in 1963 (5), but its use in microwave radar holograms was discussed several years earlier (6). Both papers discussed the separation of the images in terms of spatial frequencies, and both referred to the real and virtual twin images as "positive and negative frequency components". Thus the earlier paper stated: "If the complex function is shifted in frequency ... the spectral display of the optical system is such that the positive and negative frequency components are independently available". In like manner, the later, hologram paper stated: "Since one is on a positive carrier, and the other is on a negative carrier, they can be separated by spatial filtering". Perhaps it was this relating the offset procedure to a frequency shift away from a spatial carrier frequency, that let some to the conclusion that the original Gabor holograms were incomplete records. Thus it was stated in one publication (7) that the first holograms of Gabor failed to record properly

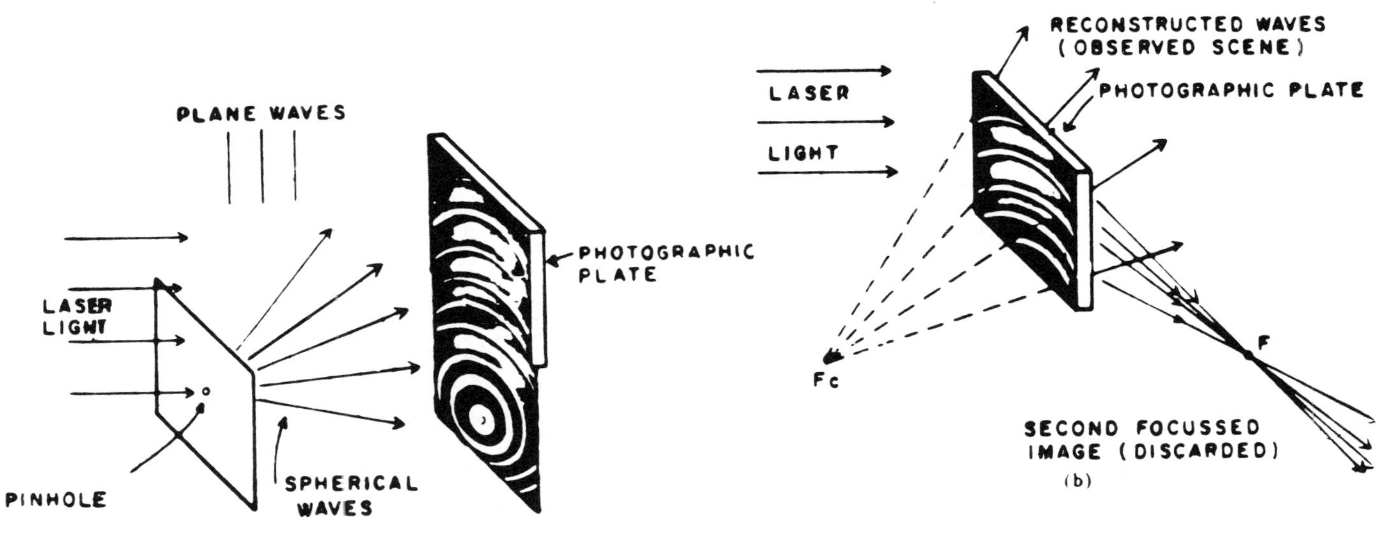

FIG. 10 In making a Gabor hologram, the plane reference waves passing around the object to be recorded (in this case the object is a pinhole in an opaque screen) are permitted to interfere with the spherical waves issuing from the object. The interference pattern thus generated corresponds to the pattern of a two-dimensional zone plate, which, when photographically recorded and reilluminated with coherent light, causes waves to be reconstructed which appear to come from the original pinhole. A more complicated, three-dimensional scene can be considered as many point sources of light, each of which generates its own zone plate when the hologram is made, and each of which is later reconstructed in its original three-dimensional position.

the phase of the incident light.

When, however, the offset procedure is looked upon as a diffraction process, it is readily recognized that all of the information needed for separation of the images is indeed fully available in the original Gabor technique. Fig. 10 shows the process in which only one single beam of laser light is used to form the hologram. It strikes the object (a pinhole) from behind and also acts as the reference wave.

Along the same line, it was not too long ago that the hologram aspects of coherent radar were questioned by some. Thus, in October, 1967, the journal I.E.E.E. Spectrum returned a paper constituting the author's 1967 Tokyo presentation with the comment: "Our reviewers do not feel that anything worthwhile is gained by relating synthetic antennas to holograms[8]. Subsequently, the I.E.E.E. Proceedings did publish a report on that presentation[1], and shortly thereafter Leith and Ingalls published similar conclusions on the same subject[9]. In their paper they noted that synthetic aperture radar procedures are truly a form of holography and further stated that "the holographic viewpoint appears to be more flexible than the communications theory or cross-correlation viewpoint, and has led to designs which are not easily explicable from the latter viewpoint".

However, even in that paper, and in a still more recent paper by Brown and Porcello[10], the Doppler aspects of side-looking radar were stressed. Thus Leith and Ingalls[9] stated: "The synthetic antenna originated as a technique for resolving targets by filtering of the Doppler frequencies present in the radar returns". In true holography, there is no Doppler, so that once the holographic viewpoint is embraced there is no need to consider Doppler effects. Then the way is immediately open for consideration of other forms of coherent radar, such as stationary (Doppler-free) radars (and sonars) and others.

Stationary Doppler-free Coherent Radar

In holography, as seen in Fig. 10, point sources of light are recorded photographically as interference patterns formed between the emitted waves and a set of reference waves. Such recordings of point sources are zone plates, which, when later illuminated, reconstruct the original point sources.

Because the usual optical imaging process employs lenses or paraboloidal reflectors, only one plane section of the image field can be recorded in truly sharp focus; all other planes being out of focus in varying degrees. Ordinary radars obey similar optical laws.

This focusing problem is completely absent in both holography and side-looking radar. Each point in the hologram scene (or each reflecting point in the radar field) forms its own zone plate, and each such recorded zone plate then causes coherent light to be reconstructed (i. e. focused) at the proper (equivalent) point in space.

Thus far the only form of coherent (hologram) radar has involved a moving aircraft, but, as in true holography, much broader uses of coherent radar concepts can be made, particularly when large-apertures are involved. Such radar antennas could take the form of extremely long linear arrays of independent receivers. Other forms include crossed arrays (Mills Crosses), and square arrays. In holography even a small portion of the hologram is able to reconstruct the full image. Similarly, in stationary coherent radars, retaining only the end sections of the long linear arrays, or the four corner sections of square arrays (thereby maintaining maximum resolution) should be satisfactory in many situations.

Perhaps the simplest form of a stationary coherent radar (or sonar) antenna would consist of a very long linear array of elements, each element being identical to the antenna carried by the aircraft in a side-looking radar. In operation all elements would first transmit, simultaneously and in synchronism, a short pulse of coherent microwaves, following which all units would, as in an ordinary radar, act as receiving elements. A certain amount of the coherent microwave signal would continually be fed to all of the elements, thereby providing the holographic reference signal. The returning, reflected echo signals would interfere at each receiver unit with this reference wave, so that along the entire array a wave interference pattern would exist. This pattern could be photographically recorded, for example, through the use of a cathode ray tube having, instead of a single beam, a large number of beams (the number corresponding to the number of elements in the receiver array), with the beams intersecting the luminous face of the tube in a single horizontal line, and all moving upward together, and the intensity of all beams would, as in that figure, be modulated by the output signal from each elemental array antenna. As in Fig. 2 (and in the radar record of Fig. 6), range would be plotted vertically, but instead of the vertical lines being recorded sequentially, by one beam, all would be recorded simultaneously by the many, upward moving beams. A photo made of the picture generated on the face of the tube would thus be completely equivalent to the radar record of Fig. 6. When processed in the same manner as the radar record, all reflecting objects would similarly generate one-dimensional zone-plate interference patterns, and the reconstruction would similarly provide a picture of the area located in the field of the radar.

We consider now the resolving capability of this arrangement. The length of the recorded zone plates will depend upon the range of the reflecting object which generates them. This effect is illustrated in Fig. 11. In the figure, Δa is the minimum resolvable fringe spacing, and this is determined by (i. e. it is closely equal to) horizontal dimension (horizontal aperture) of the individual receiving antennas (as in the case of the aircraft's side-looking radar antenna). It is seen that since $a^2 \approx 2rn\lambda$, and $a \approx r\lambda/\Delta a$, when $\lambda = 0.1$ ft. and $\Delta a = 20\lambda$, the half zone plate length (the half synthetic aperture, a) is 5,000 ft. at a 100,000 ft. range, but only 50 ft. at a 1,000 ft. range.

This variation of synthetic aperture size with range causes the resolution of such radars to be independent of range. This is illustrated in Fig. 12, showing the focusing action of a zone plate. The heavy lines indicate the envelope (to the first null) of the wave energy which is focussed (diffracted) by a zone plate Z having an aperture 2a. (In Fig. 11 only the half aperture a was discussed.) Assuming that the energy concentration is diffraction limited, the azimuthal resolution R_a is approximately equal to $\lambda r/a$ and the range resolution R_r is approximately equal to $\lambda r^2/a^2$. These results show that when $r = 10,000$ ft., $\lambda = 0.1$ ft., and $\Delta a = 20\lambda$, R_a is only 2 ft. (i. e. it is equal to the antenna size). Thus, for these parameters, the azimuthal resolution is 20 times better than the range resolution. Also, when the range r is increased (or decreased), the zone plate Z of Fig. 11 becomes larger and moves to the left (or becomes smaller and moves to the right), with the angle θ and also the values of R_a and R_r remaining unchanged. We note that for the example picked earlier, where the half synthetic

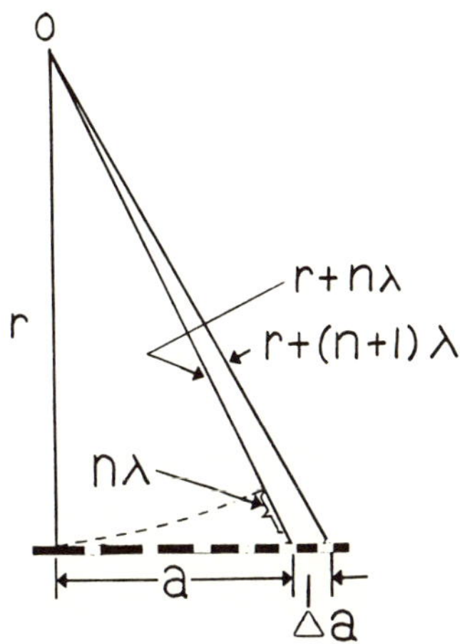

FIG. 11 Geometry of a half-zone plate.

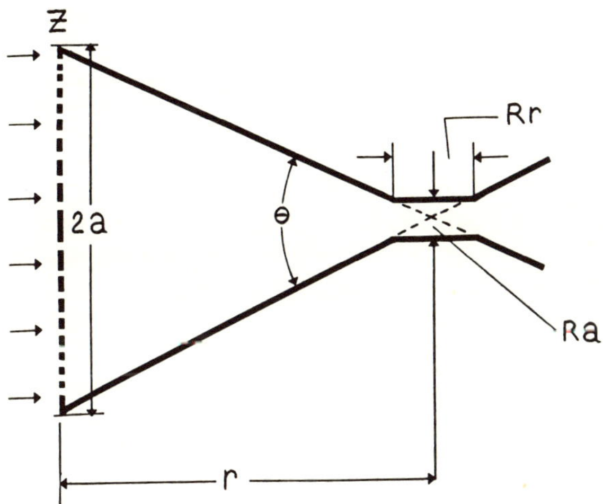

FIG. 12 Focusing action of a zone plate.

aperture, a, was 5,000 ft., the ratio of the length of the zone plate aperture (10,000 ft.) to the azimuth resolution (2 ft.) is 5,000. For the situation where Δ a extends over only one wavelength, the range and azimuth resolutions become equal. Under these conditions good range resolution would be acquirable without the use of pulses, i. e. through the action of the superimposed zone plates alone.

In the arrangement just described, no vertical motion of the large number of cathode ray tube spots would be required. They would remain stationary, generating a single, horizontal line pattern. The procedure would result in all of the (horizontal) one-dimensional zone plates (i. e. all of the horizontal lines of Fig. 6) in the radar record being collapsed and superimposed to form one horizontal line. (In the usual radar record, superposition of the one-dimensional zone plates occurs only for reflectors at the same range; in this situation superposition of zone plates occurs for all reflecting points.) As in a true hologram, however (for which this superposition of zone plates of reflecting objects at all ranges also occurs), reconstruction of the original scene is not hampered by this superposition.

As seen in Fig. 12, the azimuth resolution is not affected by this collapsing process. However, range resolution can be. In the pulsed technique, i. e. in Fig. 2, range resolution is determined by the pulse length, whereas range resolution in the hologram procedure is set as R_r (Fig. 12). It is of interest to note that if the resolution R_r (obtained by holographic diffraction only) is considered satisfactory, and if the minimum range requirement is not severe, then a very long pulse (equal to twice the minimum range required, divided by the velocity of propagation) could be employed, thus yielding the same range resolution R_r, but a greatly enhanced signal to noise ratio.

Forward Scatter and Bistatic Coherent Radar

Hologram concepts can be extended to a forward scatter radar or sonar; i. e. to a coherent radar or sonar modeled after the original Gabor hologram arrangement[11]. In this arrangement, the area or targets of interest would be located between the transmitter and a receiving array; i. e. the transmitter would be placed at 180° from the receiver (relative to the illuminated area of interest). Such an arrangement would take advantage of the rather high forward-scatter signal diffracted by objects located between the transmitter and receiver[12]. As discussed earlier, the receiver could again be either one long linear array of many elemental receivers, or two end sections of that array. In the hypothetical case where only one point-scatterer is located between the coherent source and the receiving array, the interference pattern between the scattered signal and the coherent background signal would again be a zone plate, and the receiving array would intercept a linear section of it (a one-dimensional zone plate). By photographically recording the individual outputs of all of the receiving array elements as one photographic line (comparable to one line of a side-looking coherent radar record), the position both in range and azimuth of this scatterer could be determined by processing the one-dimensional photographic zone plate with laser light (as in the radar case).

In the usual case in which many scatterers are present, a multiplicity of zone plates will be generated, and these would be superimposed on the photographic record. As in a hologram, however, positional information on all of the scatterers could still be retrieved through illumination of the photographically recorded, single-line, composite interference pattern with laser light. To make the photographic line record, the receiver element outputs could be sampled sequentially, and the resulting signal could intensity modulate a synchronously moving cathode ray tube beam with the light signal then being recorded photographically. In the purely forward scatter case the scattered signal is strongest; i. e. when the transmitter is placed at exactly 180° toward the receiving array, maximum signal-to-noise results. Also, the interference fringe size is the greatest (permitting the fewest elements to be needed in the array), and the coherence length requirements on the signal source are minimal. However, for the 180° case, Gabor's problem of twin images exists, and it would accordingly be preferable probably to place the coherent transmitting source at some off-axis bistatic angle (say 135°). In such bistatic arrangements the transmitter could operate continuously (i. e. it would not be pulsed). Range and bearing (azimuth) would both be acquired by the array as indicated in Fig. 12. It is interesting to note that the bistatic, continuous wave form of coherent radar was proposed in 1966 by G. L. Tyler[13]. In his article, he stated: "When radars with pulsed wave forms are used for mapping, it is easy to visualize how resolution in range and azimuth are obtained. Just how this works when continuous waveforms are used may not be quite so obvious. It can be explained by an optical analogy" ... "The hologram provides a photographic analog of bistatic radar ... the entire process of making the hologram and playing it back ... corresponds to the bistatic radar mapping scheme".

Passive (Cooperative) Radar Concepts

The coherent radar concept can also be applied to passive or cooperative forms. We consider again a long multi-element receiving array and assume that there is an aircraft present which is radiating a strong, single frequency, highly coherent, microwave signal. We assume further that one of the array elements receives this signal, amplifies it, and feeds it as a reference signal to all of the other elements of the array. These elements will, however, also be receiving the radiated signal from the aircraft directly, so that an interference pattern will again be generated along the array. This pattern can be photographically recorded and used to reconstruct the location of the plane in azimuth and range. In this arrangement there is no radiation from the ground, so that it is a form of passive radar, as opposed to a transmitting or active radar. The plane has thus "cooperated" in providing the coherent radiation.

A variant of this is to have a coherent transmitter located on the ground, sending signals to all aircraft in the vicinity. They are equipped with receivers which pick up the coherent signal, amplify it, and re-radiate it * , preferably after shifting the frequency slightly. The ground transmitter signal is now also shifted in frequency by an equal amount and piped to all of the elemental receivers of a long array so as to act as the reference wave. The receivers also pick up signals from all aircraft, so that many superimposed zone plates are generated, as in the case of previously discussed active systems. For this case either a continuously operating version or a pulsed version could be considered.

Far Field Hologram Radar

Because holography and coherent radar are both superior to lens optics and ordinary radar in providing detailed informa-

* This procedure was suggested to the author by E. Grogan Shelor of the Bendix Communications Division, Towson, Maryland.

tion concerning objects located in the near field, one might be tempted to dismiss consideration of the far field performance of coherent radar as unimportant, since ordinary radar performs this task well. On the other hand, hologram techniques may provide simpler ways of achieving performance equivalent to the usual far field radars or sonars.

Consider, for example, a multi-element, square array, radar, having 100 elements on a side and hence 100 x 100 or 10,000 total elements, positioned $\lambda/2$ apart. The far field, defined as beginning at a distance $2a^2/\lambda$ out from the array, beginning for this case at 5,000 λ. For microwaves of one-inch wavelength this far field begins at a distance of only 5,000 inches or a little over 400 ft. from the radar. Such a radar would thus usually be employed as a "far-field" radar. We now suppose further that the array itself is used only for receiving, and that the transmitted signal, radiated by another, high power antenna, illuminates a conical volume of sizeable angular extent. The array is to acquire, to the best of its ability, information regarding reflecting objects located within the radiation cone of the transmitter.

One way of accomplishing the desired function of the array would be establish thousands of "pre-formed" beams, whereby, for any single beam direction, delays would be provided as necessary for each of the 10,000 array elements, so that all elements would contribute properly to the receiving beam for that direction. Because the array is 50 wavelengths on a side, the width of each of the preformed beams will be approximately 1 degree ($51\lambda/a$), so if a 90° pyramid of coverage ($\pm 45^{\circ}$) is desired, approximately 90 x 90 or 8,100 preformed beams would be required. Whether the required delays (they total, for 10,000 elements and 8,100 beams, approximately 80,000,000 delays) are provided individually (by analog methods) or by digital computer techniques, the task is obviously not a simple one.

Consider now providing these same far field beams by holographic, i. e. by stationary coherent radar, procedures. The transmitted signal is then coherent, provided by a highly stable oscillator. A small amount of this oscillator signal would be used as a reference signal and would be supplied continuously to each element of the receiving array (possibly by direct line to the element) with the phase adjusted so as to

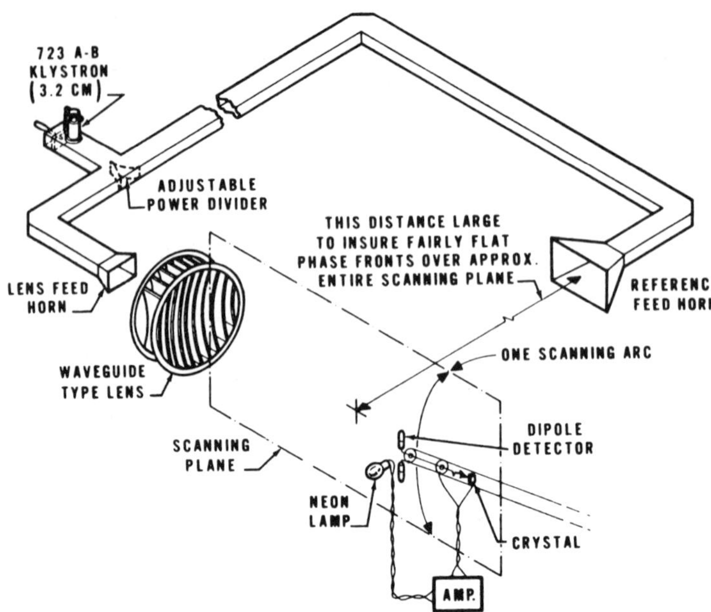

FIG. 13 Procedure for making hologram of Fig. 14. Reference waves from feed horn interfere with waves emanating from the lens. Interference patterns are detected in the scanning plane by the dipole, amplified to light the lamp and photographed.

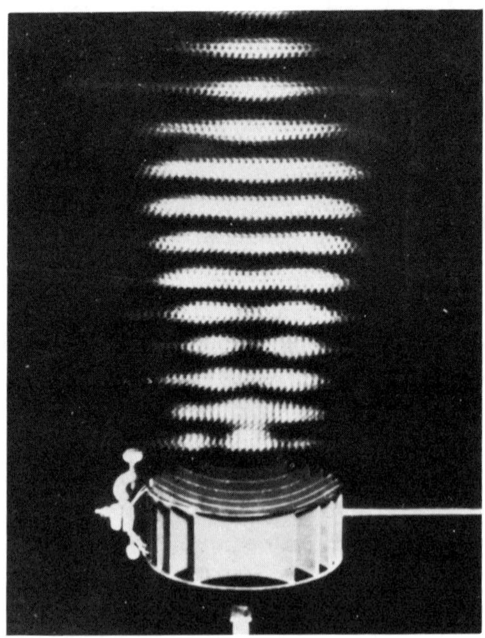

FIG. 14 Microwave hologram as recorded at Bell Telephone Laboratories in 1950.

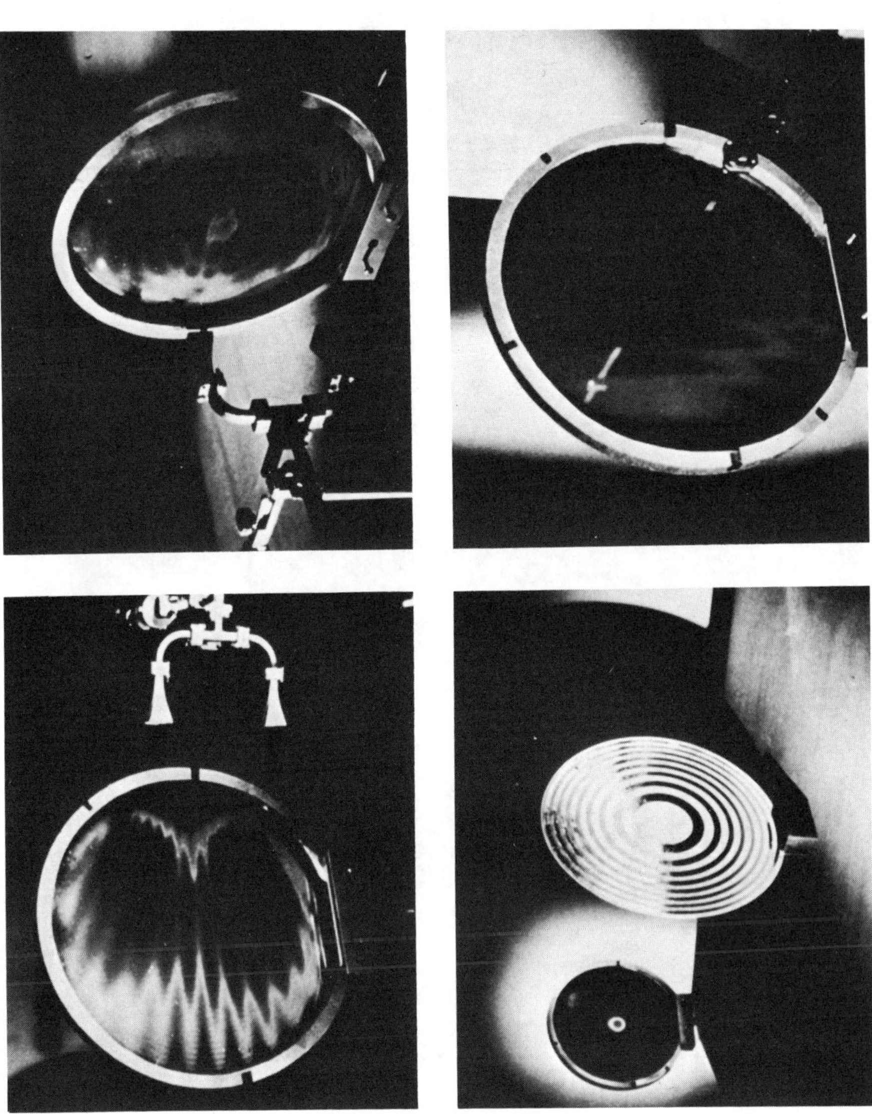

FIG. 15 Liquid Crystal microwave Patterns.

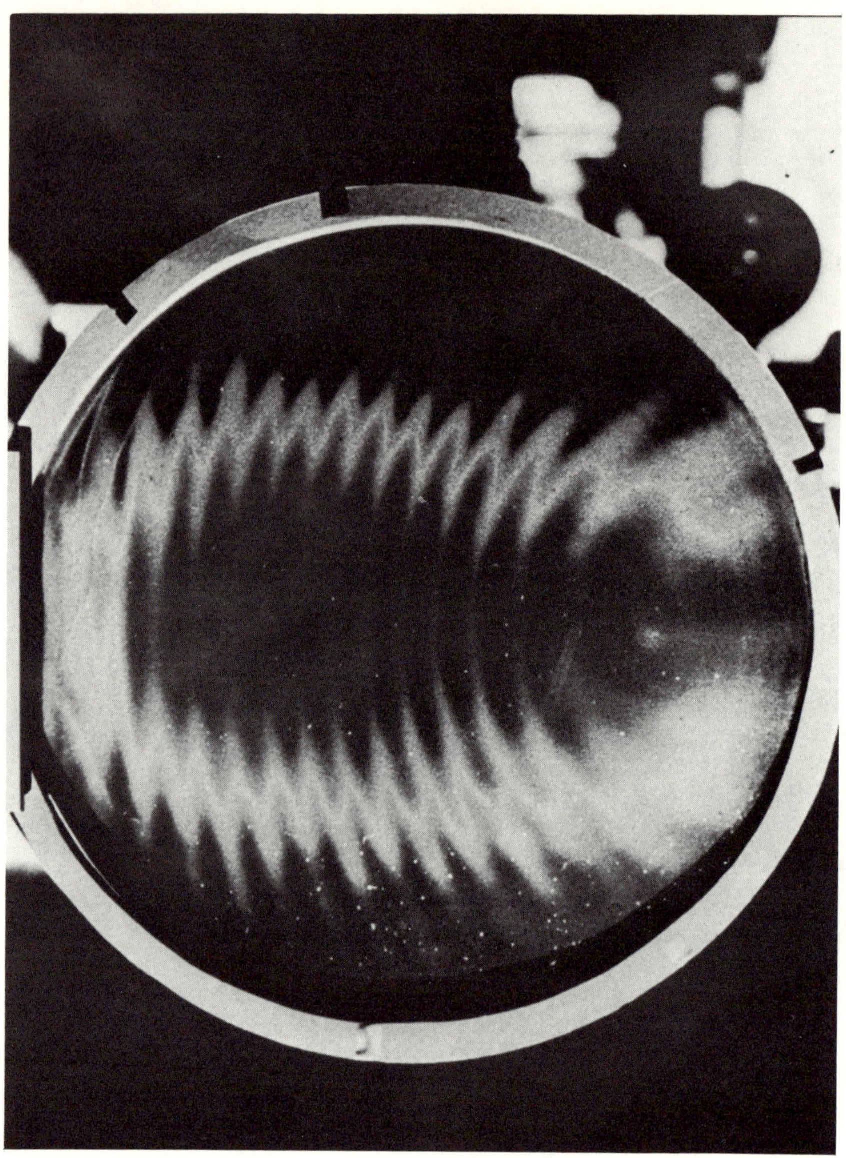

FIG. 16 A liquid crystal pattern formed between a point microwave source and a plane wave source. This pattern, a microwave hologram, is a zone plate, which when greatly reduced in size and illuminated with laser light, can reconstruct virtual and real images of the original point microwave source. Because it is offset, the straight-through (zero-order) component and the two images can be separated.

simulate a plane wave of this frequency impinging (preferably at some angle) on the entire array. Reflecting objects located within the cone of the transmitter will reflect some of the transmitted signal back to the array, thereby generating, in combination with the reference signal, a holographic interference pattern which is "sampled" by all the elements. The resultant values at all elements are then recorded photographically, and, after development of the photographic record, the far field reflecting objects are reconstructed by coherent light.

At first glance it would appear that the first procedure using delays, which combines all 10,000 elements into a preformed beam, thereby achieving an extremely high directivity gain (40 decibels or so), must be superior to the procedure in which each single isolated unit of the 10,000 elements is made to affect, by itself, the exposure of the photographic plate at that point. Actually, the coherence of the signals generating the interference pattern on the array and the similar coherence of the laser reconstructing beam, causes each recorded point to combine coherently (during the reconstruction process) with all the rest of the points, thus providing, holographically, a comparable array gain for each of the 8,100 directions.

Microwave Holograms Using Liquid Crystal Displays

In an early technique for making microwave holograms[(14)], the microwave interference or fringe pattern was photographically recorded by scanning the field with the combination of a microwave probe and light source (Fig. 13). Fig. 14 shows such a photographically recorded microwave fringe pattern (hologram). In these two figures, the wave source of interest was the radiation pattern of a waveguide metal lens, and a source of plane reference waves acted as the reference signal. Recently the use of liquid crystals, whose colors are determined by the strength of the microwave field, was investigated for recording microwave space patterns[(15)]. Fig. 15 shows several records of such interference patterns. Fig. 16 shows a similar photographically recorded microwave interference pattern formed by two coherent microwave sources. One of these was a point source and one approximated a plane wave source, and the two wave sets were caused to interfere at the plane of the liquid crystal device[(16)]. The resulting pattern is identical to the pattern of an offset microwave zone plate. Reconstruction of an image (either real or virtual) of the original point

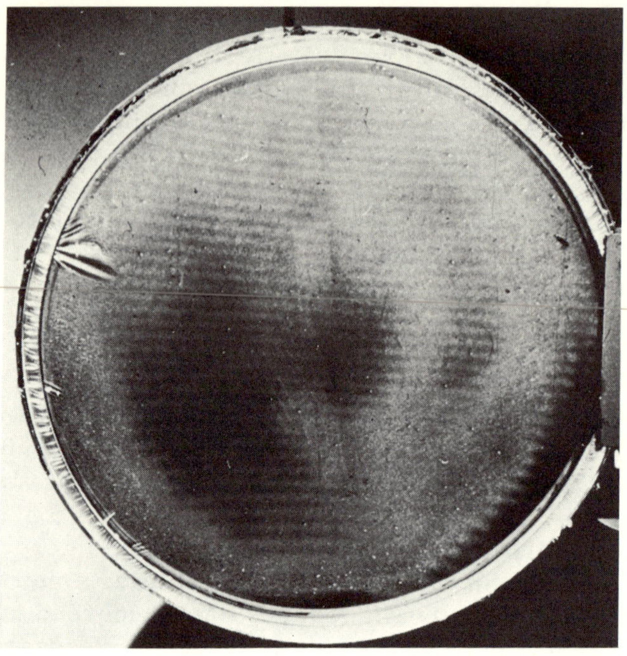

FIG. 17 Microwave hologram of the Letter L

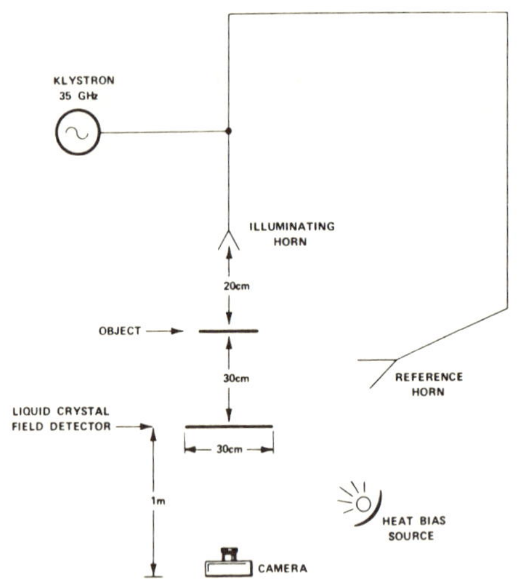

FIG. 18 Procedure for Making the Hologram of FIG. 17

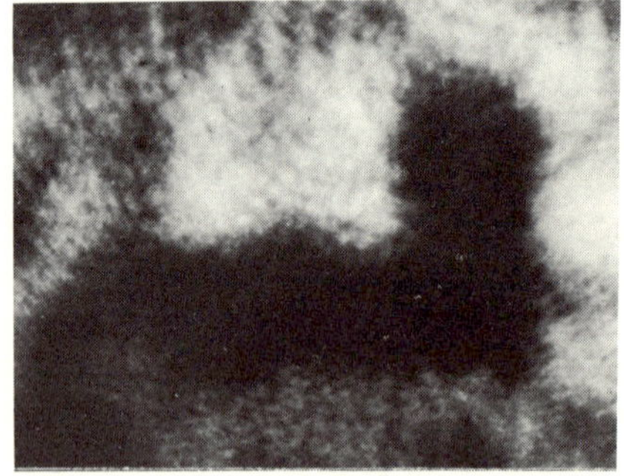

FIG. 19 Reconstruction of the Hologram of FIG. 17

microwave source can be accomplished optically by photographically reducing the record and illuminating it with coherent (laser) light. Fig. 17 is a liquid crystal microwave hologram of the letter L. It was made[17] as shown in Fig. 18. Reconstruction of this hologram is shown in Fig. 19.

Although very convenient for pictorially representing strong microwave fields, the liquid crystal technique requires, at the present stage of development, approximately 1 milliwatt of microwave power per square inch to create a well-defined pattern. This field strength is not easy to generate for those microwave holograms which are usually of interest, namely those involving the reflection of microwave energy from rather diffusely reflecting scenes. However, with high-power microwave illumination, or as the sensitivity of the liquid crystals increases, the liquid crystal technique may become very significant in microwave holography.

Holographic Processing of Radar Signals

An interesting application of holography to the processing of radar signals was recently described[18]. The radar is of the pulse-doppler type, transmitting a coherent train of pulses. It is able to measure, with excellent resolution, both the range and range rate of each of many targets. In the radar processor a sequence of radar returns is stored as a sequence of holograms. Reconstruction of these holograms then yields the desired range and range rate information.

In any radar, good resolution in range, which is measured by the round-trip time taken by the pulse upon reflection from a target, requires short-duration pulses. On the other hand, good range rate resolution, determined by measuring the doppler frequency shift of the signal returning from a target, requires a long-duration pulse. To satisfy these two opposing requirements, a long train of many very short pulses is employed as the transmitted radar "pulse" signal. The signal remains coherent over the length of the pulse train. As shown in Fig. 20, the burst consists of a train of N pulses equally spaced T seconds apart, with each pulse having a duration of t sec. The pulse duration t determines the range resolution, and the burst duration NT determines the doppler frequency-shift resolution (equal to 1/NT hertz).

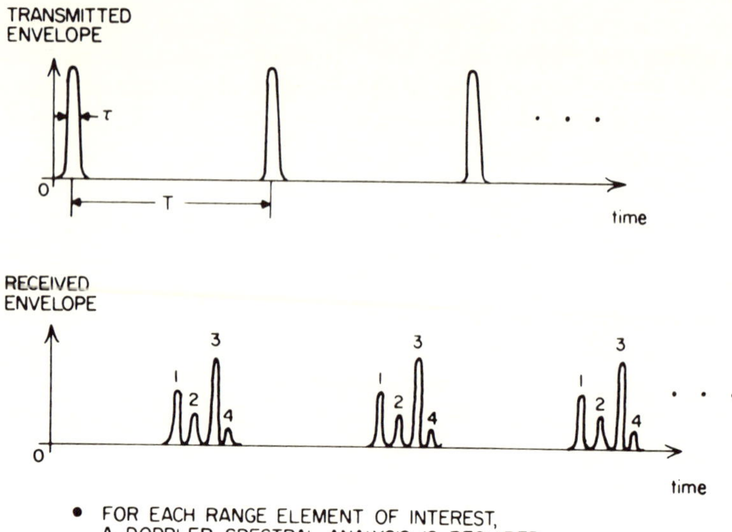

FIG. 20 Outgoing radar signal comprising a train of short pulses (top) and, at bottom, received signal with four targets present.

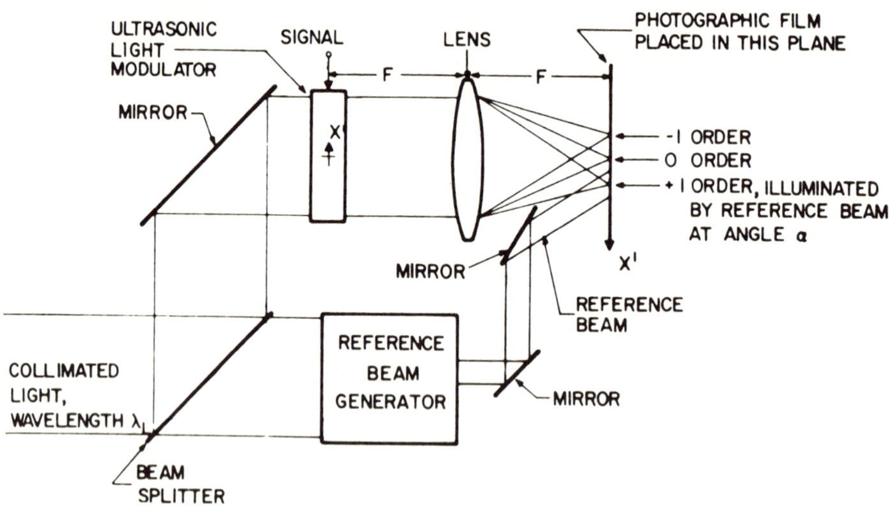

FIG. 21 Procedure for generating hologram for radar signal processing.

In the next line of Fig. 20 it is seen that each radar transmission elicits a return from each of the four targets. The delay between a transmitted pulse and the time of the corresponding return from one of the targets establishes the target's range with the required resolution. The range <u>rate</u> of each target is obtained by feeding the entire sequence of N returns from the target into a spectrum analyzer and noting the doppler frequency shift (the bottom line of Fig. 20). Since such a spectral analysis must be performed for each element of range of interest, many spectrum analyzers are normally needed. In the new technique, simple holographic procedures accomplish the same task, and thus eliminate the need for the many analyzers. The holograms, acting as optical gratings, in effect are the spectrum analyzers.

In the first step in holographic processing, the radar returns are stored holographically, as shown in Fig. 21. A collimated beam of laser light is split into reference and signal components, and these are later re-combined to form an interference pattern on a surface of photographic film. In the upper path the signal beam is generated by passing laser light through an ultrasonic light modulator in which the radar return is travelling (at the velocity of sound). The lower path directs a second tilted plane-wave laser reference beam onto the film surface. The signal in the aperture of the ultrasonic light modulator is the reflected signal obtained from the several targets that are interrogated by one transmitted radar pulse of the train. When the entire radar return has moved into the ultrasonic light modulator, the laser light is briefly turned on to provide the required hologram exposure time. This one hologram occupies one narrow vertical strip on the film. After each individual pulse of the pulse train has thus been photographically recorded, the film is moved sideways, i. e. perpendicular to the plane of the figure, so that a sequence of N radar returns is recorded as N holograms side by side on the photographic film. This series of holograms forms, for each target, an optical grating whose tilt is a function of the range rate of that target. After the film sheet of many side-by-side holograms is developed, it is again illuminated by the original reference wave. The holographically reconstructed configuration is shown in Fig. 22. It shows two information-bearing areas outlined in the output or frequency plane. Either of the two areas contains all the range-doppler information.

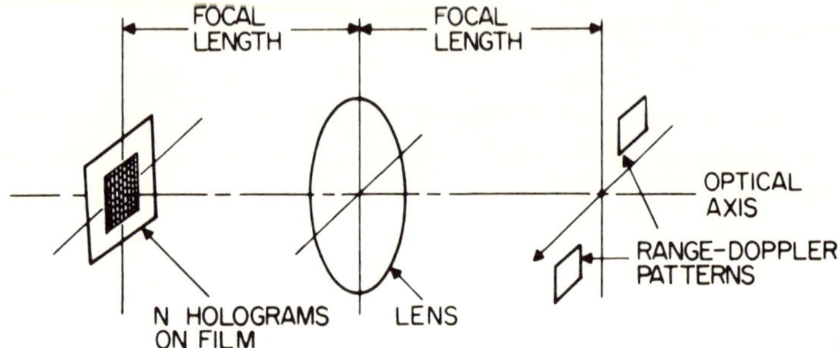

FIG. 22 Reconstructing the hologram made as shown in FIG. 21.

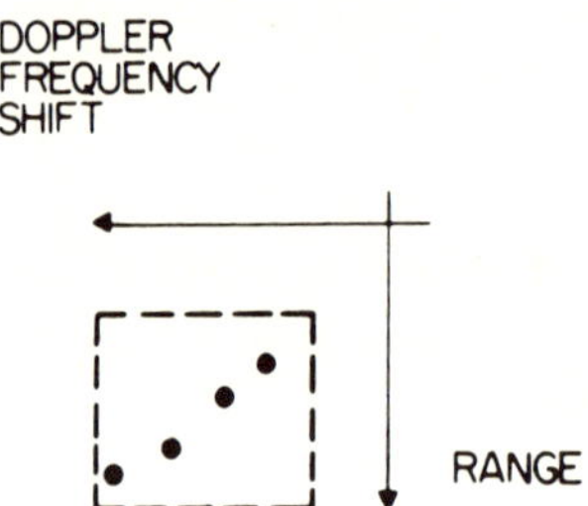

FIG. 23 Enlarged view of reconstructed hologram.

An expanded view of one of these areas is shown in Fig. 23. Each of the four bright spots of light corresponds to one of the targets that was interrogated by the N transmitted radar pulses. The horizontal position of the spot is indicative of the target's range, and the vertical is indicative of the target's range rate.

Holographic Pulse Compression

Pulse compression techniques are employed in radar and sonar to permit long transmitted pulses to be converted, upon reception, into shorter pulses. The power in the transmitted pulse can thus be made greater (because it is longer), without increasing the peak power, and the shortening procedure still yields the good range resolution, which can otherwise be achieved only with very short pulses. One variety, called chirp, varies the frequency during the duration of the pulse(19). Upon reception, a matched filter (such as a transmission line which causes signals of different frequencies to travel at different velocities, thereby permitting the later, more rapidly travelling, portion of the long pulse to catch up to the slower, earlier generated portion) converts the long pulse to a short one.

In a recently described procedure(20) the amplitude of the long pulse is varied instead of the frequency. Such a procedure dispenses with the matched filters required in the chirp process. It can be looked upon as a holographic procedure, as it uses similar zone plate reconstruction techniques.

In Fig. 7 there is a prominent one-dimensional zone plate located near the bottom of the blank area. As we saw, the optical processing procedures of side-looking radar retrieve from that zone plate the azimuthal position of the strongly reflecting point in the terrain which generated the zone plate pattern. We recall that the accuracy of determination of the position of that reflecting point (in azimuth) is about equal to the separation of the last resolvable dots in the zone plate series. The ratio of the zone plate length to this resolution width is practically unlimited; an exemplary ratio of 5,000 to 1 was discussed earlier.

In the amplitude pulse compression procedure, the total length, in time, of the original pulse would correspond to the total length of the zone plate in that figure, i.e. the long, continuous pulse would be given an amplitude modulation pattern corresponding to that of the one-dimensional zone plate pattern in the figure. The received echoes from a reflecting point would again be photographically recorded. The film would be moved sideways very rapidly to record the incoming, reflected signals, which, in the radar case, are returning at the speed of light, and in the sonar case at the speed of sound. The echo from a single, isolated target would generate a replica of the outgoing pulse, i.e. a photographically recorded pattern similar to the zone plate of Fig. 7. In processing this recorded pattern, laser light would be diffracted (focussed) by the zone plate into a very tiny area, thereby providing very accurately the range of the reflecting point. This accuracy would be equal to the final dot (fringe) spacing, so that the ratio of 5,000 to 1 discussed earlier would here correspond to a pulse compression ratio of 5,000 to 1. In that example, the minimum dot spacing (fringe spacing) was assumed to be 20 wavelengths; this would be equivalent here to a minimum amplitude modulation length of 20 cycles of the carrier and would result in a quite modest bandwidth requirement (approximately ± 2.5%).

For the more usual case of a multitude of reflecting points (targets) at various ranges, many zone plates will be generated in the reflected signal, and these will be superimposed on the photographic record. This line record will then resemble one of the other, more complicated, line records of figures 6 or 7, in which numerous superimposed zone plates, generated by many reflecting points at identical ranges (but at different azimuths) are recorded in the radar record. As in the radar record, processing this record with laser light will cause the individual targets to be resolved in range.

The above-described amplitude modulation procedure should be particuarly adaptable to side-looking (synthetic aperture) radar, where optical processing techniques are already employed. Leith has described[(21)] a pulse compression procedure using optical processing of frequency modulated (chirp) signals in such radars, and the amplitude modulation described here provides an alternative method. To increase the energy in the outgoing pulse in such radars, it would be made quite long and be given a zone plate envelope. This will result in the thin, one-dimensional zone plate of the figure acquiring an extended vertical dimension, thereby becoming a sort of two-dimensional zone plate. Optical processing procedures then would be used to collapse this pattern into a one-dimensional pattern, as shown, thus retrieving accurate range information. Following this, the usual radar processing optical procedure would be employed to retrieve azimuth information. For the actual case, where many targets at different ranges and azimuthal positions exist, zone plate superposition would occur in both horizontal and vertical directions. However, as in the horizontal only case, the optical processing procedures would still permit the information to be fully retrievable.

In conclusion it would appear that the technologies of radar and sonar can benefit from optical holography and optical processing. Similarly, perhaps the technology of coherent radar can be advantageously extended to the visible, coherent light techniques.

REFERENCES

1. W. E. Kock, "Side-looking radar, holography, and doppler-free coherent radar", Proc. IEEE, vol. 56, no. 2, February, 1968, pp. 238-239.

2. W. E. Kock, "Microwave Holography", Microwaves, vol. 7, no. 11, 46 (1968).

3. W. E. Kock, "Stationary coherent (hologram) radar and sonar", Proc. IEEE, vol. 56, no. 12, December, 1968, pp. 2180-2181.

4. L. J. Cutrona, E. N. Leith, L. J. Porcello, and W. E. Vivian, "On the application of coherent optical processing techniques to synthetic aperture radar", Proc. IEEE, vol. 54, pp. 1026-1032, August, 1966.

5. E. N. Leith and J. Upatnieks, "Wavelength Reconstruction with Continuous Tone Transparencies", Journal of the Optical Society of America, pp. 1377-1381 (1963).

6. L. J. Cutrona, E. N. Leith, C. J. Palermo, and L. J. Porcello, "Optical Data Processing and Filtering", IRE Trans. Prof. Group on Infor. Theory, vol. IT6, no. 3, pp. 386-400.

7. E. N. Leith and J. Upatnieks, "Wavefront Reconstruction Photography", Physics Today, pp. 25-32, August, 1965.

8. Discussion in a Forum letter: W. E. Kock, IEEE Spectrum, vol. 6, no. 12, p. 9, December, 1969.

9. E. N. Leith and A. L. Ingalls, "Synthetic antenna data processing by wavefront reconstruction", Applied Optics, vol. 7, pp. 539-544, March, 1968.

10. W. M. Brown and L. J. Porcello, "An introduction to synthetic aperture radar", IEEE Spectrum, vol. 6, no. 9, pp. 52-62, September, 1969.

11. W. E. Kock, "A hologram form of bistatic radar or sonar", Proc. IEEE, vol. 57, no. 1, January, 1969, pp. 100.

12. W. E. Kock, J. L. Stone, J. E. Clark, and W. D. Friedle, "Forward scatter of electromagnetic waves by spheres", 1958 WESCON Conv. Rec., pt. 1, p. 86; W. E. Kock, "Related experiments with sound waves and electromagnetic waves", Proc. IRE, vol. 47, pp. 1192-1201, July, 1959.

13. G. L. Tyler, "The bistatic, continuous -wave radar method for the study of planetary surfaces", Journal of Geophysics Research, vol. 71, no. 6, pp. 1559-1567, March 15, 1966.

14. W. E. Kock and F. K. Harvey, "Sound wave and microwaves space patterns", Bell System Technical Journal, vol. 20, no. 7, pp. 564-587, July, 1951.

15. C. F. Augustine, "Field detector works in real time", Electronics, vol. 41, pp. 118-121, June 24, 1968.

16. C. F. Augustine and W. E. Kock, "Microwave holograms using liquid crystal displays", Proc. IEEE, vol. 57, no. 3, pp. 354-355, March, 1969.

17. C. F. Augustine, C. Deutsch, D. Fritzler, Emanuel Marom, "Microwave holography using liquid crystal area detectors", Proc. IEEE, vol. 57, no. 7, pp. 1333-1334, July, 1969.

18. Marvin King, "Holographic processing of radar signals", Laser Focus, pp. 43-45, August, 1969; also, M. Arm and M. King, "Holographic storage of electric signals", Applied Optics, vol. 8, pp. 1413-1419, July, 1969.

19. J. R. Klauder, A. C. Price, S. Darlington, and W. J. Albersheim, "The theory and design of chirp radar", Bell System Technical Journal, vol. 39, p. 745.

20. F. Tuttle and W. E. Kock, "A holographic pulse compression technique employing amplitude modulation", Proc. IEEE, vol. 58, no. 1, January, 1970, p. 170.

21. E. N. Leith, "Optical processing techniques for simultaneous pulse compression and beam sharpening", IEEE Trans. on Aerospace & Electronic Systems, vol. AES-4, no. 6, pp. 879-885, November, 1968.

ACOUSTICAL HOLOGRAPHY*

Fredrick L. Thurstone, Ph. D.

Division of Biomedical Engineering,

Duke University, Durham, North Carolina, 27706

It would be very good to be able to report that acoustical holography, in the few short years since its inception, had reached the point of being a practical and useful technique for the three-dimensional visualization of opaque structures, particularly for the visualization of biological tissue structures for diagnostic purposes. Unfortunately, this is not the case. There remain two primary limitations on the development of such a practical holographic visualization technique. The fundamental limitation remains that of the inability to detect an acoustical energy distribution in a two-dimensional field with sufficient sensitivity for diagnostic use, and a secondary limitation is imposed upon the image presentation procedure because of the image anomalies associated with the great difference in wavelength between the investigating acoustic energy and the reconstructing electromagnetic energy in the visible spectrum. There is, however, reason to be optimistic about the future development and application of holographic imaging using acoustical energy. The reasons for optimism include the very considerable amount of research interest and endeavor on the part of numerous research groups and secondly, the very broad spectrum of acoustical frequencies under investigation for different applications. These frequencies range from subaudible frequencies for geologic application to gigahertz frequencies in limited volumes in solids.

Because of the author's limited experience and specific interests in the possible medical diagnostic use of acoustical holography, the remainder of this report will be directed specifically toward this application. Some aspects of the problem may be much more

*This work supported in part by USPHS Grants GM 15892, HE 41131, HE 12715, HE 5716.

generally applicable but clearly, such a great spectrum of frequencies as are involved in acoustical holography cannot be encompassed by a single set of techniques.

In order to put into perspective the possibilities and limitations of holographic imaging of biologic tissues using ultrasound, consider that in the frequency range of 1 - 10 MHz which propagates well in soft tissue structures, the acoustical wavelengths vary from 1.5 - 0.15 mm. Thus the number of image elements may well be on the order of 1000 x 1000 or 10^6 elements. Adding amplitude and/or phase brings the amount of image information to 10^7 bits or more. It is clear that the detection and recording of this amount of information is not a trivial problem.

Holographic imaging offers some distinct advantages over the more conventional direct imaging techniques. The primary advantage would appear to be the recording of information from a three-dimensional volume in a single recording with resolution capability approximating the acoustic wavelength thus obviating the limited depth of focus or lack of depth information inherent in direct imaging procedures. A second and certainly very attractive advantage is the possibility of reconstructing three-dimensional images which may have greater subjective or diagnostic value.

There are also distinct disadvantages associated with holographic imaging. First, the amount of data that must be recorded is in general greater than the amount necessary to provide the same image resolution using direct imaging procedures. Second, because of the coherence requirements, the speckle effect degrades the image quality. And third, the complexity of recording, processing, and reconstructing the images is in general greater than simple, direct techniques. It should also be recognized that the subjective character of three-dimensional visualization has not yet been achieved using acoustical holography.

There are a number of problems associated with the development of a practical imaging system using ultrasound whether holographic or more conventional in nature. Although there may not be universal agreement among workers in this field, the primary problem is thought to be that of signal detection or the conversion of acoustical energy into some other usable form or record. Numerous techniques are available and have been investigated including thermal, chemical, mechanical, and electrical techniques. None of the available techniques, however, are capable of recording the necessary amount of information at a sufficient data acquisition rate and with sufficient sensitivity to produce high quality diagnostic images. Ideally, a detection system should be sensitive to acoustic intensities on the order of a few milliwatts per square centimeter, be fast enough to record in a fraction of a second, have a resolution of a wavelength or less, and be capable of reasonably large physical and numerical aperture. In brief, what is needed is a good ultrasound "film."

A second major problem is the very great dynamic range of the acoustical information that is present in the desired image. This dynamic range may easily exceed 60 db. This range is produced in part by the range in size and nature of the various subjects within the field of view to be imaged simultaneously, and also in part due to the highly specular nature of acoustical reflection from biological tissue interfaces. This great dynamic range leads to problems both in detection and in image presentation. Most display devices and even film have a much more limited range. How or where to compress the data is a problem that has not been studied sufficiently. Most of the direct imaging systems developed to date for diagnostic use have not approached this problem. In fact, most have used storage type oscilloscopes for image presentation. Such a display is essentially a one bit binary display yielding extremely high contrast or black-white images.

There are other problems associated with the development of an imaging system such as image resolution and image registration errors. However, in general, these problems can be overcome with a systematic approach to the design and fabrication of the imaging system.

Signal detection then remains as the primary limitation on the development of a practical system, and this limitation must be overcome by the development of new procedures. Effort should continue to be directed toward the search for such a detection scheme and should also be directed toward the improvement of image processing and presentation techniques. There can be no question that ultrasound fields of relatively short wavelengths should yield reasonably good visual images of opaque tissue structures. With proper processing and presentation, these images will undoubtedly have great diagnostic value.

There is another area that is important in the development of an imaging system, and that is the area of the psychometrics of visualization. For example, the optimum compression procedure for reducing the dynamic range of acoustical information is not easily determined by objective criteria. The real question is what is the use to be made of the image -- that is, what decision is to be made or what value to be recieved? Only after answering these questions can an image presentation scheme be optimized. This procedure may actually be easier in diagnostic imaging, where the image is usually desired for relatively specific reasons.

It should also be pointed out that frequently the most desirable information is dynamic in nature, because the most significant physiologic indicators are those which are undergoing change. This, in turn, predicates higher data acquisition rates than for imaging static subjects. Additionally, the need for multiple or sequential images imposes limitations on the amount of image processing that can be accomplished in a practical system.

A number of techniques have been used or proposed for recording ultrasound holograms, and the capabilities and limitations of the various procedures are quite different. The earliest demonstrated and perhaps most widely used procedure involves mechanically scanning a small point detecting transducer through a two-dimensional plane. The resulting electrical signal is processed and used to modulate a light source which produces a hologram in the form of a film transparency. Although numerous minor problems are inherent in this method, the primary limitation is the time required to obtain the required amount of data. For high resolution images with low sampling rates, the time required may be on the order of hours. This procedure does have the advantage that the object signal itself is detected, allowing an arbitrary reference phase or wavefront to be synthesized, and additionally any arbitrary combination of signal amplitude and phase may be recorded in any arbitrary range of optical density in the hologram recording.

A logical extension of the single point detector are various schemes of detector arrays. These include both one and two- dimensional arrays which may be sampled or scanned either mechanically or electrically. Although faster than the single detector system, these arrays, in general, cannot provide as much image information because of limitations on the number and uniformity of the elements of the array.

Another technique which has been investigated by a number of groups involves the use of the Sokolov ultrasound image converter tube for recording an ultrasound hologram either directly or with a synthesized reference signal. This system utilizes a single, large, piezoelectric crystal transducer which is scanned by an electron beam. Although quite rapid in recording due to the electronic scanning, this system is severely limited in the amount of data that can be recorded. This limitation is imposed by the limited size and resolution of the piezoelectric crystal, and also because of the very limited angle through which incident acoustic energy is detected.

Probably the best visual images reconstructed to date have been obtained by the use of a liquid-air interface to produce an ultrasound hologram. Both object and reference fields are made to impinge on the interface as in normal optical holography. The varying radiation pressure in the resulting interference pattern then produces varying levitation of the liquid surface. This pattern can be photographed, or alternatively, the surface can be illuminated with the reconstructing light field for real-time visualization of the resulting images. Obvious attributes of this technique are the rapid response and on-line operation. Limitations on the method include streaming effects which require pulsed operation and mechanical stability requirements to reduce unwanted surface deformations. The primary limitation for diagnostic use, however, is probably the lack of sensitivity of the method. Maximum sensitivity requires approximately 100 mw/cm^2 incident energy from the object. Assuming appreciable attenuation in the tissue under study, it is clear that only relatively thin specimens can be examined if the incident energy is to be kept below acceptable levels.

Several other schemes have been investigated or proposed which involve scanning a laser beam or using pulsed lasers. Most of these procedures involve the detection of the motion of a surface or membrane, and thus are subject to the same characteristics as the liquid surface levitation procedure just described.

Except for the liquid surface levitation technique which can be visualized directly or with a video system, essentially all ultrasound holography has utilized a film transparency as the ultimate recording medium, and in turn, reconstructed images from these using laser light for visualization. Frequently the object under study has been geometrically within the reference beam, either real or synthetic. In these cases a stop is usually employed to remove the undiffracted light from the area of the reconstructed image. Since three-dimensional visualization has not been accomplished, either of the conjugate images may be recorded photographically or viewed directly.

Because the wavelength of the visible light field used for reconstructing the images differs from the wavelength of the acoustic field used for recording the information, there are a number of anomalies present in the reconstructed image. These anomalies can be made to disappear if and only if the hologram recording is linearly scaled by the ratio of these wavelengths, and the recording geometry is exactly reproduced in the reconstruction process. Under this condition, the image size is reduced by this wavelength ratio, and is, in general, too small for direct visualization. If the resulting image is magnified for visualization, or if the hologram is scaled by a factor of less than the wavelength ratio, there results a primary longitudinal or axial distortion in the image. This primary distortion is equal to the wavelength ratio reduced by the amount of linear reduction in the hologram or the reduction in lateral magnification. Since the wavelength ratio is on the order of 500:1, this first order longitudinal distortion has precluded the subjective visualization of three-dimensional images.

Since the realization of three-dimensional visualization is one of the advantages of holography over direct imaging procedures, considerable effort has been directed toward the reduction of the longitudinal distortion factor. Several schemes have been demonstrated which reduce the distortion factor, and although no single method has eliminated the distortion, it may be that a combination of procedures will allow three-dimensional visualization. The fact that linear scaling of the hologram reduces the distortion has already been pointed out. To the extent that a reduction in image size does not reduce the resolution beyond the capability of the unaided eye, this technique is appropriate for reducing the distortion.

If a scanning technique is used to record the hologram, it is frequently possible to scan either the detecting transducer or the illuminating transducer. If both transducers are scanned simultaneously, the resultant hologram will be produced with twice the amount of phase change and will be as if recorded at a shorter wavelength by as much as a factor of two. As a result, the distortion is also reduced by a factor of two.

In some cases it is possible to utilize a higher order of diffracted energy in the reconstruction process. Although some other image anomalies are also affected, the primary longitudinal distortion factor is reduced by the order of the diffracted image.

Another method of reducing the distortion factor has been proposed and demonstrated. This method involves a non-linear scaling of the hologram or fractionation. If the hologram is fractionated into a large number of elements, and then each element is linearly reduced without changing position relative to the other elements, then the distortion factor is reduced by the amount that each element is reduced. This technique has been demonstrated, however, it is doubtful if a reduction by more than a factor of five or at most ten can be accomplished.

It has been proposed to synthesize a three-dimensional display, perhaps with a synthesized hologram, using individual image planes of the reconstructed distorted image. In this procedure, individual image planes might be recorded and subsequently reproduced with a lesser spacing between planes. In this way the longitudinal distortion might be reduced or eliminated. A primary limitation on this procedure aside from its complexity and the amount of processing required is that in the recording of individual image planes, all other image information appears out of focus and hence is equivalent to noise superimposed on the image. For very high contrast simple objects this might not be objectionable, but for complex subjects with great dynamic range, this would be a very severe limitation. It is possible in the recording process to isolate an individual object plane by means of range gating or by FM or chirp methods. However, multiple recordings would then be necessary in order to synthesize a three-dimensional image.

Still another method of reducing the longitudinal distortion would involve the synthesis of a stereohologram or integral photograph in which the multiple perspectives employed in recording the synthesized record would be modified in a manner which would make the resultant views similar to those which would be obtained from an undistorted object. A limitation on this technique is imposed by the limited resolution and depth of field in recording the individual perspective views.

Ultimately, the limited amount of information contained in an ultrasound hologram will provide a limitation on the amount of subjective three-dimensional character that can be produced. This is because of the limited resolution that can be obtained with long wavelength investigating fields.

In conclusion, perhaps it would be well to inject a tone of optimism. Acoustical holography is a relatively new imaging procedure. With the great interest that has developed in this field, new and better detection methods will undoubtedly be developed, and improved image presentation procedures will be found. Hopefully, in the relatively near future, these advances will allow the development of practical imaging systems for a variety of applications including medical diagnosis.

PROPOSED APPLICATIONS OF HOLOGRAPHIC TECHNIQUES TO THE OPTICS OF THE EYE AND VISION RESEARCH

Hitoshi Ohzu*

Waseda University

Shinjuku-ku, Tokyo, Japan

INTRODUCTION

Many kinds of optical measuring instruments in ophtalmology have been developed in order to determine the optical constants of the eye. Because the object of study is the living eye, one always finds some difficulties in making measurements, and also some limitations on the accuracy of the data. For the further study of processing visual information in the living human eye, we surely need more precise information about the optical constants and the configuration of the eye including the retinal receptors.

In this paper we wish to discuss the possibility and some of the merits of the application of holographic techniques to this field of study.

*Now at Washington University, St. Louis, Missouri

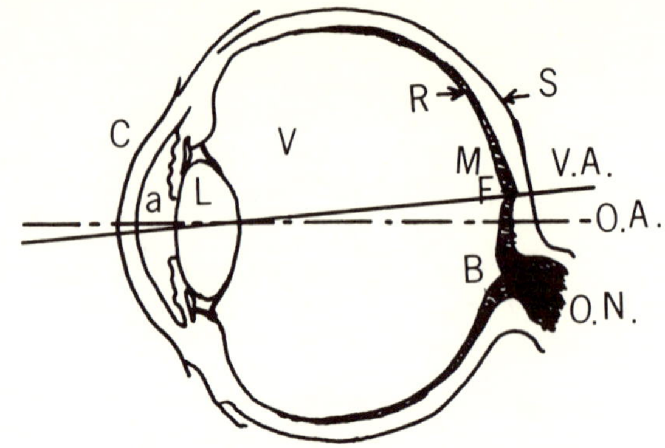

Fig. 1 Schematic Diagram of the Eye

C:Cornea a: Anterior chamber
L: Eye Lens V:vitreous humor
R:retina M:fovea
F:central fovea S:sclera
B:pappilla nervi optici
V.A.:visual axis
O.A.:optical axis
O.N.:visual nerve

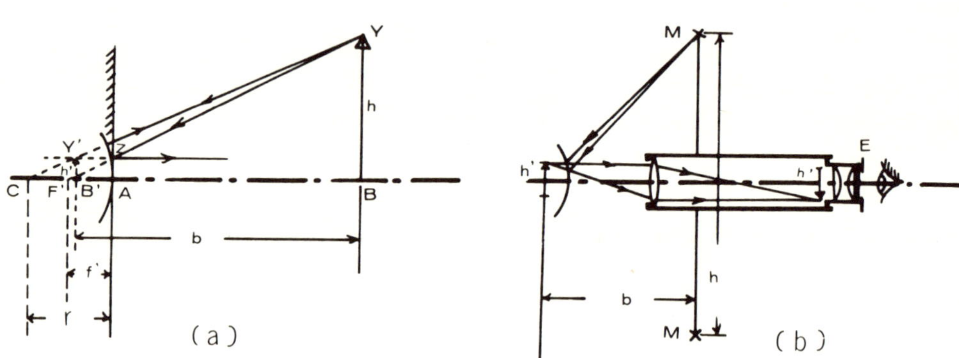

Fig.2 An Example of an Ophthalmometer

(a) The principle of the measurement of r (the first Purkinje-Sanson image P)
(b) The principle of an ophthalmometer

REVIEW OF OPTICAL INSTRUMENTS USED IN OPHTHALMOLOGY

Let us consider the ophthalmometer, ophthalmophakometer and fundus camera. The former two instruments are used mainly to measure the optical properties of the cornea, the eye lens, the aqueous humor, and the vitreous humor which together constitute the optical imaging system of the eye. The ophthalmophakometer is rarely used by clinicians. The fundus camera has been developed to observe the retina which contains not only the light receptors, but also acts to process partially visual information in the eye (Fig. 1).

Ophthalmometer (Keratometer)

The ophthalmometer is used to measure the radius of curvature of the front corneal surface. The principle of the measurements is to observe the first Purkinje-Sanson image, h' which is a virtual image of the illuminating point source made by the front surface of the cornea. In Fig. 2a, the virtual image of the point source Y by the front surface of cornea will occur in Y', where F' and C are the focal point and the center of curvature respectively. Then

$$\frac{ZA}{YB} = \frac{AF'}{BF'}, \quad \text{and if} \quad b \gg f'$$

$$\frac{h'}{h} = \frac{f'}{b} = \frac{r}{2b}, \quad \therefore \quad r = \frac{2b}{h}\, h'$$

Thus we can find r through the measurement of h'. In the real apparatus, a telemicroscope system of shorter focal length and two illuminating targets are usually used as shown in Fig. 2b. In order to avoid the effect of eye movements, some improvements, such as a fixed doubling method, have been developed.

The curvature of the back surface of cornea is also measured by using the Purkinje-Sanson's second image and the refractive index of the cornea. After measuring the distance h_2 between two images of the target sources which are set symmetrically about the optical axis of the eye by a photographic method, one may calculate the curvature r_2 by the following equation:

$$r_2 = \frac{n_c}{D_c - 1 + \dfrac{1}{d'_c + r'_2}} - d_c$$

where $$r'_2 = \frac{h_2}{h_1}\, r_1$$

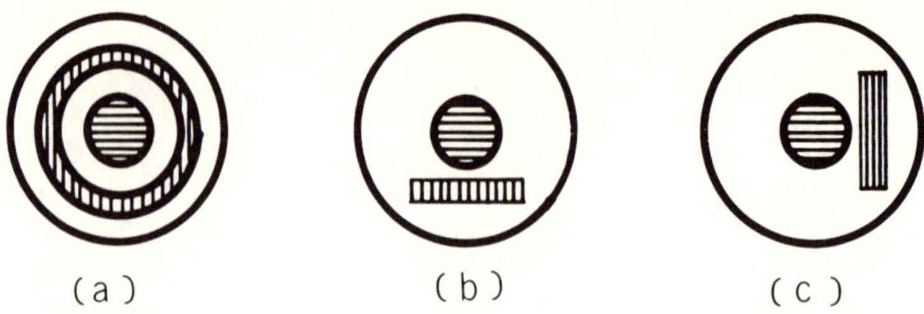

Fig. 3 Various type of illuminating light (vertical hatches) and observing light (horizontal hatches) in fundus camera

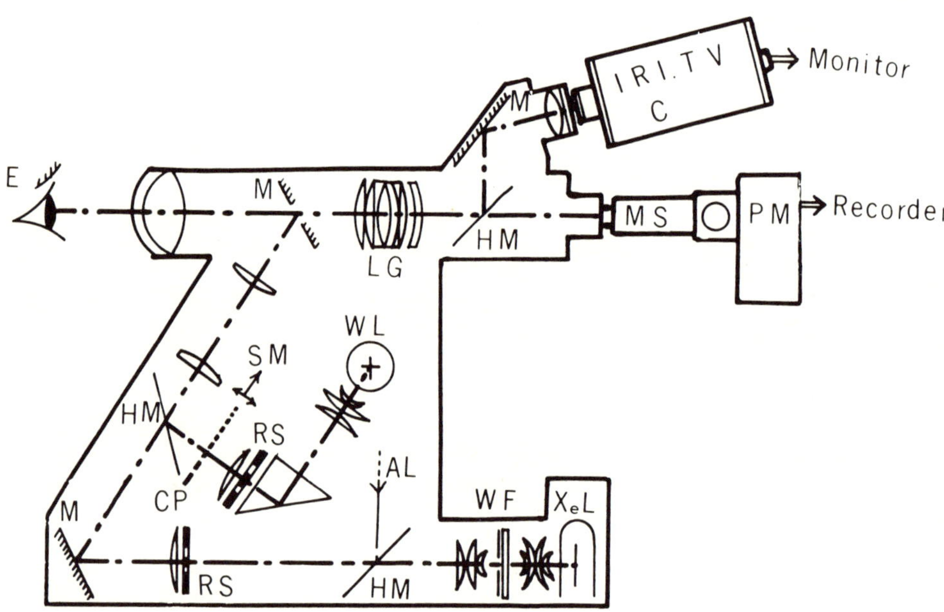

Fig. 4 Fundus camera which is modified for use in the measurement of the optical transfer function of the living eye and fundus reflectometry in our laboratory

X_eL:xenon flash lamp XL:white light source RS:ring slit
WF:interference filter of wedge type CP:frequency chart of grating
AL:additional flicker light source MS:microscope objective lens
SM:servo motor to drive the test chart PM:photomultiplier
C:camera or IRV TV camera HM: half mirror M:mirror

n_c: refractive index of cornea

D_c-1: refractive power of front surface of cornea

d_c: thickness of cornea

d'_c: $d_c \times n_c$

The thickenss of the anterior chamber may be measured also by an ophthalmometer-type device. However, the refraction of the back surface of the cornea is ignored and the refractive indexes of the cornea and aqueous humor are assumed to be the same in such measurements.

Ophthalmophakometer

The Purkinje-Sanson's third and fourth images may be used to measure the front and back surfaces of the eye lens. From these images the approximate curvature of these lens surfaces may be calculated. The apparatus for this purpose is usually called ophthalmophakometer (Tscherning). The refractive index is independently measured by using excised eyes. The length of the optical and visual axis to the fundus is measured by x-ray or ultrasonic methods.

In any case, those measurements are based on first order geometrical optics, and many approximations and statistical manipulations are made on the values measured (particularly in estimates of the refractive properties of the eye lens). If one wants more precise information, for example, the change of curvature during accommodation, the difference from spherical surface, or the difference of data between the optical axis and the visual axis etc., better techniques are needed. In order to advance our understanding of the imaging ability of the eye, we need more precise knowledge concerning the optical aberrations of the eye, not only for paraxial imagery and the infinite viewing state, but also for each accommodative state and for off axis viewing. It is also very important to know what the merits of the eye lens are during accommodation, when matching the optics of the eye with industrial compound systems of lenses used in conjunction with the eye.

Fundus Camera

In the case of fundus observation, large amounts of reflected light from the cornea and the eye lens limit the contrast of the fundus image. The optical paths of the illuminating and observing light are usually separated in fundus camera (Fig. 3), in order to minimize this problem. A non-spherical objective lens is usually used in order to compensate for the aberrations of the eye. An example of

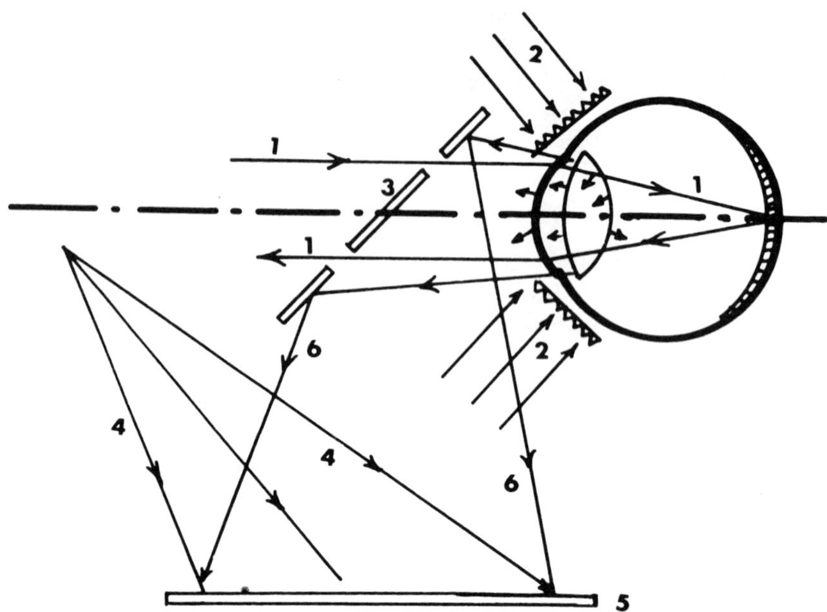

Fig. 5 Schematic Diagram of the recording procedure for holography of the living eye

1:collimated laser beam of small bundle 4:reference beam
2:diffuser for laser illumination 5:photographic plate (hologram)
3:Mirror 6:diffracted light from the eye

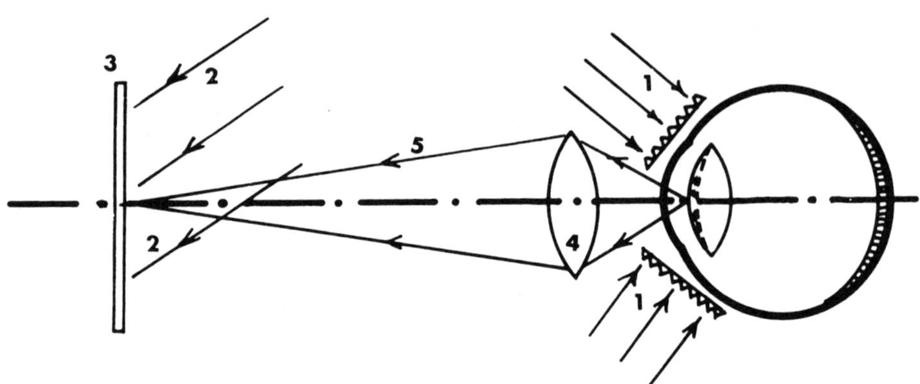

Fig. 6 Schematic diagram for holographic interferometry on the eye lens surface

1:diffuser for laser illumination 4:objective lens
2:reference beam 5:reflected light from the eye lens surface
3:photographic plate (hologram)

a fundus camera which has been somewhat modified to measure the optical transfer function of the eye is shown in Fig. 4. The magnification and the field of view are limited. Stereoscopic observation using a binocular system is also available.

APPLICATION OF HOLOGRAPHIC TECHNIQUES

We are now going to discuss the merits and possibilities of holographic techniques for the study of the optics of the eye. It is convenient to classify the following three groups of problems:

(1) Holographic observation of the optical constants of the eye and the light path through the cornea and eye lens.

(2) Holographic interferometric measurements of the configuration changes which accompany accommodation or changes in intraocular pressure of the eye.

(3) Three-dimensional observation of the fundus retina.

Holographic Observation of the Optical Constants of the Eye and the Light Path Through the Eye Lens

The reflectance from the front surface of the cornea is about 2.5% and the other refracting surfaces have about 0.2 - 0.7% reflectance. These media are transparent to visible light but always exhibit some scattering. It might be possible to separate the reflected light contributed by each of the refracting surfaces and the scattered light from each eye medium by holographic techniques. The mean value of the refractive index of the cornea, aqueous humor, eye lens and vitreous humor are listed in the Gullstrant schematic eye as 1.376, 1.336, 1.386 (lens cortex), 1.409(lens nucleus), and 1.336 respectively. If the eye is illuminated from the front surface of the cornea by a diffused laser beam and if at the same time a small bundle of collimated laser light is incident at the eye (Fig. 5), the hologram is composed of the reflected light from each of the eye surfaces and by the scattered light from each of the media. The penetration of the small laser beam will serve to analyze the phase change in the lens media in the reconstructed image. The eye lens is constructed of many layers of differing refractive index. This distribution of index varies with accommodation. We may also find the real length of the optical and the visual axis of eye, if we can record in the hologram the scattered light originating from the small laser beam returning from the retina. Most of the data measured by ophthalmometry and/or phakometry provides us with information relative to the optical axis of the eye (approximate), and not the visual axis. Calculations used to determine the optical constants are usually made with the assumption of perfect spherical surfaces and uniform distribution of refractive index of each medium.

If we can find the real light path through the living eye and its three-dimensional configuration in the reconstructed image of the hologram, it will be a great advance. The problem is that the reflected light might be too limited to form an adequate hologram. Of course, one has to limit the laser beam energy to prevent retinal damage in the living eye. I recommend that a monitored pulsed laser be employed. A good diffuser is needed. Again, to prevent retinal damage it is probably wise to have the laser beam penetrating the eye out of focus at the retina.

Holographic Interforometric Measurements of the Configuration Changes which Accompany Accommodation of the Eye

The eye itself has many aberrations. However, the living eye has a great advantage since it has a controlling servo system. The eye lens is pliable. The curvature of both surfaces varies with accommodation, but most changes occur in the front lens surface. Of necessity, the distribution of the refractive indexes in the lens must also change. It is very interesting to know how it varies. Again this is important information that is needed for the design of commercial lens systems applied to the eye. For this study, I propose to use a holographic technique such as double exposure inteferometric method. It would be desirable to measure the change of curvature of the eye lens surface in two different states of accommoation. It will also be valuable to see if this factor is altered as a function of age. For this purpose, a configuration such as that in Fig. 6 is used. The laser light is diffused and enters the eye lens through the sclera. The fraunhofer hologram is made by using a lens of short depth of focus compared to the thickness of the eye lens (5mm thickness), e.g. a microscope objective. The same procedure is repeated for both states of accommodation, i.e. a doubly exposed hologram is made in real time. Then an interference pattern will be obtained in the reconstructed image. From this pattern one may calculate the change in eye lens curvature. The living eye continually undergoes eye movements. Therefore, again it is convenient to use a pulsed laser, or very short exposures of a CW laser. The usual optical measurements of the curvature give us only a mean value of the lens surface and assume it to be spherical. The precise measurement of the configuration of the eye lens in the living state is very important in order to estimate its aberrations, its optical transfer function, etc.

Three-Dimensional Observation of Fundus Retina

It is virtually impossible to resolve the individual receptor cells of the living retina by available ophthalmological techniques such as the fundus camera. Our ability to do so is limited by the aberrations of the eye, vibrations and movements of the body, and the

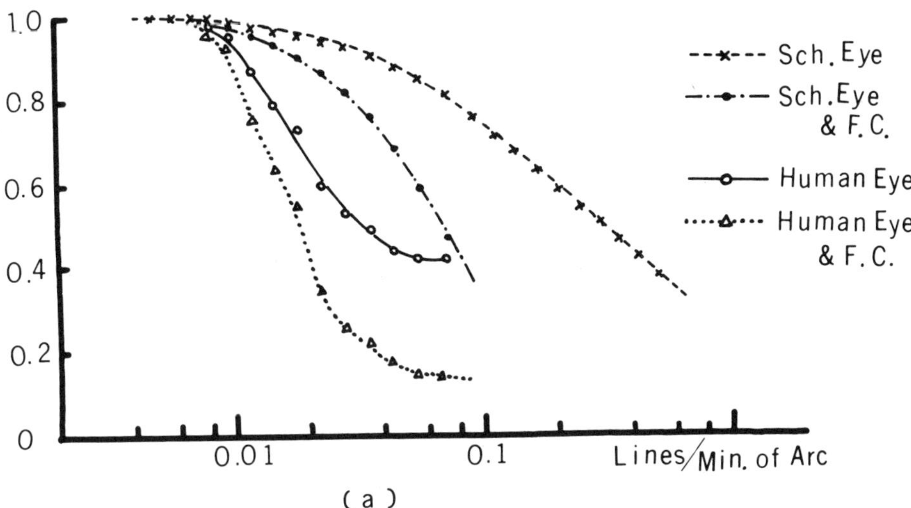

(a)

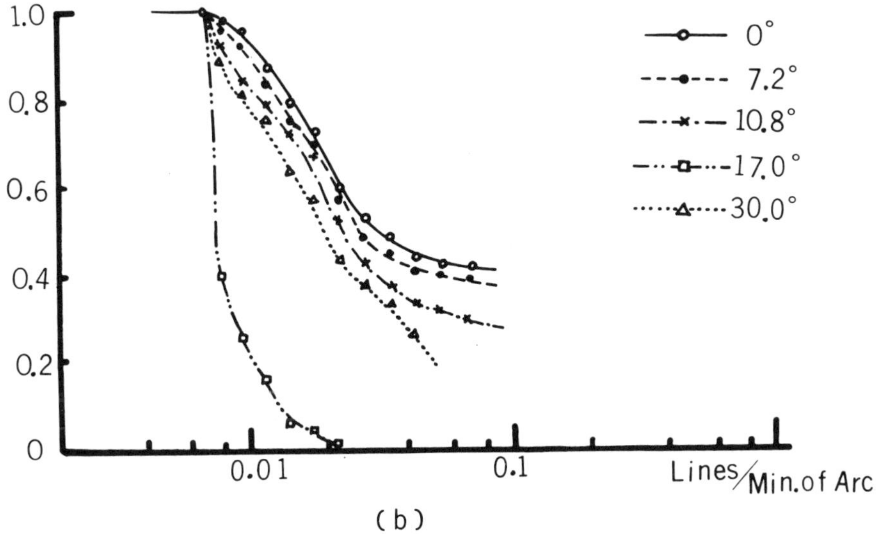

(b)

Fig. 7 Optical transfer functions of the human living eye (measured)

(a) Comparisons with the schematic eye
(b) Optical transfer functions of several view points

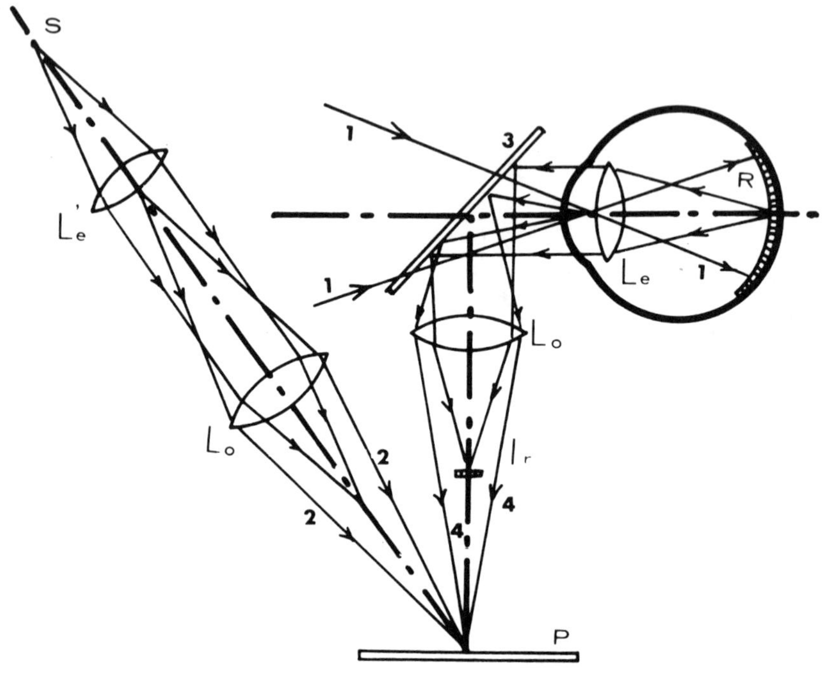

Fig. 8 Schematic diagram for holographic aberration free observation of the retina

S:point source of laser
R:retina
Le: eye lens
Le':equivalent lens to the eye
Lo:relay lens
Ir:Image of retina through eye lens
P:photographic plate (hologram)

1:illuminating laser beam in Maxwellian view
2:reference beam
3:half mirror
4:imaging light path of the wave front through the eye lens

movements of the eye. It is very important to study the optical properties of the retinal receptors and the processing mechanisms for visual information of the retina. A precise knowledge of the real three-dimensional configuration of the fundus of the living eye is also desired in clinical ophthalmology. Is it really impossible to resolve a single cell (cones or rods) through the living eye? The smallest cone has a diameter of about 1.5μ in the central fovea, the maximum limiting diameter of the pupil of the eye is about 8 mm and the focal length of ca. 22 mm in the medium of refractive index of 1.336. It is easy to estimate that the cut off frequency is about 1000 lines/mm in the retinal plane when the lens is aberration free. This means, then, that we have the possibility of resolving single cones in the retina through the eye, if we could eliminate the aberrations of eye lens; at least, cells in parafovea might be resolved. We also need much more knowledge regarding spatial distribution, that is, the array of receptor cells in the living state, for the study of vision. For example, the Stiles-Crawford effect and color problems are influenced to some extent by the optics of the retinal receptors. For this purpose, holography might also be applied. The optical transfer functions of the living eye lens are shown in Fig. 7. This data was recently obtained in our laboratory. We were not, however, satisfied with values measured at the higher frequencies. If we know the optical transfer function of the eye lens, we may reconstruct the fundus image with higher resolving power. The principle of the procedure is shown in Fig. 8. The fundus is illuminated by laser light in a maxwellian view. The image I_1 of the retina will be focused at the focal plane of lens L_o through the eye lens L_e and a relay lens L_o.

The photographic plate is placed in a plane conjugate to the pupil of the eye lens so that the wave front from the fundus through the eye lens is recorded on the plate P.

On the other side, we chose a lens L_e' which has the same OTF as the eye lens. The wave front through this lens is used as the reference beam and is recorded at the same time on the photographic plate P. Now, when the point spread function of the eye lens L_e, and L_e' are denoted as K(x,y), then the amplitude distribution of the retinal image is

$$T_r(x,y) = \iint R(x',y') \cdot K(x-x', y-y')\, dx' \cdot dy'$$

The point image through L_e' is described as

$$T_s(x,y) = \iint \delta(x'-a, y'-b) \cdot K(x-x', y-y')\, dx' \cdot dy'$$

$$= K(x-a, y-b)$$

The intensity distribution on the photographic plate is (Stroke)

$$I_h = |t_r|^2 + |t_s|^2 + t_r \cdot t_s^* + t_r^* \cdot t_s$$

where t_r and t_s are the fourier spectra of T_r and T_s. Because the amplitude transmission is proportional to I_h, the spectrum of the reconstructed image becomes $t_r^* \cdot t_s$, and $t_r \cdot t_s^*$. If we denote the fourier transform of $R(x,y)$ and $K(x,y)$ as $r(u,v)$ and $k(u,v)$ respectively, then t_r and t_s are equal to $r(u,v) \cdot k(u,v)$ and $k(u,v)$.

In conclusion, the reconstructed image has the spectrum $r(u,v) \cdot |k(u,v)|^2$ and $r^*(u,v) \cdot |k(u,v)|^2$ Now, as we are dealing with a coherent optical system, $k(u,v)$ will be expressed only as a phase function, and then $|k(u,v)|^2 = 1$, that is the reconstructed image is independent of the aberrations of eye lens. By this procedure, we need a lens which has the same optical transfer function as the eye lens. The several schematic eyes cannot be used for this purpose, because their optical constants are determined only to simulate the geometric optics of the living eye, and have not been considered from the point of view of wave front aberrations, etc.

Thus, it will be necessary to perform the experiments mentioned above first in order to deal with this task.

Because our object here is a living body, and may be subject to damage by high energy laser irradiation, caution is advised. We have now developed the apparatus to measure the optical transfer functions of the human living eye, and we hope the experiments applying holographic techniques will be initiated in the near future.

The author is pleased to acknowledge the considerable assistance of Dr. J. M. Enoch of the Department of Opthalmology, Washington University.

A HIGH CAPACITY HOLOGRAPHIC STORAGE SYSTEM

James Lipp, Jerry L. Reynolds

IBM SDD

Poughkeepsie, New York

ABSTRACT

This paper describes a potential storage system that would provide 2^{24} bytes of data with a random access of 2μsec, and a data rate of 50 megabytes/sec. The system discussed is organized as a block oriented optically accessed memory and utilizes holography as the storage mechanism. The information is stored holographically.

The holograms are recorded in a silver halide emulsion on a glass substrate providing a nonvolatile storage medium. This technology combines high bit densities with large positional tolerances. Each holographic block consists of 512 bytes of data which have been word organized. The plate is accessed by means of a laser beam which transfers, in parallel, an entire data block to a detector array.

The system consists of the following elements: a frequency doubled Nd:YAG laser operating at 5300A; a 15-stage digital light deflector subsystem which provides 32,768 randomly accessed positions, an information storage plate, means for changing plates, a silicon integrated photodetector array that converts the 512 bytes (64 x 72 bits) of optical data into electrical signals at a nominal 1

A need exists for high capacity, low cost storage systems for use in such applications as system residence, archival storage, program library, and storage of tables. This paper describes a potential storage system that could be used in these applications. This system would provide 2^{24} bytes of data with a random access of 2μsec, and a data rate of 50 megabytes/sec.

As computing systems become more economical, large data base systems find more applications. In order to utilize a large data base system, a high capacity storage device becomes a primary requirement.

The use of optical technology for storing data is a solution to the high capacity storage problem. Optical memories can be classified as bit, byte, or block oriented. These memories can be further classified into conventional photographic records, spot patterns generated on substrates, or holographic storage systems. The first two approaches are generally bit oriented and have the same data throughput characteristics provided by tapes and discs. If one were required to store 10^6 bits per square inch, then severe mechanical tolerance problems would be presented. The positional accuracy in a system having a 10^6 bits per square inch (25 micron bits) with 20 percent positional tolerance is 5 microns, a rather difficult goal to achieve and one which is eliminated with a holographic storage system. Furthermore, the redundancy of holograms eliminates the substrate stability problems.

In. Fig. 1, a block oriented experimental holographic storage system is shown. This system uses a frequency doubled Nd:YAG laser operating at 5300A, followed by a set of beam forming optics which spatially filters and converges the beam. This beam is deflected through a digital light deflector.* This deflector is driven by solid state circuits and consists of a 15-stage binary arrangement which provides 32,768 randomly accessed positions. A set of beam expanding optics follows next, which magnifies the optical output to agree with the requirements of the holographic storage plate. The holograms are recorded in a silver halide emulsion supported upon glass plates which provide a nonvoltatile storage medium and combines high bit densities with large positional tolerances. This plate when illuminated by a laser beam transfers a block of data in parallel to the detector array as shown. A channel is defined as the beam expander, information plate, and detector array combination.

The system shown in this slide is limited by the size of the detector array. By this we mean, that the state of the art limits the silicon ship size and the number of associated chips that make up an array.

---

*W.Kulcke et al "Convergent Beam Digital Light Deflector" Chap. 23 of J. T. Tippett et al Optical and Electro-Optical Information Processing, MIT Press, Cambridge, 1965.

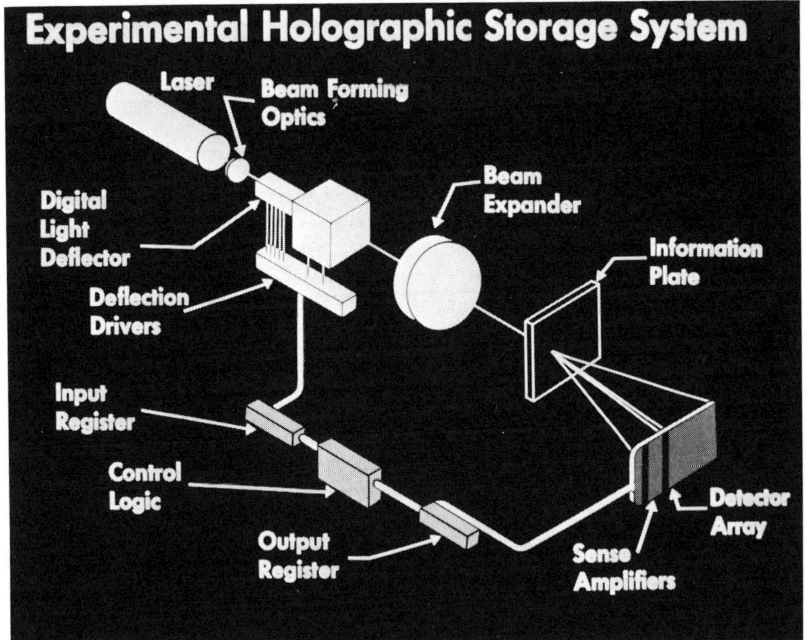
Experimental Holographic Storage System
Laser
Beam Forming Optics
Beam Expander
Digital Light Deflector
Information Plate
Deflection Drivers
Input Register
Control Logic
Output Register
Sense Amplifiers
Detector Array

FIG. 1

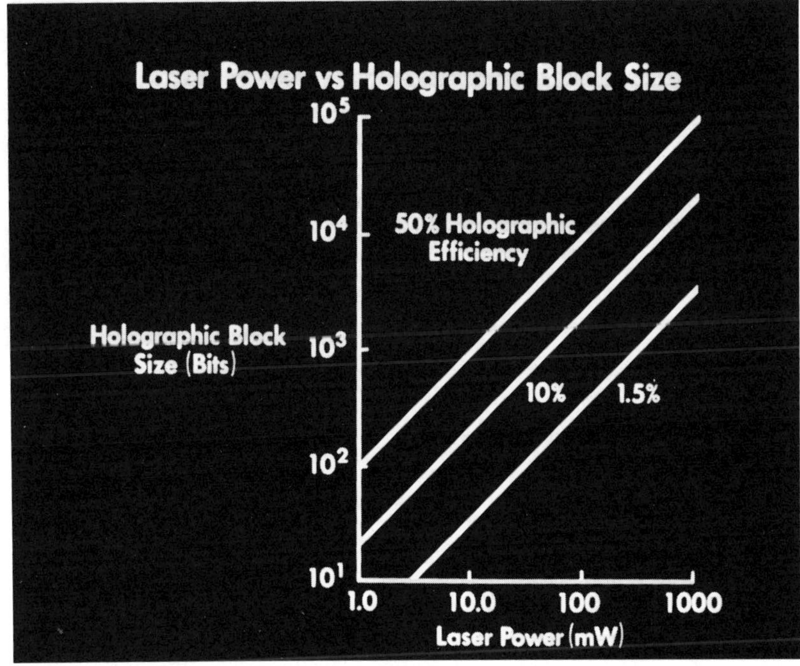
Laser Power vs Holographic Block Size
10^5
10^4
10^3
10^2
10^1
50% Holographic Efficiency
10%
1.5%
Holographic Block Size (Bits)
1.0
10.0
100
1000
Laser Power (mW)

FIG. 2

Fig. 2 is a plot of the number of holographic bits that can be stored in a holographic block versus the system laser source power requirement for several holographic reconstruction efficiencies. For this graph it was assumed the power to a detector may be maintained at 1μw, and the system transmission was assumed to be 20%. One can see that for an absorption hologram of 1.5% efficiency, a 512 byte block (4608 bits) would require a laser with more than one watt output. If one were to use phase holograms, the power requirement would be reduced to approximately 180 mw for a 10% hologram and 36 mw for a 50% phase hologram. With these considerations in mind one can construct a system consisting of several channels.

Table I
HOLOGRAPHIC STORAGE SYSTEM SPECIFICATIONS

Capacity	16×10^6 Bytes
Access	2μseconds - Random
Block Size	512 Bytes
Cycle Time	10μseconds
Data Rate	51.2 megabytes per second

With the 1μw per bit specification previously mentioned and a 512 byte detector array for each of the eight channels a 200 mw laser would be more than adequate. The digital light deflector that would be necessary for this system would be a 15 stage device giving 32,768 positions. Eight storage plates would be required with each plate storing 4096 hologram blocks of 512 bytes each. A holographic block in this system would be 1.5 mm in size making the overall storage area 3.8 inches on a side.

Table II
DETECTOR ARRAY SPECIFICATIONS

Size	512 bytes 64 x 72 bits per array
Technology	Integrated Silicon Photodetectors
Detector Spacing	250μ
Sensitive Area	200μ
Word Cycle	50 nseconds
Radiation Rate	10^7 Photons per Second
Signal to Noise Ratio	10 db

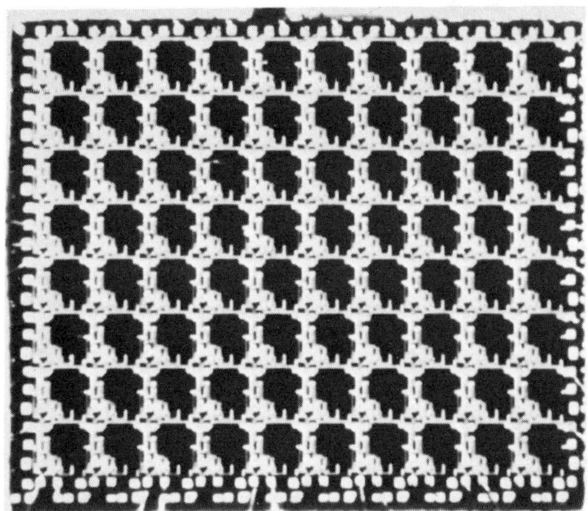

FIG. 3 Integrated detector array chip with an 8 x 10 array of silicon photodetectors

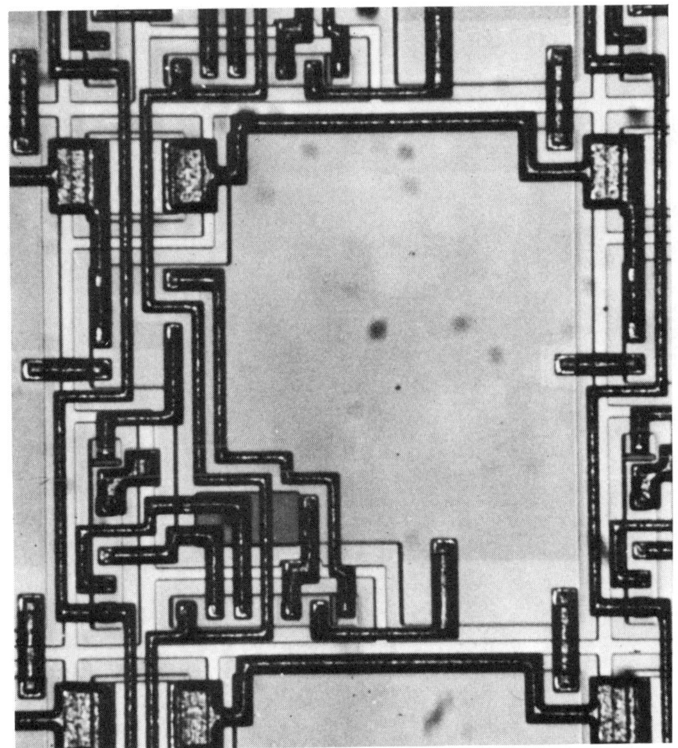

FIG. 4 One cell of the array of FIG. 3

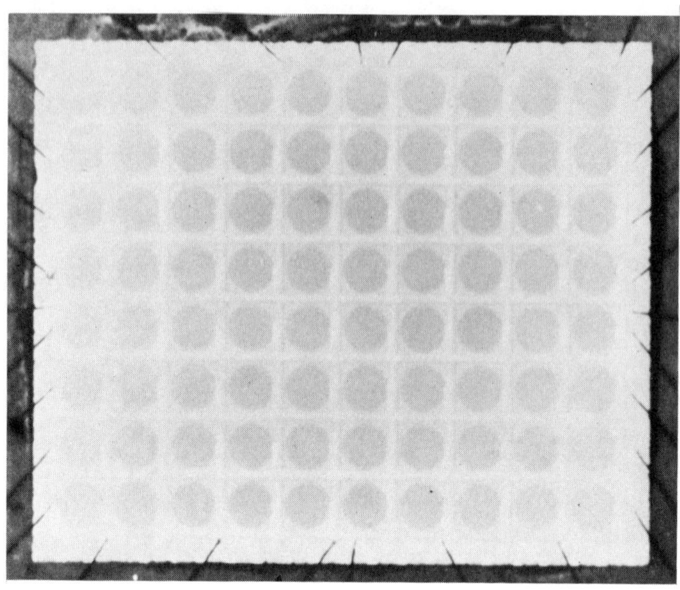

FIG. 5 Integrated detector array chip with an 8 x 10 array of silicon photodetectors and FET amplifiers

FIG. 6 A unit cell of the array in FIG. 5

The parameters of a detector array for this proposed system that would detect, amplify, and transmit to an output register is shown in Table 2. This detector array would be 64 x 72 bits, that is, 64, eight byte words plus error correcting codes.

The detector array would be an integrated array using either FET's or bipolar transistors as amplifying elements. To meet the 5μsec system cycle time the word cycle time for this array is 50 nsec. producing a read-out time of 3.2 μsec. The sensitivity of this array would necessarily be 10^7 photons per bit, which is 1 μwatt for a 500 nsec. integration time at 5300A. The descrimination level of this array would be 10 dB.

To obtain a minimum detector array size, consistant with system tolerance considerations a 200 micron detector bit size was selected resulting in a 250 micron detector spacing.

If one were to construct an array with these considerations in mind the array would consist of 64 chips, each chip containing 72 bits arranged in a 8 x 9 matrix. Allowing one bit spacing between chips yields an overall detector array size of approximately 17 mm x 19 mm.

Fig. 3 shows a detector array chip. The chip is an 8 x 10 array of integrated silicon photodetectors with associated transistor amplifing circuits.

The word lines are shown horizontal and are 10 bits wide to accommodate error correcting codes. The 8 bit lines are shown vertical. This chip would be representative of one of the 64 chips required for a completed array that would sense a 512 byte holographic block.

Fig. 4 shows one cell or bit of the chip of Fig. 3. The large area in the center of the cell is in the silicon photodetector. The cell is defined as the area bounded by the isolation diffusion which appears as lightly colored areas. This cell is 25 mils square and the detector area is approximately 17 x 19 mils. Further details of this array and circuitry will be given in a future paper.

Fig. 5 shows another detector array chip. The chip differs from Fig. 3 in that the amplifying devices are FET's. Again, this is an 8 x 10 matrix of silicon photodetectors. The detector spacing is 25 mils. The detector array for this system will consist of 16 such chips giving an overall array size of 32 x 50 bits. This represents 32, 4 byte words plus error correcting codes.

In Fig. 6 we have a close-up of the array chip. The photodetector area can be seen as the round area in the center. This area is 20

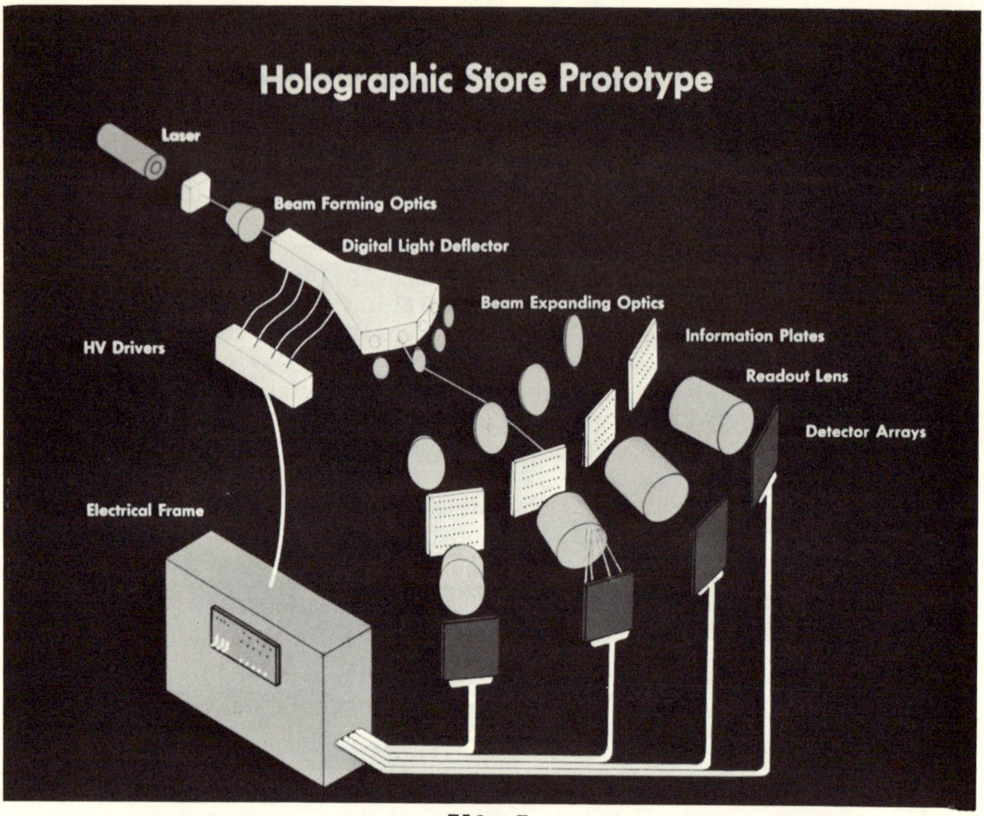

Holographic Store Prototype
Laser
Beam Forming Optics
Digital Light Deflector
Beam Expanding Optics
Information Plates
HV Drivers
Readout Lens
Detector Arrays
Electrical Frame

FIG. 7

Input Optics
Digital Light Deflection System
Argon Laser
EOS(KD*P)
S A Deflection Section
$2^4_x \cdot 2^4_y$
TIR
Deflection Section
$2^2_x \cdot 2^2_y$
TIR
Tree
D
C
B
A
EOS
Output: $2^6_y \times 2^8_x$
16,384 Positions

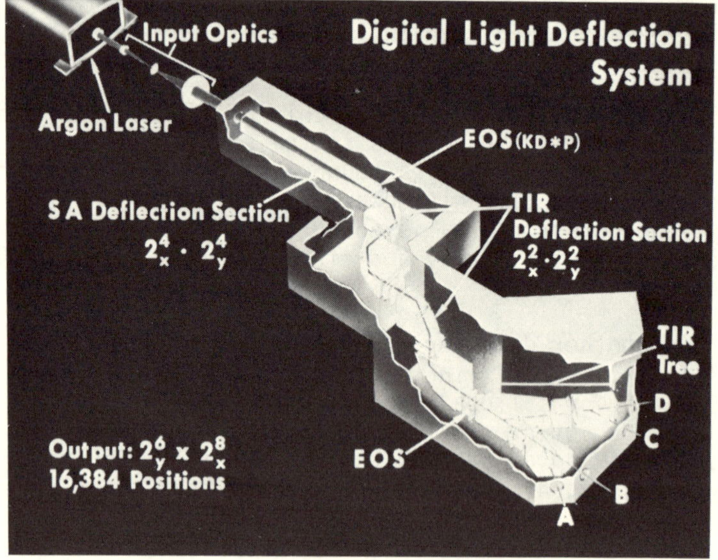

FIG. 8

mils in diameter. The FET device can be seen as the small source spot at the right of the detector.

We are presently fabricating a model of a holographic store to prove the concepts of the system we have just described. This model will have a capacity of 2 megabytes stored upon 4 storage plates. The access of 128 byte block of data will be in 4 μsec. and a memory cycle time will be 8 μsec. This produces a data rate of 16 Mbytes per sec.

An artist's conception of the system is shown in Fig. 7. It consists of the following elements: an argon laser as the light source followed by a set of beam forming optics containing a spatial filter, (this produces a convergent light beam which passes through the digital light deflector), the 14 stage digital light deflector to randomly position the beam to the required address in the holographic storage plates, and four plates, each containing a 64 x 64 matrix of recorded holograms, which provide 16,384 blocks of information.

The reconstructed image produced from each of thee holograms, consists of a matrix of bits with either a light spot present or absent at each matrix position. This image is reconstructed on a solid state photodetector array and is sensed by the photodiodes. The signal is then sensed by the sense amplifiers and sent to the output register. Each one of the holograms is recorded on the photographic plate, such that in the reconstruction, the matrix image contained in all holograms is exactly registered on a single detector array.

This reconstruction is provided by the readout lens. This lens takes the inverse transform of the hologram and registers the reconstructed images upon the detector array. Image registration can also be accomplished by moving the fourier writing lens during the writing process.

For readout, the laser beam is required to have the same convergence, and incidence angles on the holographic plate, as the reference beam had in the recording process. The readout of the desired block of information is accomplished by deflecting the laser beam, to the appropriate hologram, and detecting the presence or absence of light at the detector array.

A three-dimensional schematic of the digital light deflector is shown in Fig. 8. It consists of 14 binary deflection stages. Each binary stage consists of an electro-optic switch and a birefringent deflection element. Two types of deflection stages are used in this deflection system. The initial small deflections in two dimensions are produced with split angle elements. Calcite and potassium dihydrogen phosphate are used in fabricating these split angle elements. This type of deflection element is used for small deflection, because of its simplicity, and economic

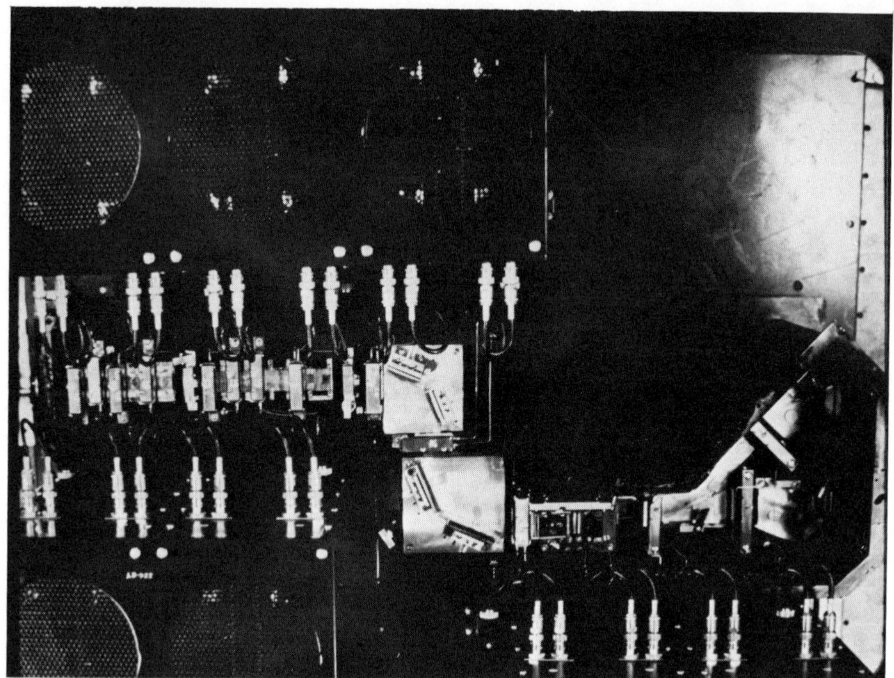

FIG. 9 Four stage digital deflector

FIG. 10 A partially filled storage plate

use of the birefringent material. The larger deflections in both axes are produced with self-compensated total internal reflection (TIR) elements. The self-compensated TIR elements are used because a) they are free of optical aberrations, b) deflection can be adjusted to precisely the desired value, and c) the deflected outputs can be adjusted to be precisely parallel. The maximum aperture required in any stage of the deflector is 0.75 x 0.75 inches. The two last stages of TIR deflection are used in one dimension to produce four separate output fields. The Electro-optic crystal switches are fabricated with KD*P.

The deflector and its high voltage drive circuits arranged on either side of the housing are shown in Fig. 9. There are reflections of the laser rays at some of the stages due to scattering at some of the optical interfaces.

The estimated light transmission for this deflection system was calculated to be 40 percent, however, initial measurements have indicated a figure of 30 percent for any required output position.

We are currently investigating the difference in these two figures which are believed to be losses due to absorption and scattering at the electrode interfaces in the electro-optic switches. The tank is filled with index matching oil and is highly transparent as evidenced by this picture.

This entire digital light deflector assembly is approximately 30" long. Although the entire housing is filled with oil, the optical path through any channel contains less than 1 1/2" of total oil path in order to limit thermal focusing. This was accomplished by the use of glass filler pieces of suitable optical index.

The digital light deflection technology makes use of solid state devices throughout. It does not use any moving parts and it is completely digital. Once constructed, this system preserves its built-in accuracy with high reliability.

Fig. 10 shows a partially filled storage plate. Each holographic block is 38.5 mils in diameter and is placed on 40 mil centers horizontally and on 38.5 mil centers vertically. This difference is due to the fact that the storage plate is placed at an angle of 15° in the read station. When fully populated this plate would contain 4096 blocks or 0.5 megabytes of data. This plate is a Kodak 649F micro flat glass. The overall dimensions of the plate are 4" x 5". The storage area dimensions are 2.56" x 2.58". The storage plate is shown in a carrier which remains with the plate through the writing, processing and reading steps. This carrier provides protection and a means of registration.

The writing sequence for our system is shown in Fig. 11. The user brings data to be stored to an IBM 360/50 machine for formatting. This data is then sent to an 1130. The 1130 controls both a mask generation station and the holographic write station. In order to write one may make up to 16,384 masks containing 128 bytes each. These masks are fabricated in the mask generation station. The masks are then used as object masks in the write station.

Several potential applications for a holographic storage system can be envisaged. It could be used as a system residence device. It is attractive when the cost per bit of the holographic store becomes less than that of the main storage. Another application would make use of reading programs stored on a holographic plate which could be changed for system update. Storage of large fixed tables is attractive since with this system the size of the data base can be easily increased. Other applications include secure read-only storage and archival storage making use of the large capacity and nonvolitale nature of this storage medium.

In summary, we have shown a possible configuration of a high capacity holographic storage system. We have also discussed design parameters and have reduced to practice sections of a model of this system. This model will enable us to evaluate absorption and phase holograms for the most procatical system application.

The work in computer generated holograms and kinoforms as reported by the IBM Scientific Computer Center in Houston could also possible be used to generate the holographic blocks of data for use in this system.

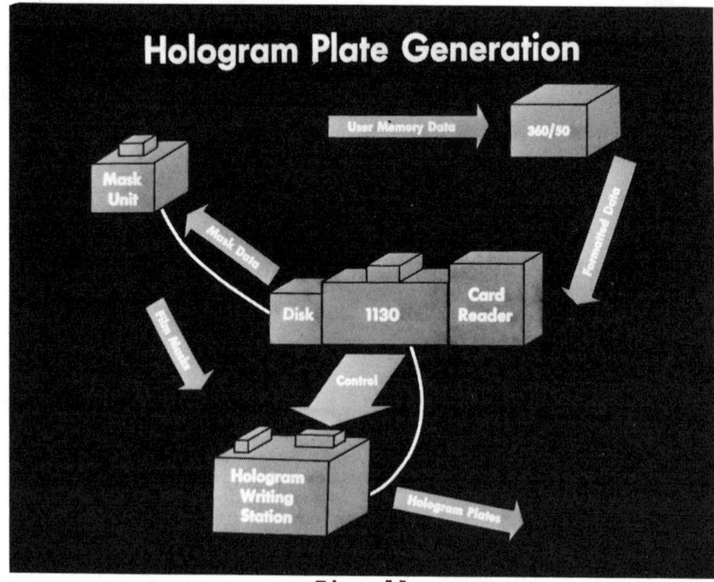

Fig. 11

AUTHOR INDEX

SUBJECT INDEX